CRC Handbook of High Resolution Infrared Laboratory Spectra of Atmospheric Interest

Editors

David G. Murcray
Department of Physics
University of Denver
Denver, Colorado

Aaron Goldman
Department of Physics
University of Denver
Denver, Colorado

CRC Press, Inc.
Boca Raton, Florida

Library of Congress Cataloging in Publication Data

Main entry under title:

CRC handbook of high resolution infrared laboratory
spectra of atmospheric interest.

Bibliography: p.
Includes indexes.
1. Atmospheric chemistry--Handbooks, manuals,
etc. 2. Infra-red spectrometry--Handbooks, manuals,
etc. 3. Molecular spectra--Handbooks, manuals,
etc. 4. Absorption spectra--Handbooks, manuals,
etc. I. Murcray, D. B. II. Goldman, Aaron.
III. Title: Handbook of high resolution infrared
laboratory spectra of atmospheric interest.
QC877.6.C17 551.6 80-23134

ISBN 0-8493-2950-7

This book represents information obtained from authentic and highly regarded sources. Reprinted material is quoted with permission, and sources are indicated. A wide variety of references are listed. Every reasonable effort has been made to give reliable data and information, but the author and the publisher cannot assume responsibility for the validity of all materials or for the consequences of their use.

Direct all inquiries to CRC Press, Inc., 2000 N.W. 24th Street, Boca Raton, Florida, 33431.

International Standard Book Number 0-8493-2950-7

Library of Congress Card Number 80-23134
Printed in the United States

PREFACE

Part of the atmospheric spectroscopy research at the University of Denver consists of detailed identification of the atmospheric absorption features present in high resolution infrared solar spectra obtained from various altitudes. This study revealed a lack of laboratory data for many of the compounds of interest with sufficiently high resolution to be used for identification. Therefore, a laboratory study was undertaken to obtain high resolution spectra of many of the species of interest. The spectra presented in this Handbook are the result of this study.

A number of individuals participated in the preparation of this handbook, most notably F. S. Bonomo, C. M. Bradford, G. R. Cook, F. H. Murcray, F. J. Murcray, and J. W. VanAllen. The research was supported by the Chemical Manufacturers Association and the National Science Foundation.

THE EDITORS

Professor David G. Murcray has been engaged in atmospheric research with particular emphasis on atmospheric spectroscopy for over 20 years. He received his B.S. from the University of Denver, his M.S. from Oklahoma State University, and his Ph.D. from the University of Denver.

Professor Murcray began his career as a teaching fellow at Harvard University from 1948 to 1949 and at Oklahoma State University from 1949 to 1950. He was an assistant instructor at Kansas University from 1950 to 1951 and a research mathematician for Phillips Petroleum Company the following year. Since 1952 he has been at the University of Denver, where he is currently Professor of Physics.

His atmospheric research has included measurements of water vapor and other minor constituents and trace constituents in the region from 13 km to 40 km, using spectroscopic techniques. This research led to the initial detection of HNO_3 and NO_2 in the lower stratosphere. Included in his research interests are atmospheric spectroscopy (particularly in the infrared), infrared instrumentation, Fourier transform spectroscopy and atmospheric chemistry.

A member of Phi Beta Kappa, Sigma Pi Sigma, Sigma Xi, and a Fellow of the Optical Society, Professor Murcray has published numerous papers and journal articles concerned with atmospheric spectroscopy and atmospheric composition.

Aaron Goldman, Research Professor of Physics at the University of Denver, has done extensive research in quantitative molecular and atomic spectroscopy since receiving his M.Sc. in 1962 from the Hebrew University, Jerusalem, Israel. He received his D.Sc. in 1965 from the Technion-Israel Institute of Technology, Haifa, Israel, where he was involved in high temperature spectral emissivity infrared and ultraviolet laboratory studies of atmospheric gases. From 1966 to present, his research has been concentrated on quantitative analysis of infrared and ultraviolet absorption of solar radiation by the earth's atmospheric gases, atomic and molecular solar lines, and infrared emission by the earth's atmosphere, from high resolution spectra obtained with ground-based and air-borne spectrometers. As part of this research, Dr. Goldman demonstrated the detection and quantification of several new molecular species in the lower stratosphere. A member of Sigma Xi and the American Geophysical Union, Dr. Goldman has authored or co-authored over 100 papers and journal articles related to atmospheric and solar spectroscopy.

ADVISORY BOARD

TABLE OF CONTENTS

INTRODUCTION

Infrared spectroscopy has been used for many years as a means of identifying and quantifying the constituents present in gaseous samples. As a result, a large number of infrared spectra have been published either in the form of handbooks or index cards (many of them by the chemical industry). For most industrial applications, resolution is not crucial, and the majority of spectra presented in previous publications have moderate to low resolution (i.e., not better than 1 cm^{-1}).

During the last decade, infrared spectroscopy also became a powerful tool in atmospheric studies. Locating a source at some distance from the spectrometer with the intervening atmosphere as an absorber provides spectra which contain a wealth of information on atmospheric composition. This technique is used extensively to obtain information concerning gases present in natural and polluted atmospheres. Recent studies in atmospheric chemistry, particularly those associated with the ozone layer, have shown that very small amounts of certain compounds can act as catalysts and have a significant impact on atmospheric photochemistry and composition. As a result, there is considerable current interest in obtaining data concerning the concentration of a large number of compounds which are present in the atmosphere at very low levels. This leads to the need for an increase in sensitivity of measurement techniques, including infrared spectroscopy. Increased spectral resolution improves the sensitivity of infrared spectroscopic measurements for almost all species.

One of the major difficulties encountered in attempting to use infrared spectroscopy for atmospheric measurements is the interference from absorption features due to other atmospheric constituents. The common minor constituents (H_2O, CO_2, N_2O, CH_4, O_3, and CO) all have strong infrared absorption features and these tend to produce absorption lines that interfere with those due to the molecular species of interest. Detection and measurement of gases in atmospheric spectra must be accomplished against this interference, and increased spectral resolution becomes very important in achieving detection under such conditions.

A recent publication by Graedel[1] lists 1600 gases which are known or thought to be present in the atmosphere. Assembling high resolution spectra of all these gases is obviously beyond the scope of this Handbook. Spectra in this Handbook represent a relatively small number of these gases, the majority of which were chosen because of current interest in the stratospheric ozone layer. Most of the spectra presented are in the 5 to 15 μ region, where most of our recent atmospheric spectral studies were carried out. Many of the molecules of interest also have strong bands at shorter and longer wavelengths which are not yet included here, but which will hopefully be included in future editions of this Handbook.

The present spectra are at least one order of magnitude higher resolution than other comparably extensive compilations of spectra. The development of modern laser spectrometers allows still higher resolution over limited spectral ranges. The resolution achieved in the spectra presented here is close to the limit for which it is practical to present spectra in handbook form. While workers often exchange data on magnetic tape, it has been our experience that a great deal of preliminary work can be accomplished by using a hard copy of the appropriate spectra. It is on the basis of our own use of hard copy spectra that we felt it worthwhile to publish this Handbook.

DESCRIPTION OF THE ATLAS

The spectra presented in this Handbook were obtained for either of two objectives. The first major objective was possible detection and measurement of new constituents in the earth's atmosphere. In most cases, the expected atmospheric absorptions from such a constituent will be quite weak. Therefore, the amount of gas in the absorption cell was adjusted so that the peak absorption in the spectrum was less than 50%. However, in some cases portions of the band may be strong enough that this criterion was exceeded.

The second objective was to obtain data for some of the common minor species (such as CH_4, N_2O, etc.) with large gas amounts in the absorption cell. Such data are needed for identifying absorption features which occur in atmospheric spectra taken over long atmospheric paths, but which are not adequately documented in previous studies. In some cases we have included spectra taken with more than one gas in the optical path for spectral regions with overlapping absorption features.

All of the spectra were taken using Fourier Transform Spectrometer systems. These systems are gradually replacing grating spectrometers for wide spectral regions in most laboratories since they have a number of advantages over grating spectrometers. One of the major advantages is the precise frequency data (linearity as well as absolute frequency calibration) that can be obtained. In many cases presented here, the spectra were calibrated against suitable reference absorption lines to within 0.1 of a resolution element. In all cases the frequencies are accurate to within a resolution element. Some of the spectra were ratioed by a corresponding spectrum of an empty cell to remove envelope trends. Some of these ratioed spectra can be considered as approximate transmittance.

The spectrum for each molecule is plotted on two scales. The first is a condensed scale, typically covering a span of 75 to 300 cm^{-1} in one frame. The second scale is an expanded view covering 20 or 10 cm^{-1} per frame. This was necessary to permit presentation of an entire band, while at the same time preserving details inherent in the high resolution used. If no high resolution structure is apparent, the expanded plots are not included.

Each plot contains a title with the name of the molecule, its chemical formula, an alternate name (if appropriate), the gas pressure, cell length and estimated resolution of the spectrum. With the two different instruments used, resolutions of 0.02 cm^{-1} and 0.06 cm^{-1} are present. All of the spectra were obtained under room conditions. In addition, each condensed scale frame contains a few additional explanatory notes. These notes typically indicate the presence of residual H_2O or CO_2, filter structure, or channel spectra.

The plots are ordered in the List of Molecular Species by formula according to an alphabetical scheme (Herzberg,[2] Rao and Mathews[3]).

A complete cross-reference index, alphabetized by molecule name, is also included. This contains references to alternate names and formula presentations that may be more familiar to the reader.

ACKNOWLEDGMENTS

Acknowledgment is made to the National Center for Atmospheric Research, which is sponsored by the National Science Foundation, for computer time used in this research. The computer programs used for plotting the spectra were written by Darwin Rolens. The figures were carefully plotted and prepared by Barbra Cox.

REFERENCES

1. **Graedel, T. E.,** *Chemical Compounds in the Atmosphere,* Academic Press, New York, 1978.
2. **Herzberg, G.,** *Molecular Spectra and Molecular Structure* II. *Infrared and Raman Spectra of Polyatomic Molecules,* D. Van Nostrand Company, Princeton, New Jersey, 1962, 568.
3. **Rao, K. N. and Mathews, C. W.,** *Molecular Spectroscopy: Modern Research,* Academic Press, New York, 1972, 412.

LIST OF MOLECULAR SPECIES

CROSS INDEX

	Molecule	Plot
Name	Acetylene	
Alternate name	Ethyne	
Index reference	C_2H_2	
Alternate formula	C_2H_2	
Plots	680-840cm^{-1} 8.0 Torr 10cm R = 0.06cm^{-1}	(130)
Name	Ammonia	
Alternate name	Ammonia	
Index reference	NH_3	
Alternate formula	NH_3	
Plots	750-950cm^{-1} 0.5 Torr 10cm R = 0.06cm^{-1}	(290A)
	1070-210cm^{-1} 5.0 Torr 5cm R = 0.06cm^{-1}	(290B)
Name	Carbon disulfide	
Alternate name	Carbon disulfide	
Index reference	CS_2	
Alternate formula	CS_2	
Plots	1460-1600cm^{-1} 0.5 Torr 10cm R = 0.06cm^{-1}	(260)
Name	Carbon oxysulfide	
Alternate name	See Carbonyl sulfide	
Index reference		
Alternate formula		
Plots		
Name	Carbonyldichloride, Carbonyl chloride	
Alternate name	Phosgene, chloroformylchloride	
Index reference	$CC\ell_2O$	
Alternate formula	$COC\ell_2$	
Plots	800-880cm^{-1} 0.8 Torr 10cm R = 0.06cm^{-1}	(70)
Name	Carbonyl chlorofluoride	
Alternate name	Chloroformylfluoride	
Index reference	$CC\ell FO$	
Alternate formula	$COC\ell F$	
Plots	1070-1150cm^{-1} 0.34 Torr 10cm R = 0.06cm^{-1}	(60)
Name	Carbonyl difluoride, Carbonyl fluoride	
Alternate name	Fluoroformylfluoride	
Index reference	CF_2O	
Alternate formula	COF_2	
Plots	740-820cm^{-1} 3.1 Torr 10cm R = 0.06cm^{-1}	(100A)
	1180-1280cm^{-1} 3.1 Torr 10cm R = 0.06cm^{-1}	(100B)

CROSS INDEX (continued)

	Molecule	Plot
Name	Carbonyl sulfide	
Alternate name	Carbon oxysulfide	
Index reference	COS	
Alternate formula	OCS	
Plots	2000-2120cm^{-1} 2.0 Torr 10cm R = 0.06cm^{-1}	(250)
Name	Chlorinenitrate	
Alternate name	Chlorine nitrate	
Index reference	$C\ell ONO_2$	
Alternate formula	$C\ell ONO_2$	
Plots	770-850cm^{-1} 14.0 Torr 5cm R = 0.06cm^{-1}	(230B)
	770-850cm^{-1} 1.9 Torr 10cm R = 0.06cm^{-1}	(230A)
	1255-1355cm^{-1} 14.0 Torr 5cm R = 0.06cm^{-1}	(230C)
Name	Chlorine nitrate + nitric acid	
Alternate name	Chlorine nitrate + nitric acid	
Index reference	$C\ell ONO_3$, HNO_3	
Alternate formula	$C\ell ONO_3$, HNO_3	
Plots	1260-1340cm^{-1} 1.2 Torr 5cm R = 0.06cm^{-1}	(240)
Name	Chlorodifluoromethane	
Alternate name	F-22	
Index reference	$CHC\ell F_2$	
Alternate formula	$CHF_2C\ell$	
Plots	770-850cm^{-1} 3.0 Torr 5cm R = 0.06cm^{-1}	(200)
Name	Chloroformylchloride	
Alternate name	See Carbonyl dichloride	
Index reference		
Alternate formula		
Plots		
Name	Chloroformylfluoride	
Alternate name	See Chlorofluoride	
Index reference		
Alternate formula		
Plots		
Name	Chloromethane	
Alternate name	See Methyl chloride	
Index reference		
Alternate formula		
Plots		

CROSS INDEX (continued)

	Molecule	Plot
Name	Chlorotrifluoromethane	
Alternate name	F-13	
Index reference	$CC\ell F_3$	
Alternate formula	$CF_3C\ell$	
Plots	1070-1250cm^{-1} 0.4 Torr 10cm R = 0.06cm^{-1}	(10)
Name	Dichlorodifluoromethane	
Alternate name	F-12	
Index reference	$CC\ell_2F_2$	
Alternate formula	$CF_2C\ell_2$	
Plots	840-940cm^{-1} 4.1 Torr 5cm R = 0.06cm^{-1}	(20A)
	1070-1250cm^{-1} 0.4 Torr 10cm R = 0.06cm^{-1}	(20B)
	1080-1180cm^{-1} 4.1 Torr 5cm R = 0.06cm^{-1}	(20C)
Name	1,2-Dichloroethane	
Alternate name	Ethylene chloride	
Index reference	$C_2H_4C\ell_2$	
Alternate formula	$C_2H_4C\ell_2$	
Plots	700-760cm^{-1} 3.5 Torr 10cm R = 0.06cm^{-1}	(190A)
	1200-1300cm^{-1} 3.5 Torr 10cm R = 0.06cm^{-1}	(190B)
Name	Dichlorofluoromethane	
Alternate name	F-21	
Index reference	$CHC\ell_2F$	
Alternate formula	$CHFC\ell_2$	
Plots	700-860cm^{-1} 5.0 Torr 5cm R = 0.06cm^{-1}	(210A)
	1060-1300cm^{-1} 5.0 Torr 5cm R = 0.06cm^{-1}	(210B)
Name	Dichloromethane	
Alternate name	Methylene chloride	
Index reference	$CH_2C\ell_2$	
Alternate formula	$CH_2C\ell_2$	
Plots	720-800cm^{-1} 5.9 Torr 5cm R = 0.06cm^{-1}	(150A)
	1240-1300cm^{-1} 5.9 Torr 5cm R = 0.06cm^{-1}	(150B)
Name	1,2-Dichlorotetrafluoroethane	
Alternate name	F-114	
Index reference	$C_2C\ell_2F_4$	
Alternate formula	$C_2F_4C\ell_2$	
Plots	810-950cm^{-1} 0.9 Torr 10cm R = 0.06cm^{-1}	(40A)
	1080-1300cm^{-1} 0.9 Torr 10cm R = 0.06cm^{-1}	(40B)
Name	Dinitrogen pentoxide	
Alternate name	Nitrogen pentoxide	
Index reference	N_2O_5	
Alternate formula	N_2O_5	
Plots	1220-1270cm^{-1} 5.0 Torr 5cm R = 0.06cm^{-1}	(320)

CROSS INDEX (continued)

	Molecule	Plot
Name	Ethane	
Alternate name	Ethane	
Index reference	C_2H_6	
Alternate formula	C_2H_6	
Plots	780-880cm^{-1} 3.9 Torr 10cm R = 0.06cm^{-1}	(140A)
	2840-3100cm^{-1} 3.9 Torr 10cm R = 0.06cm^{-1}	(140B)
Name	Ethylene chloride	
Alternate name	See 1,2-Dichloroethane	
Index reference		
Alternate formula		
Plots		
Name	Ethylene trichloride	
Alternate name	See Trichloroethylene	
Index reference		
Alternate formula		
Plots		
Name	Ethyne	
Alternate name	See Acetylene	
Index reference		
Alternate formula		
Plots		
Name	Fluoroformylfluoride	
Alternate name	See Carbonyl difluoride	
Index reference		
Alternate formula		
Plots		
Name	Formic acid	
Alternate name	Methanoic acid	
Index reference	CH_2O_2	
Alternate formula	HCOOH	
Plots	1070-1170cm^{-1} 1.6 Torr 10cm R = 0.06cm^{-1}	(220)
Name	F-11	
Alternate name	See Trichlorofluoromethane	
Index reference		
Alternate formula		
Plots		

CROSS INDEX (continued)

	Molecule	Plot
Name	F-113	
Alternate name	See 1,1,2-Trichlorotrifluoroethane	
Index reference		
Alternate formula		
Plots		
Name	F-114	
Alternate name	See 1,2-Dichlorotetrafluoroethane	
Index reference		
Alternate formula		
Plots		
Name	F-12	
Alternate name	See Dichlorodifluoromethane	
Index reference		
Alternate formula		
Plots		
Name	F-13	
Alternate name	See Chlorotrifluoromethane	
Index reference		
Alternate formula		
Plots		
Name	F-14	
Alternate name	See Tetrafluoromethane	
Index reference		
Alternate formula		
Plots		
Name	F-21	
Alternate name	See Dichlorofluoromethane	
Index reference		
Alternate formula		
Plots		
Name	F-22	
Alternate name	See Chlorodifluoromethane	
Index reference		
Alternate formula		
Plots		
Name	Hydrogen nitrate	
Alternate name	See Nitric acid	
Index reference		
Alternate formula		
Plots		

CROSS INDEX (continued)

	Molecule	Plot
Name	Hydrogen peroxide	
Alternate name	Hydrogen peroxide	
Index reference	H_2O_2	
Alternate formula	H_2O_2	
Plots	1210-1310cm^{-1} 4.0 Torr 10cm R = 0.06cm^{-1}	(280A)
	1200-1340cm^{-1} 2.5 Torr 45cm R = 0.02cm^{-1}	(280B)
Name	Methane	
Alternate name	Methane	
Index reference	CH_4	
Alternate formula	CH_4	
Plots	1200-1400cm^{-1} 10.0 Torr 45cm R = 0.02cm^{-1}	(110B)
	1200-1400cm^{-1} 80.0 Torr 45cm R = 0.02cm^{-1}	(110C)
	1400-1600cm^{-1} 80.0 Torr 45cm R = 0.02cm^{-1}	(110D)
	1600-1800cm^{-1} 80.0 Torr 45cm R = 0.02cm^{-1}	(110E)
Name	Methane	
Alternate name	Methane	
Index reference	CH_4	
Alternate formula	CH_4	
Plots	1210-1410cm^{-1} 20.0 Torr 10cm R = 0.06cm^{-1}	(110A)
Name	Methane + dinitrogen oxide	
Alternate name	Methane + nitrous oxide	
Index reference	CH_4, N_2O	
Alternate formula	CH_4, N_2O	
Plots	1120-1300cm^{-1} 76.0 Torr 5cm R = 0.06cm^{-1}	(120)
Name	Methanoic acid	
Alternate name	See Formic acid	
Index reference		
Alternate formula		
Plots		
Name	Methyl chloride	
Alternate name	Chloromethane	
Index reference	$CH_3C\ell$	
Alternate formula	$CH_3C\ell$	
Plots	670-770cm^{-1} 72.0 Torr 5cm R = 0.06cm^{-1}	(160A)
	950-1110cm^{-1} 72.0 Torr 5cm R = 0.06cm^{-1}	(160B)
Name	Methyl chloroform	
Alternate name	See 1,1,1-Trichloroethane	
Index reference		
Alternate formula		
Plots		

CROSS INDEX (continued)

	Molecule	Plot
Name	Methylene chloride	
Alternate name	See Dichloromethane	
Index reference		
Alternate formula		
Plots		
Name	Nitric acid	
Alternate name	Hydrogen nitrate	
Index reference	HNO_3	
Alternate formula	HNO_3	
Plots	1270-1370cm^{-1} 0.42 Torr 45cm R = 0.02cm^{-1}	(270D)
	1670-1750cm^{-1} 0.42 Torr 45cm R = 0.02cm^{-1}	(270E)
Name	Nitric acid	
Alternate name	Hydrogen nitrate	
Index reference	HNO_3	
Alternate formula	HNO_3	
Plots	760-920cm^{-1} 1.57 Torr 10cm R = 0.06cm^{-1}	(270A)
	840-940cm^{-1} 2.2 Torr 5cm R = 0.06cm^{-1}	(270B)
	1275-1355cm^{-1} 2.1 Torr 10cm R = 0.06cm^{-1}	(270C)
Name	Nitric oxide	
Alternate name	Mononitrogen monoxide	
Index reference	NO	
Alternate formula	NO	
Plots	1800-1960cm^{-1} 36.0 Torr 5cm R = 0.06cm^{-1}	(300)
Name	Nitrogen dioxide	
Alternate name	Nitrogen dioxide	
Index reference	NO_2	
Alternate formula	NO_2	
Plots	1500-1700cm^{-1} 2.0 Torr 45cm R = 0.02cm^{-1}	(310)
Name	Nitrogen pentoxide	
Alternate name	See Dinitrogen pentoxide	
Index reference		
Alternate formula		
Plots		
Name	Phosgene	
Alternate name	See Carbonyl chloride	
Index reference		
Alternate formula		
Plots		

CROSS INDEX (continued)

	Molecule	Plot
Name	Tetrafluoromethane	
Alternate name	F-14	
Index reference	CF_4	
Alternate formula	CF_4	
Plots	1200-1300cm^{-1} 0.075 Torr 2cm R = 0.06cm^{-1}	(80A)
	1250-1310cm^{-1} 0.04 Torr 5cm R = 0.02cm^{-1}	(80B)
Name	Tetrafluoromethane + methane + nitrous oxide	
Alternate name	F-14 + methane + nitrous oxide	
Index reference	CF_4, CH_4, N_2O	
Alternate formula	CF_4, CH_4, N_2O	
Plots	1050-1350cm^{-1} 0.04, 75 Torr 2.5cm R = 0.02cm^{-1}	(90)
Name	1,1,1-Trichloroethane	
Alternate name	Methyl chloroform	
Index reference	$C_2H_3C\ell_3$	
Alternate formula	$C_2H_3C\ell_3$	
Plots	700-760cm^{-1} 1.1 Torr 10cm R = 0.06cm^{-1}	(180)
Name	Trichloroethylene	
Alternate name	Ethylene trichloride	
Index reference	$C_2HC\ell_3$	
Alternate formula	$C_2HC\ell_3$	
Plots	740-900cm^{-1} 2.8 Torr 10cm R = 0.06cm^{-1}	(170)
Name	Trichlorofluoromethane	
Alternate name	F-11	
Index reference	$CC\ell_3F$	
Alternate formula	$CFC\ell_3$	
Plots	810-870cm^{-1} 0.07 Torr 10cm R = 0.06cm^{-1}	(30A)
	810-890cm^{-1} 2.6 Torr 5cm R = 0.06cm^{-1}	(30B)
	1060-1160cm^{-1} 2.6 Torr 5cm R = 0.06cm^{-1}	(30C)
Name	1,1,2-Trichlorotrifluoroethane	
Alternate name	F-113	
Index reference	$C_2C\ell_3F_3$	
Alternate formula	$C_2F_3C\ell_3$	
Plots	760-940cm^{-1} 2.9 Torr 10cm R = 0.06cm^{-1}	(50A)
	1090-1250cm^{-1} 2.9 Torr 10cm R = 0.06cm^{-1}	(50B)

Spectra

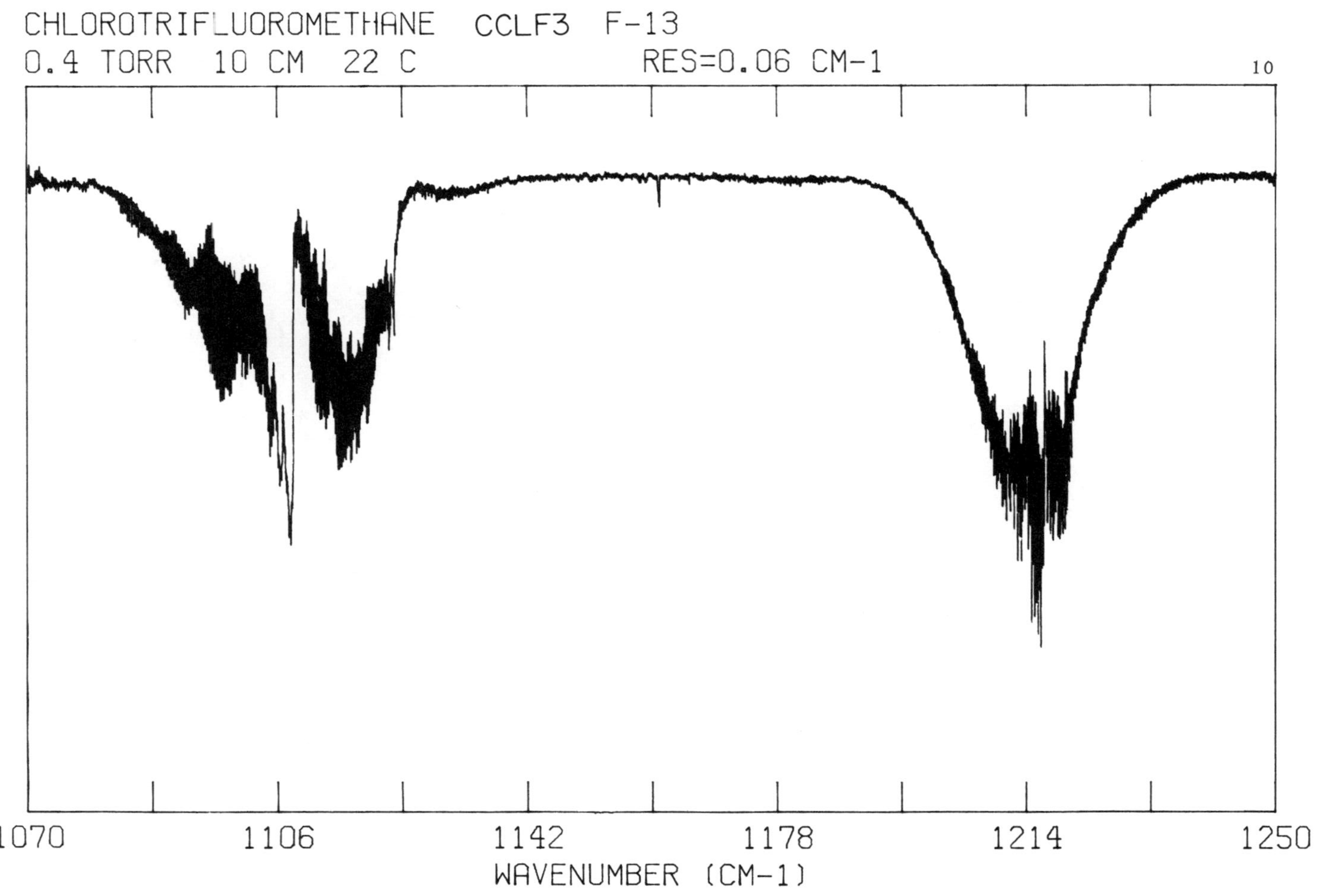
CHLOROTRIFLUOROMETHANE CCLF3 F-13
0.4 TORR 10 CM 22 C RES=0.06 CM-1
10
1070
1106
1142
1178
1214
1250
WAVENUMBER (CM-1)

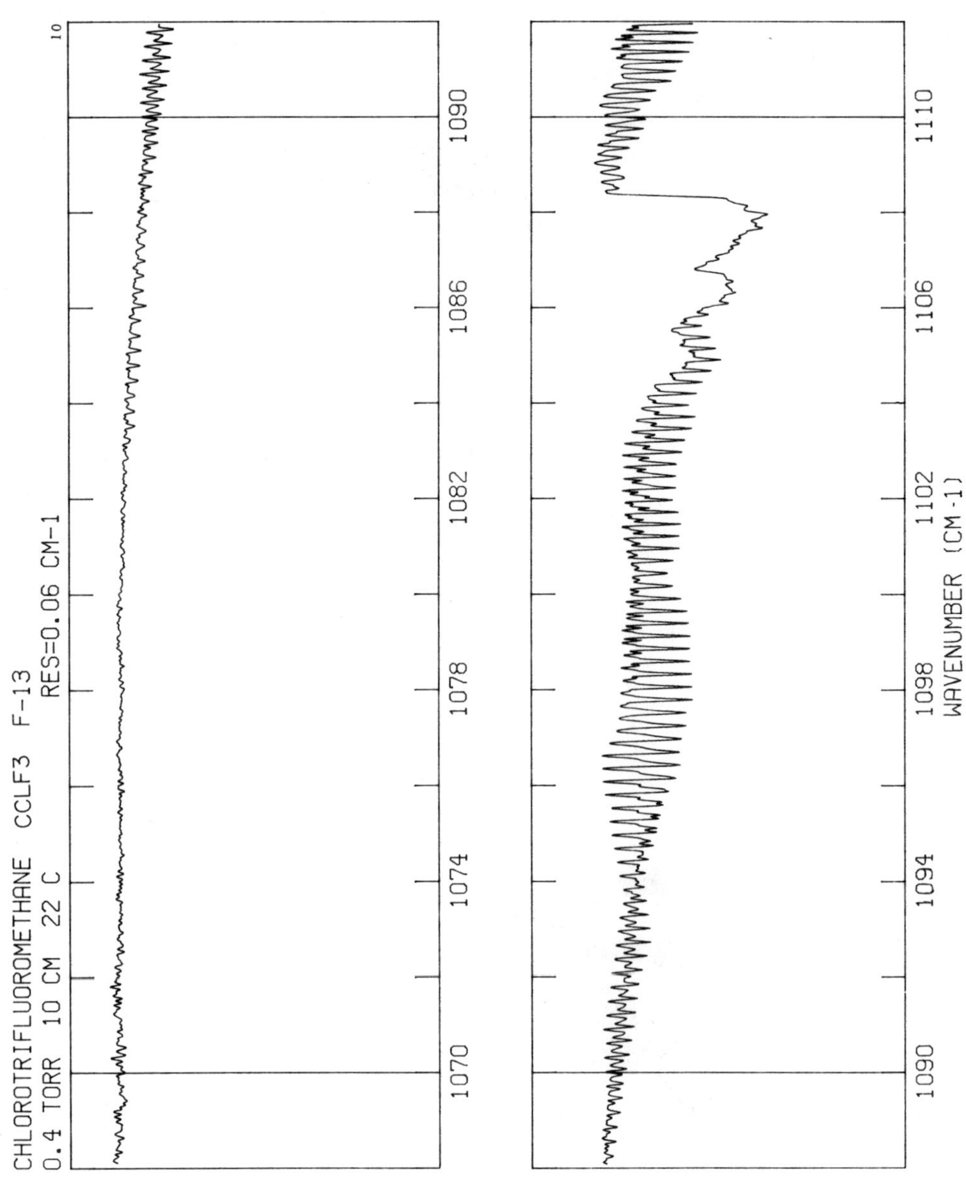
CHLOROTRIFLUOROMETHANE CCLF3 F-13
0.4 TORR 10 CM 22 C RES=0.06 CM-1
10
1070
1074
1078
1082
1086
1090
1090
1094
1098
1102
1106
1110
WAVENUMBER (CM-1)

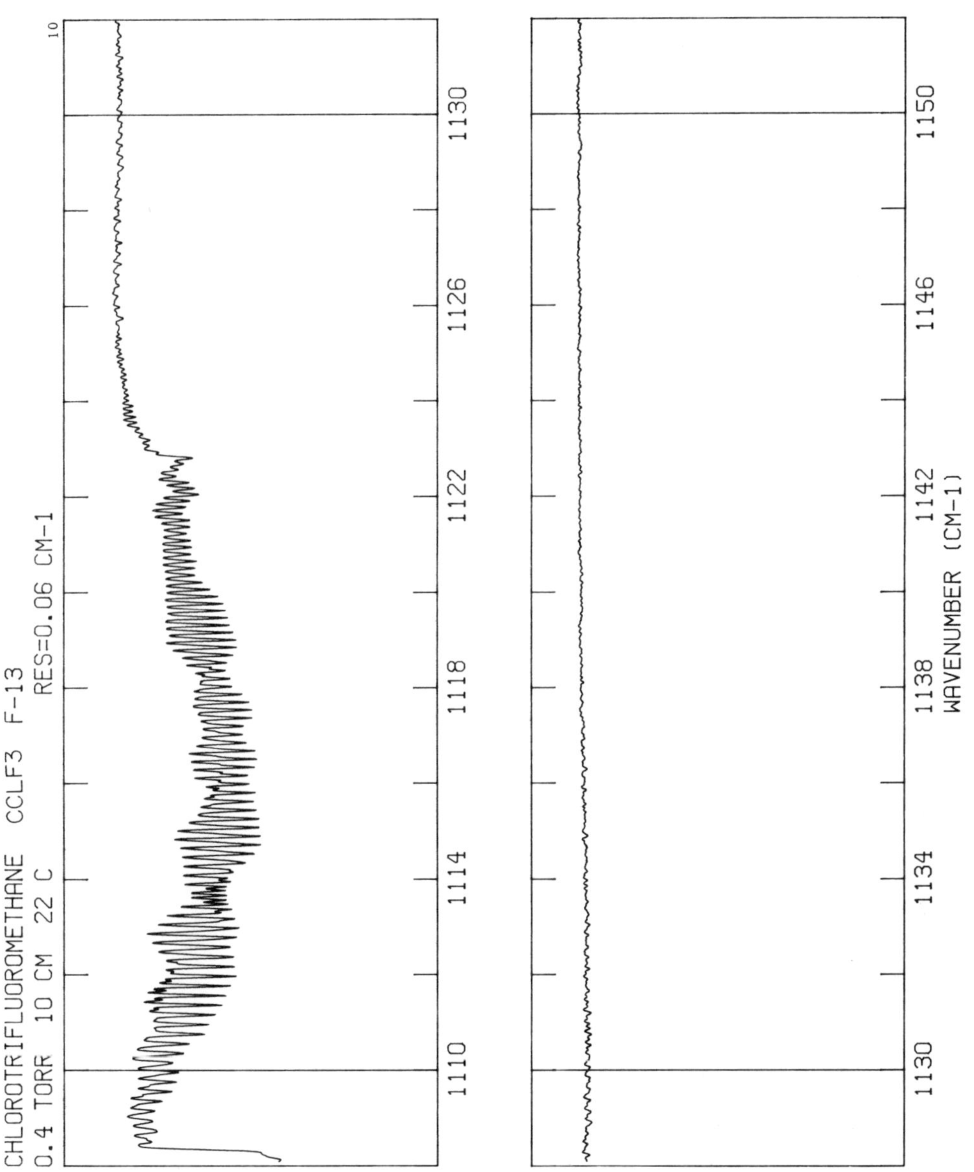
CHLOROTRIFLUOROMETHANE CCLF3 F-13
0.4 TORR 10 CM 22 C RES=0.06 CM-1
10
1110
1114
1118
1122
1126
1130
1134
1138
1142
1146
1150
WAVENUMBER (CM-1)

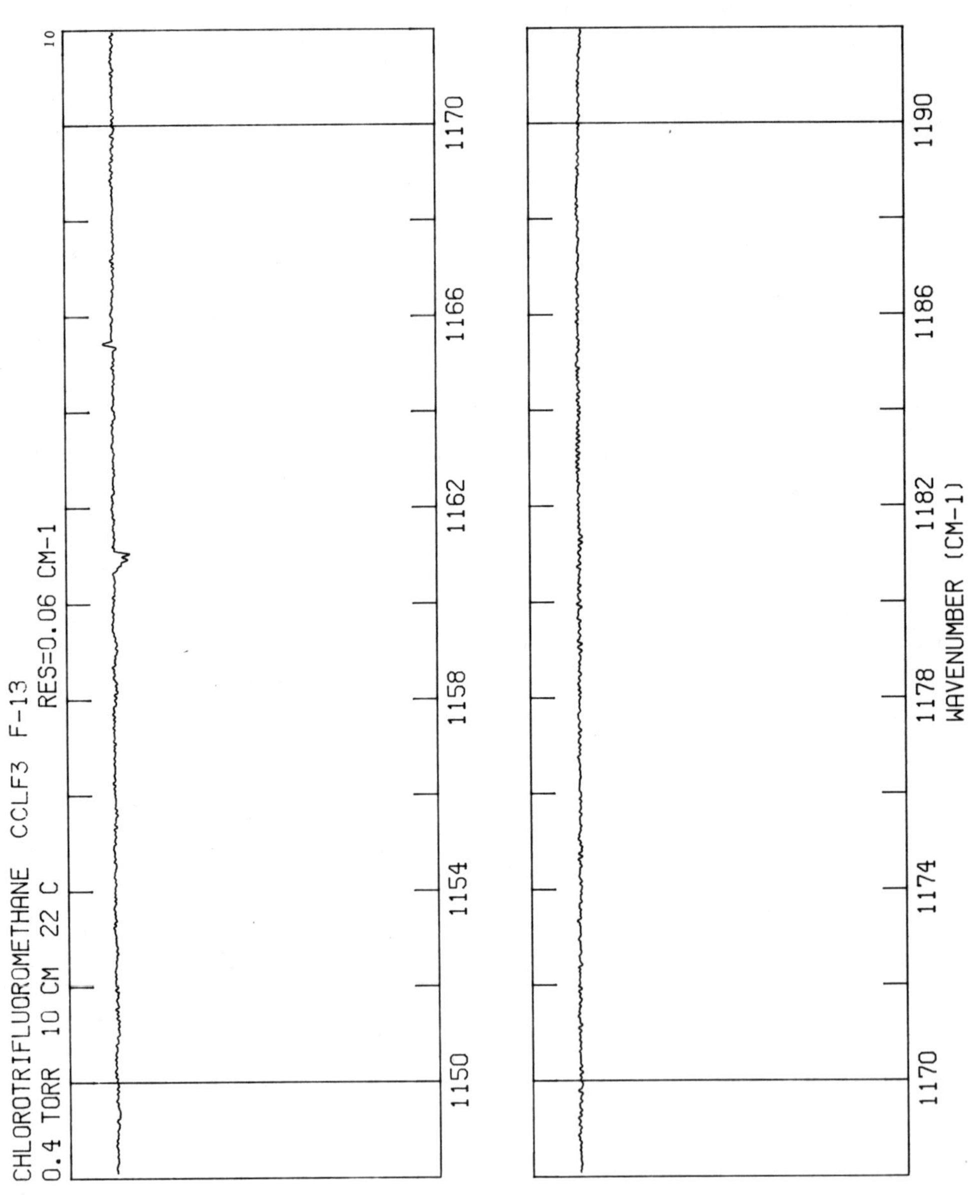
CHLOROTRIFLUOROMETHANE CCLF3 F-13
0.4 TORR 10 CM 22 C RES=0.06 CM-1
10
1150
1154
1158
1162
1166
1170
1174
1178
1182
1186
1190
WAVENUMBER (CM-1)

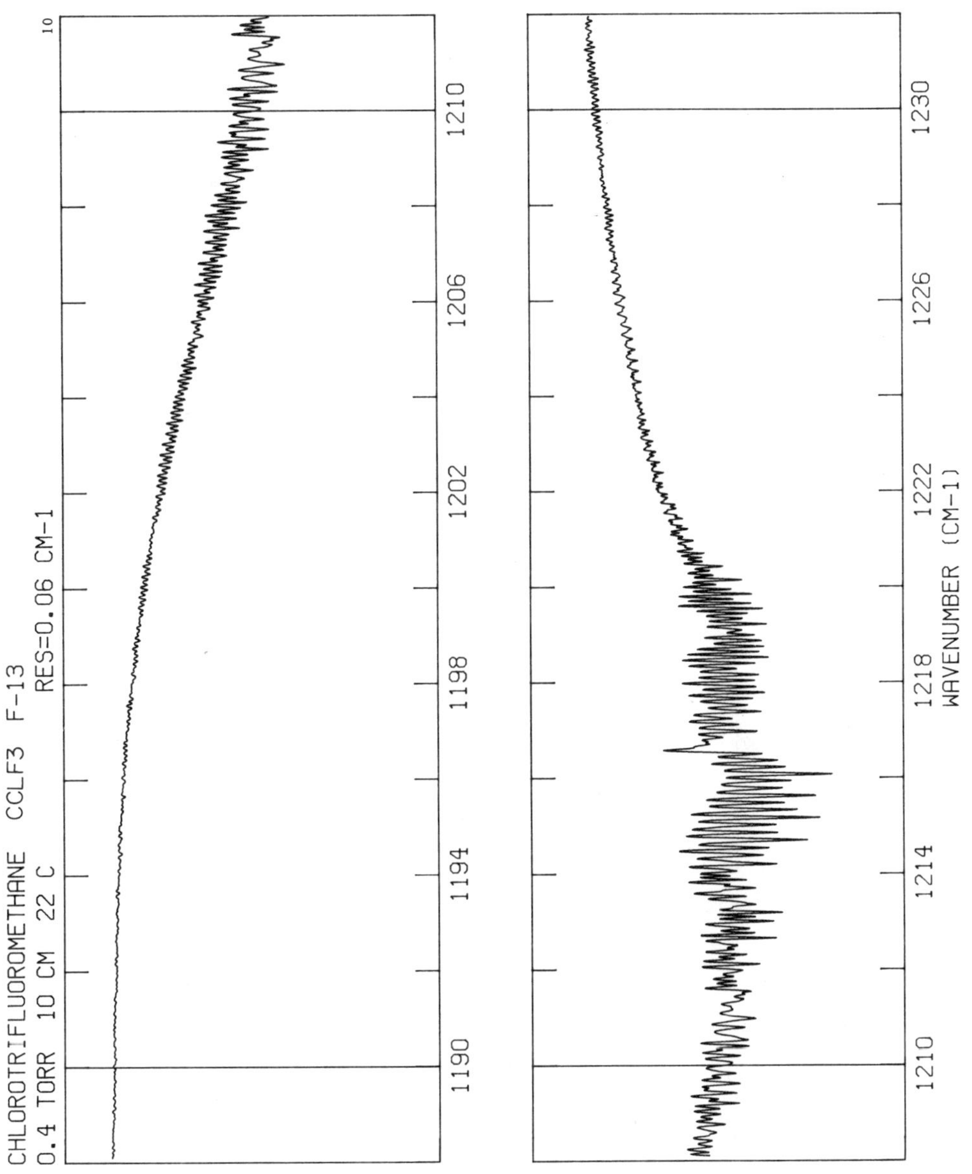

CHLOROTRIFLUOROMETHANE CCLF3 F-13
0.4 TORR 10 CM 22 C RES=0.06 CM-1
10
1190
1194
1198
1202
1206
1210
1214
1218
1222
1226
1230
WAVENUMBER (CM-1)

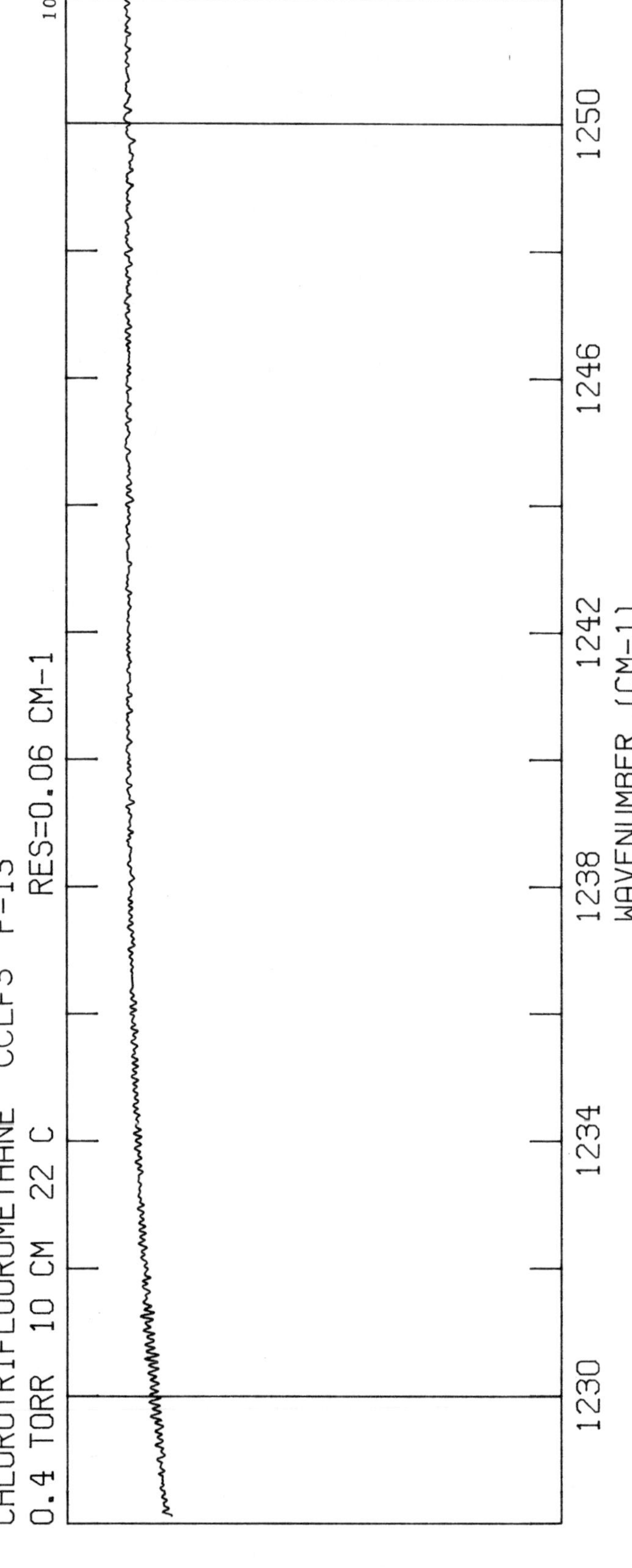
CHLOROTRIFLUOROMETHANE CCLF3 F-13
0.4 TORR 10 CM 22 C RES=0.06 CM-1
10
1230
1234
1238
1242
1246
1250
WAVENUMBER (CM-1)

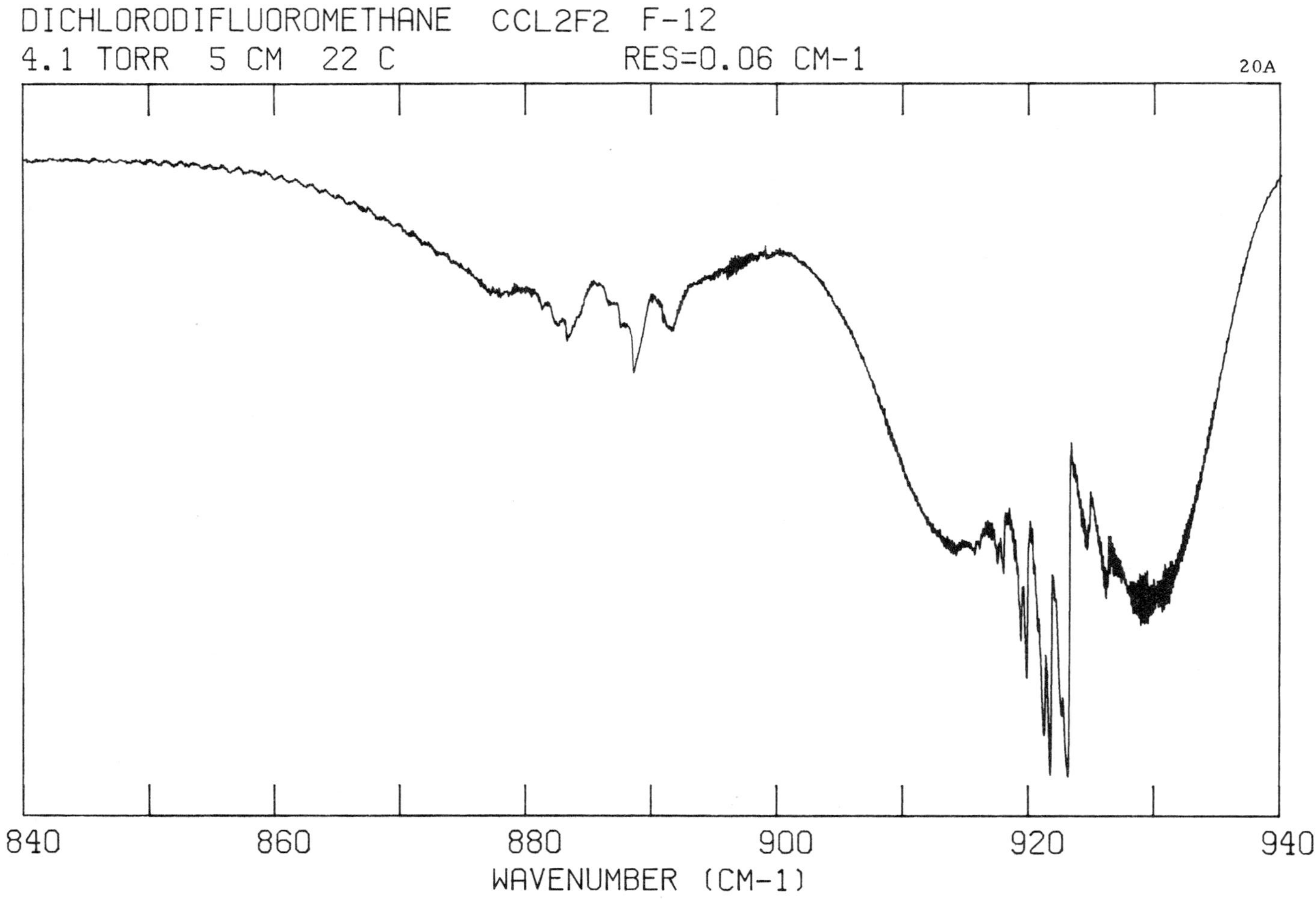
DICHLORODIFLUOROMETHANE CCL2F2 F-12
4.1 TORR 5 CM 22 C RES=0.06 CM-1
20A
840
860
880
900
920
940
WAVENUMBER (CM-1)

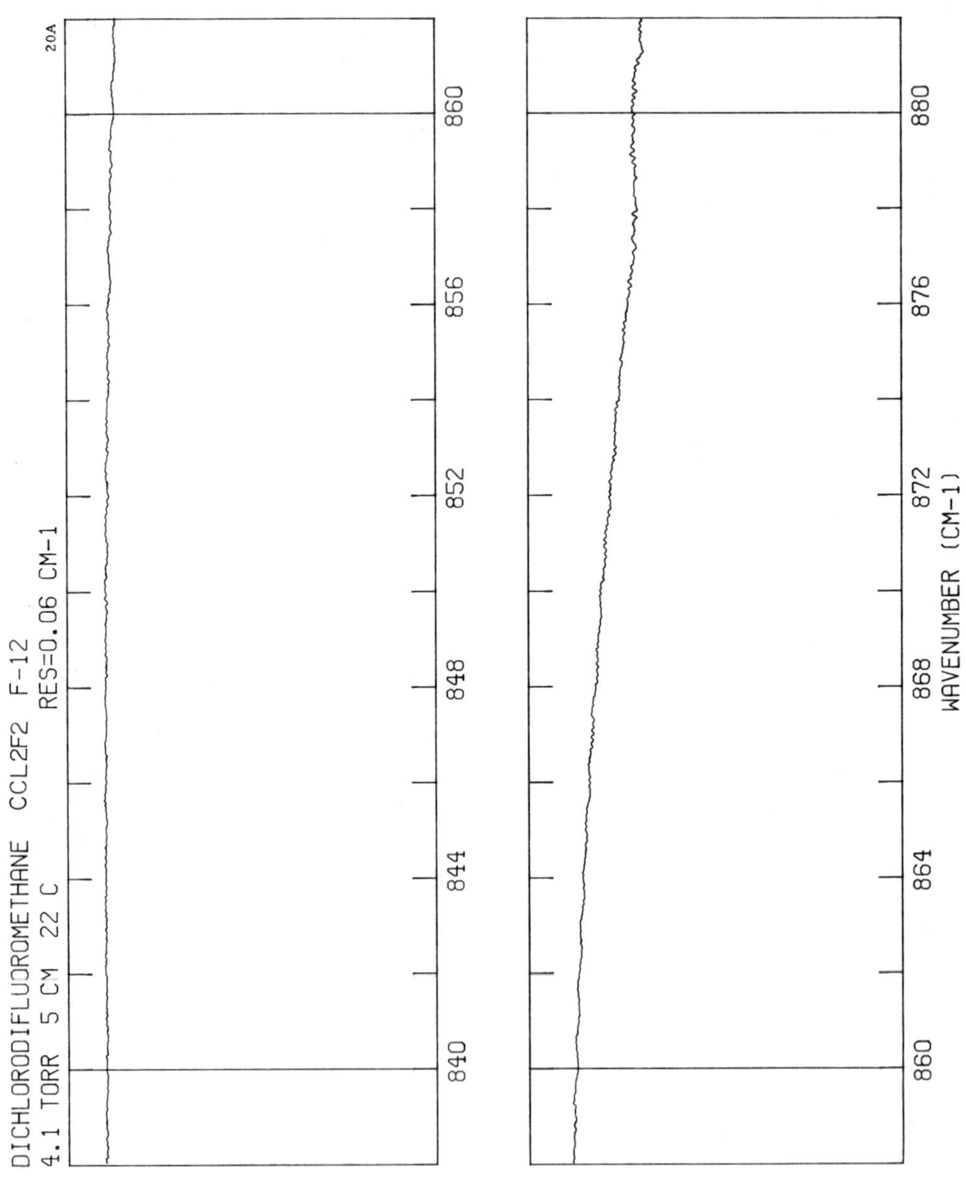
DICHLORODIFLUOROMETHANE CCL2F2 F-12
4.1 TORR 5 CM 22 C RES=0.06 CM-1
20A
840
844
848
852
856
860
860
864
868
872
876
880
WAVENUMBER (CM-1)

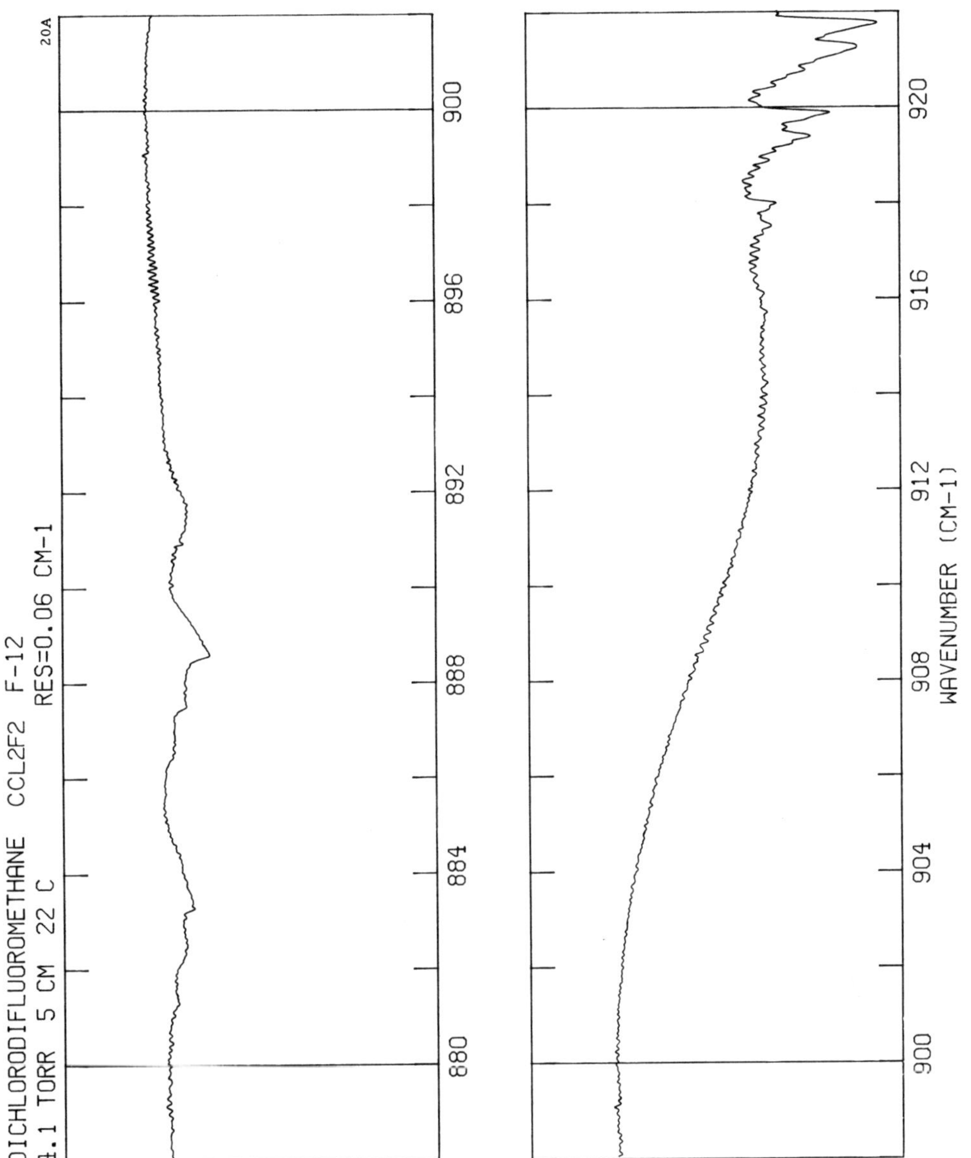
DICHLORODIFLUOROMETHANE CCL2F2 F-12
4.1 TORR 5 CM 22 C RES=0.06 CM-1
20A
880
884
888
892
896
900
900
904
908
912
916
920
WAVENUMBER (CM-1)

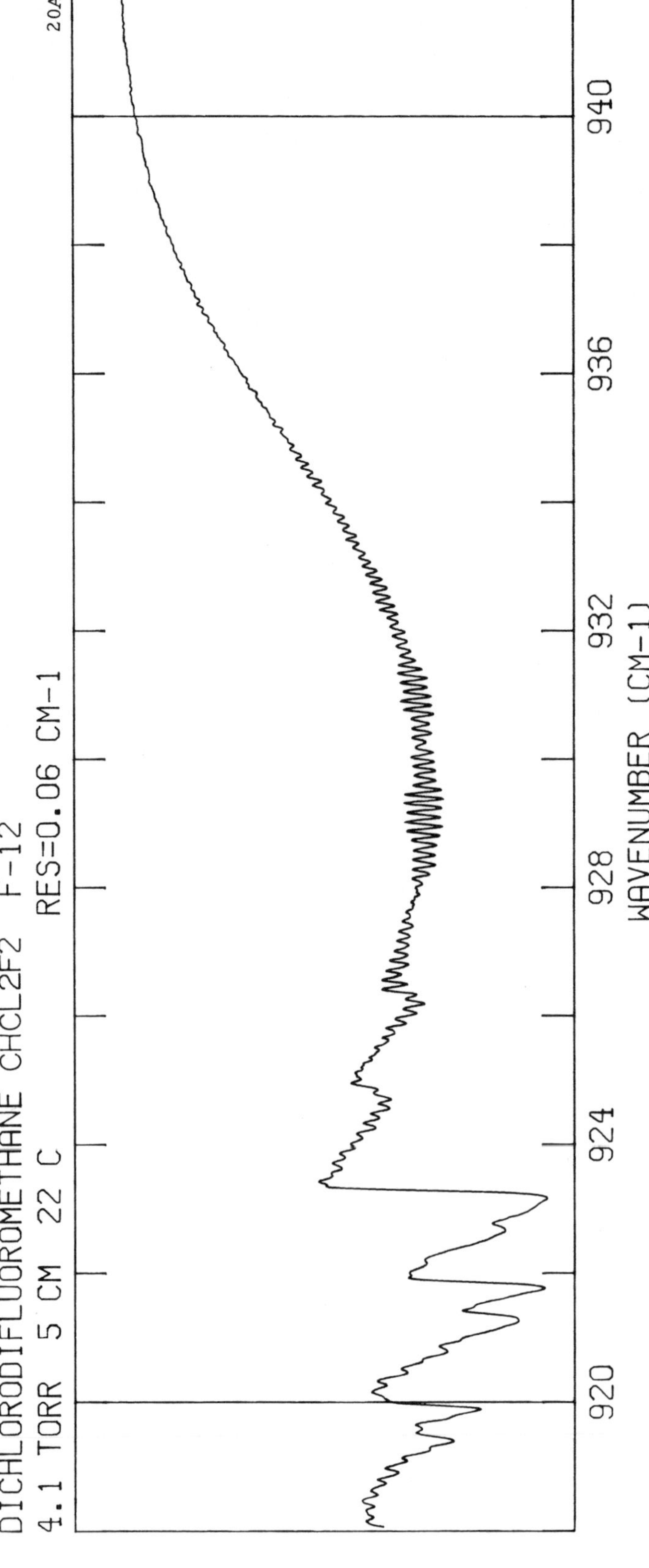
DICHLORODIFLUOROMETHANE CHCL2F2 F-12
4.1 TORR 5 CM 22 C
RES=0.06 CM-1
20A
920
924
928
932
936
940
WAVENUMBER (CM-1)

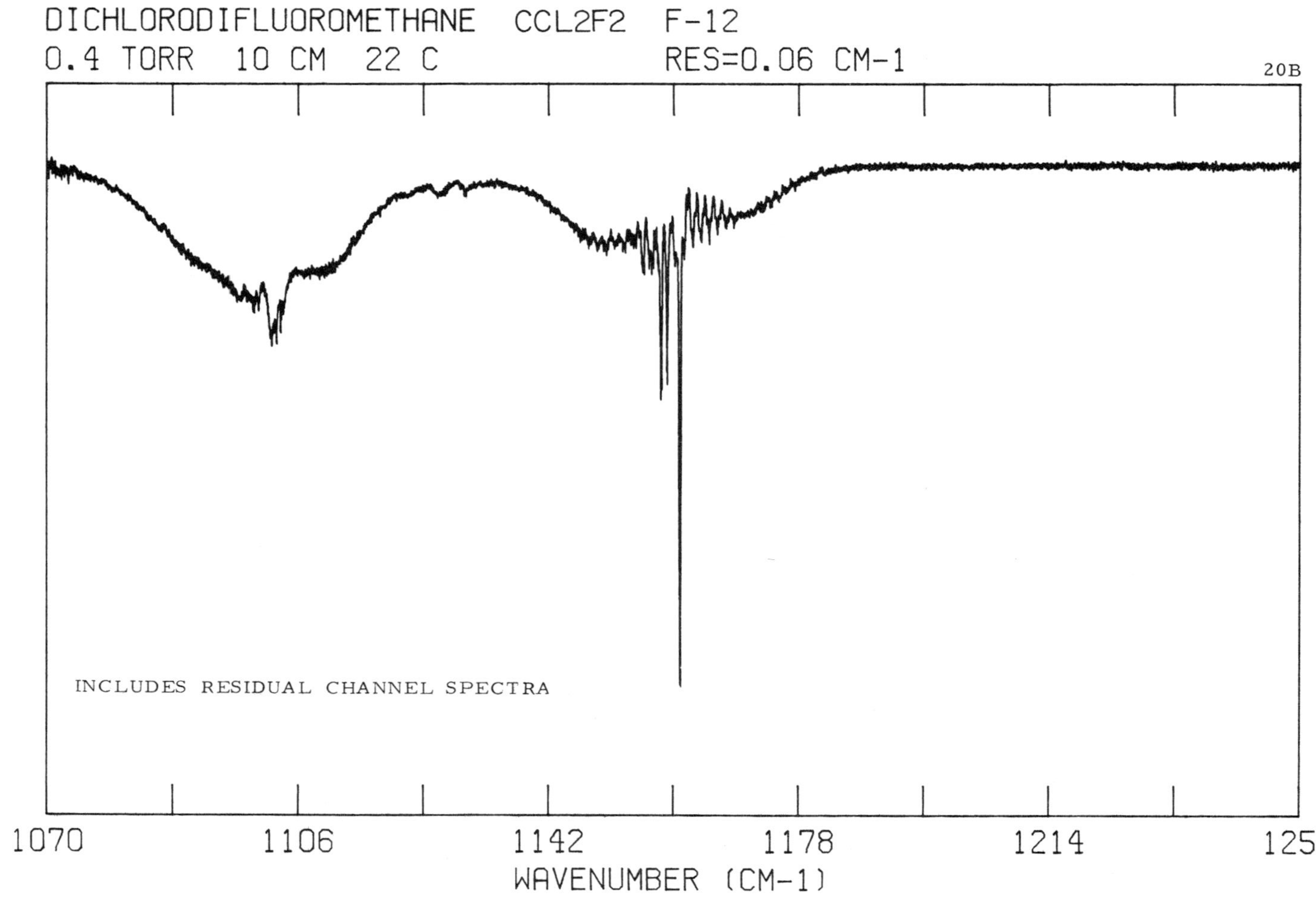

DICHLORODIFLUOROMETHANE CCL2F2 F-12
0.4 TORR 10 CM 22 C RES=0.06 CM-1
20B
INCLUDES RESIDUAL CHANNEL SPECTRA
1070
1106
1142
1178
1214
1250
WAVENUMBER (CM-1)

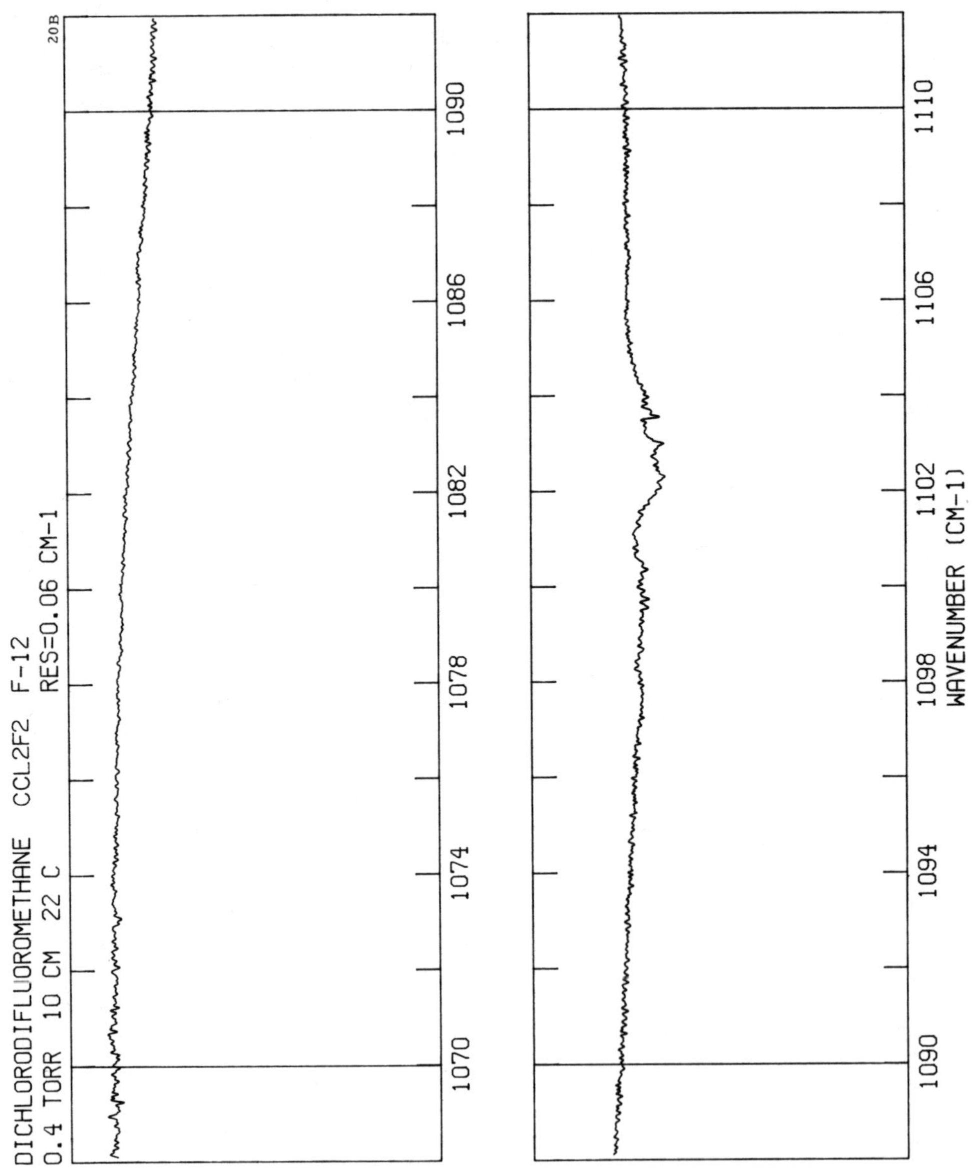
20B
DICHLORODIFLUOROMETHANE CCL2F2 F-12
0.4 TORR 10 CM 22 C RES=0.06 CM-1
1070
1074
1078
1082
1086
1090
1090
1094
1098
1102
1106
1110
WAVENUMBER (CM-1)

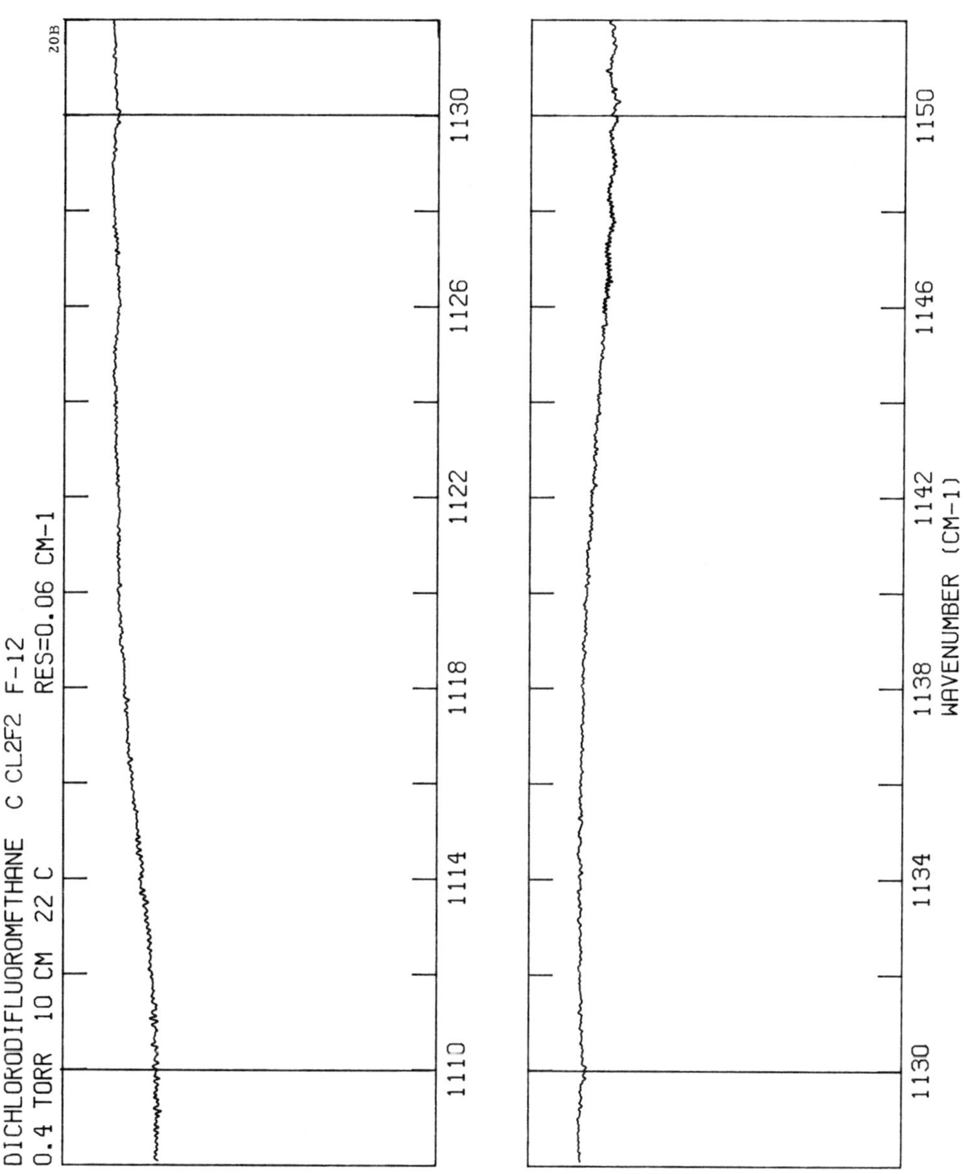
20B
DICHLORODIFLUOROMETHANE C CL2F2 F-12
0.4 TORR 10 CM 22 C RES=0.06 CM-1
1110
1114
1118
1122
1126
1130
1134
1138
1142
1146
1150
WAVENUMBER (CM-1)

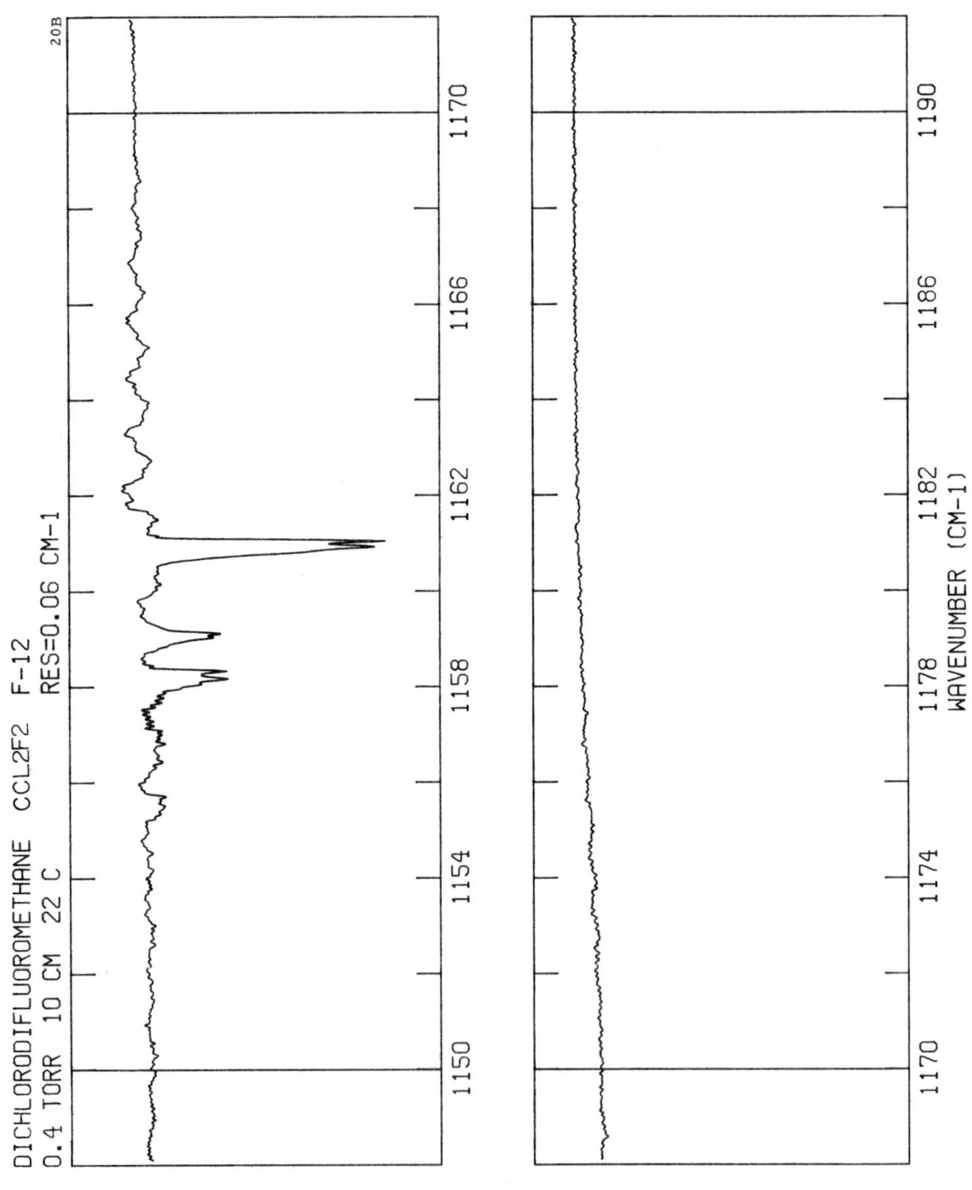
DICHLORODIFLUOROMETHANE CCL2F2 F-12
0.4 TORR 10 CM 22 C RES=0.06 CM-1
20B
1150
1154
1158
1162
1166
1170
1170
1174
1178
1182
1186
1190
WAVENUMBER (CM-1)

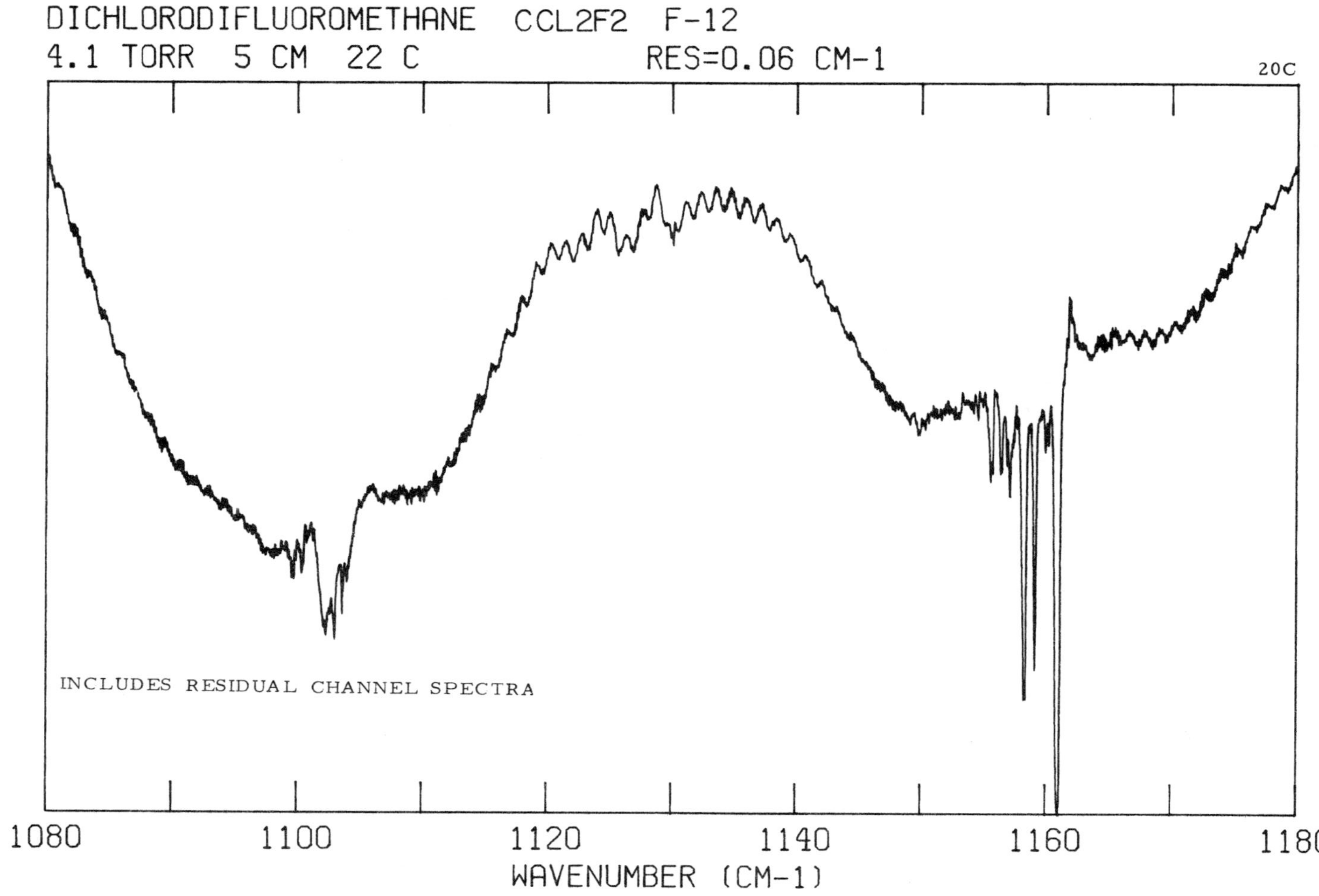
DICHLORODIFLUOROMETHANE CCL2F2 F-12
4.1 TORR 5 CM 22 C RES=0.06 CM-1
20C
INCLUDES RESIDUAL CHANNEL SPECTRA
1080
1100
1120
1140
1160
1180
WAVENUMBER (CM-1)

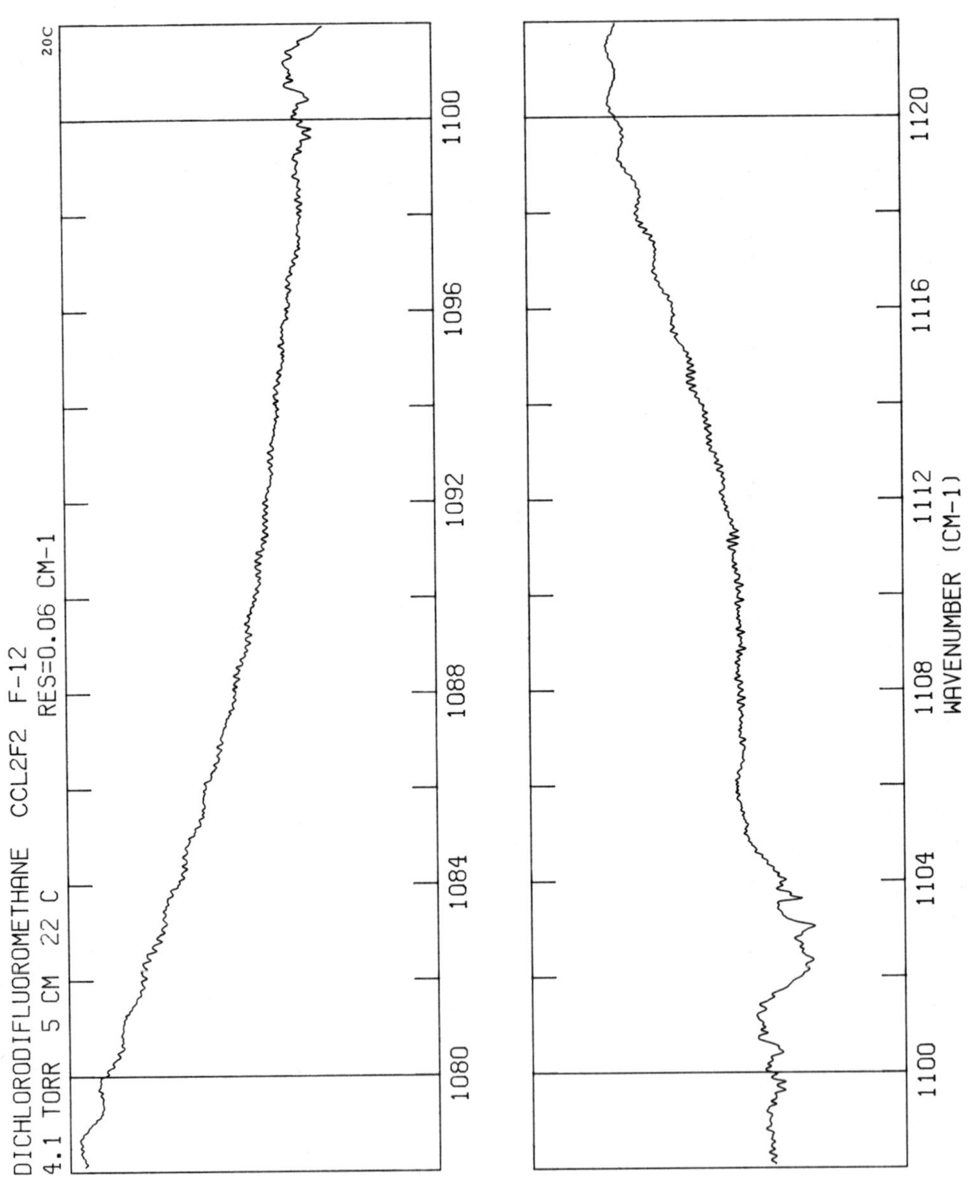
DICHLORODIFLUOROMETHANE CCL2F2 F-12
4.1 TORR 5 CM 22 C RES=0.06 CM-1
20C
1080
1084
1088
1092
1096
1100
1100
1104
1108
1112
1116
1120
WAVENUMBER (CM-1)

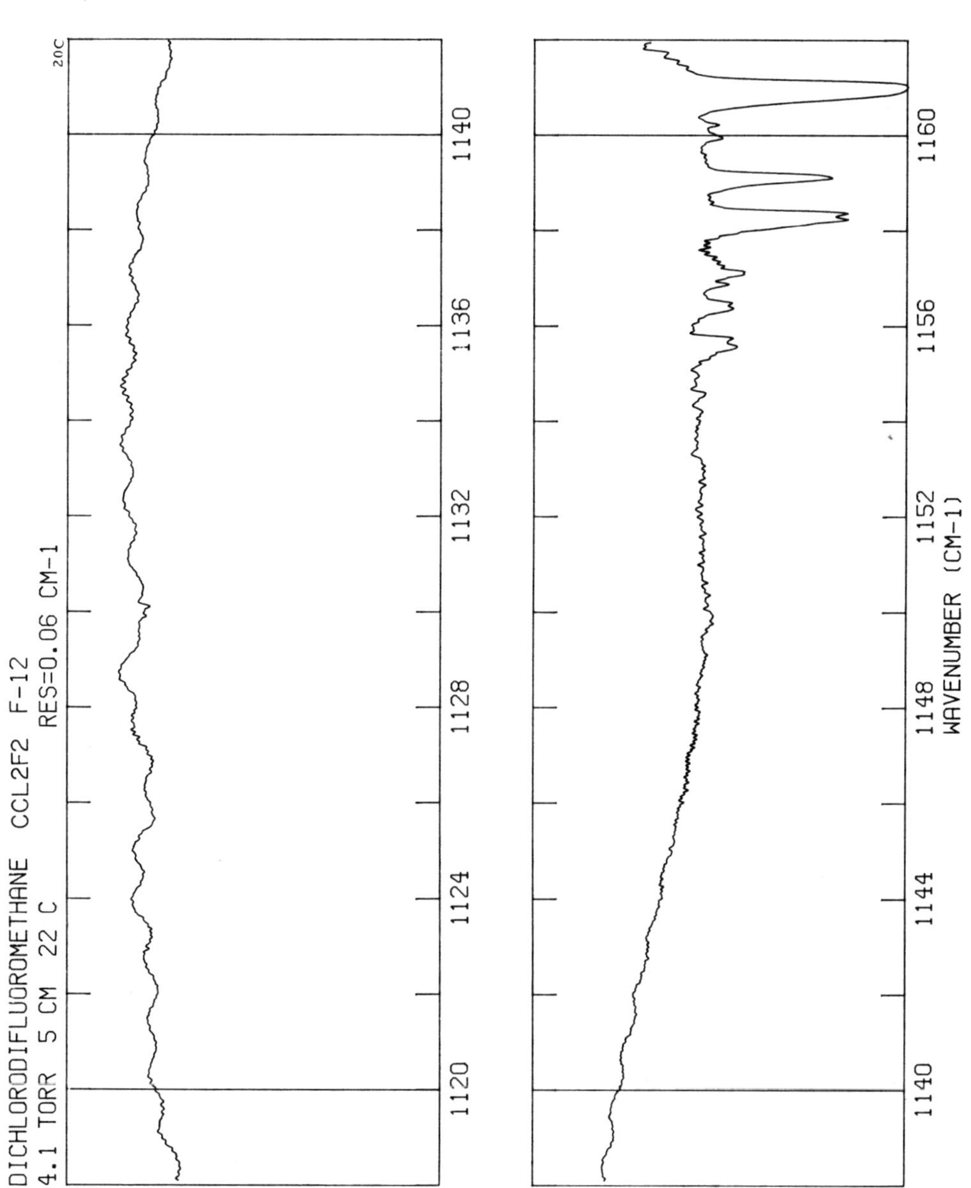
DICHLORODIFLUOROMETHANE CCL2F2 F-12
4.1 TORR 5 CM 22 C RES=0.06 CM-1
20C
1120
1124
1128
1132
1136
1140
1140
1144
1148
1152
1156
1160
WAVENUMBER (CM-1)

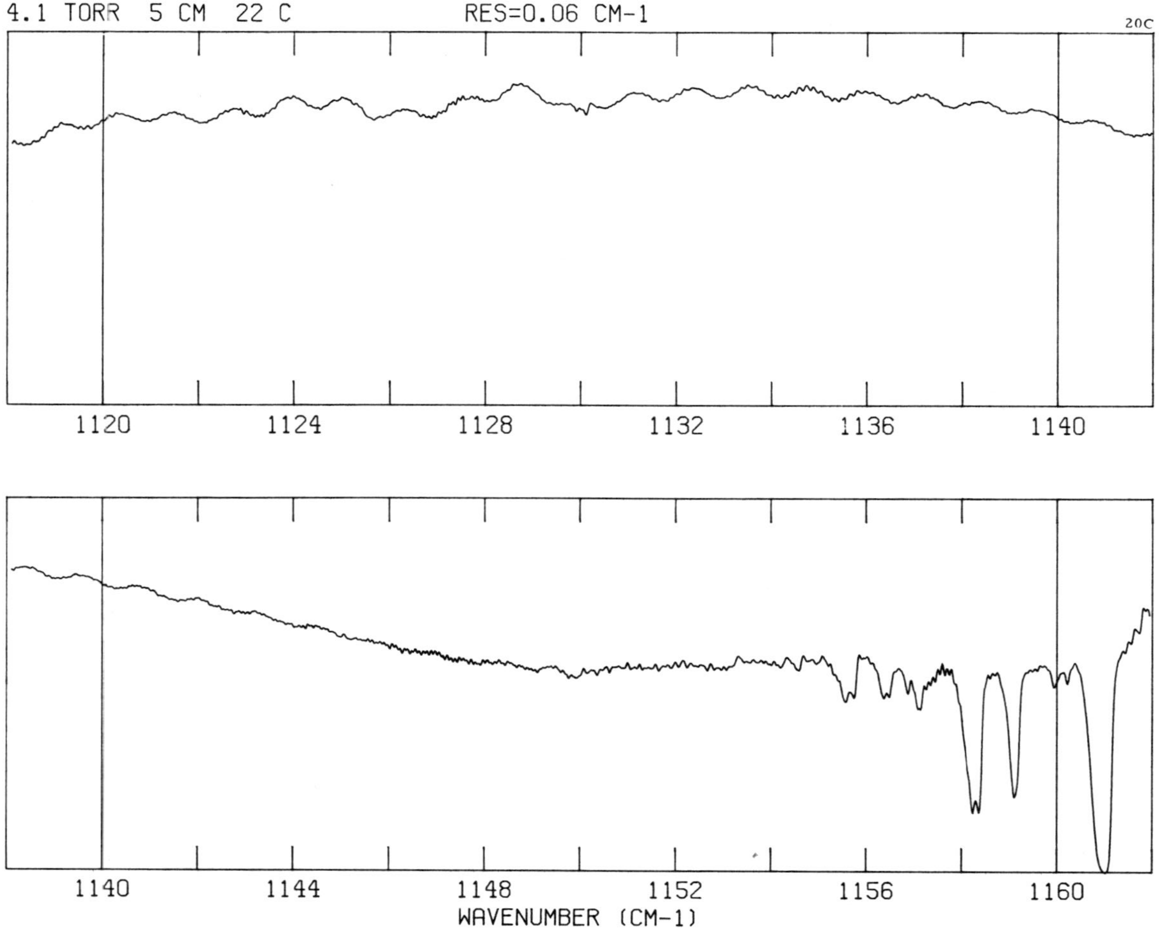

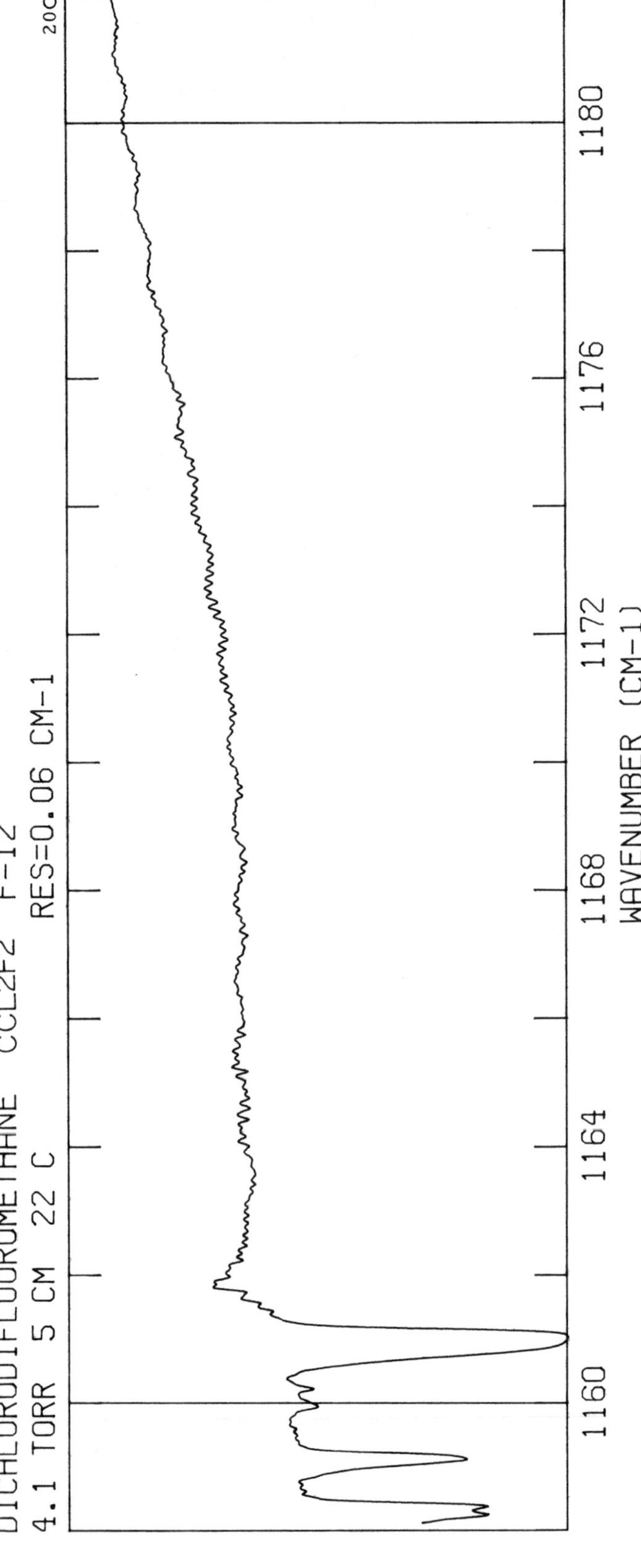
DICHLORODIFLUOROMETHANE CCL2F2 F-12
4.1 TORR 5 CM 22 C RES=0.06 CM-1
20C
1160
1164
1168
1172
1176
1180
WAVENUMBER (CM-1)

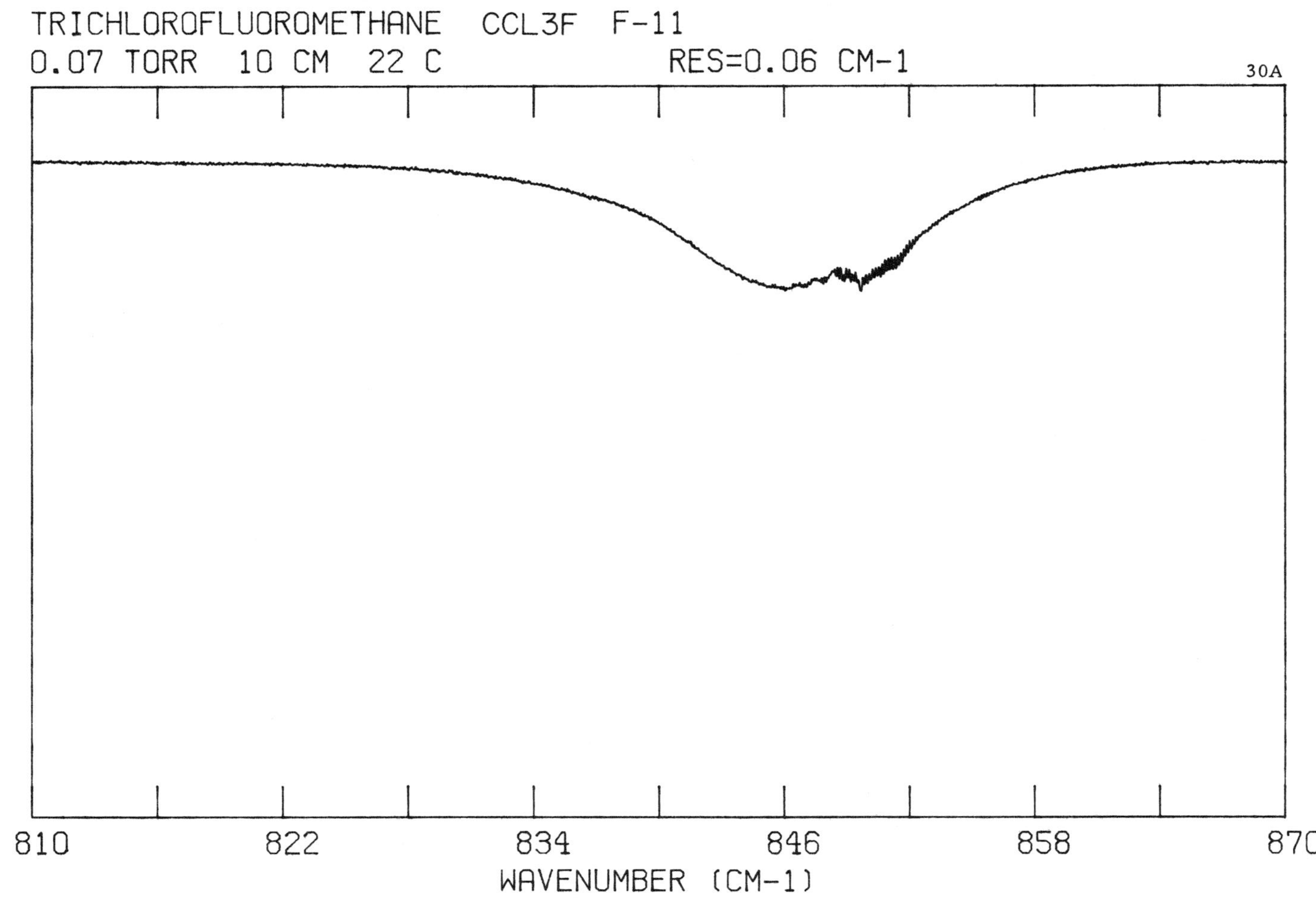
TRICHLOROFLUOROMETHANE CCL3F F-11
0.07 TORR 10 CM 22 C RES=0.06 CM-1
30A
810
822
834
846
858
870
WAVENUMBER (CM-1)

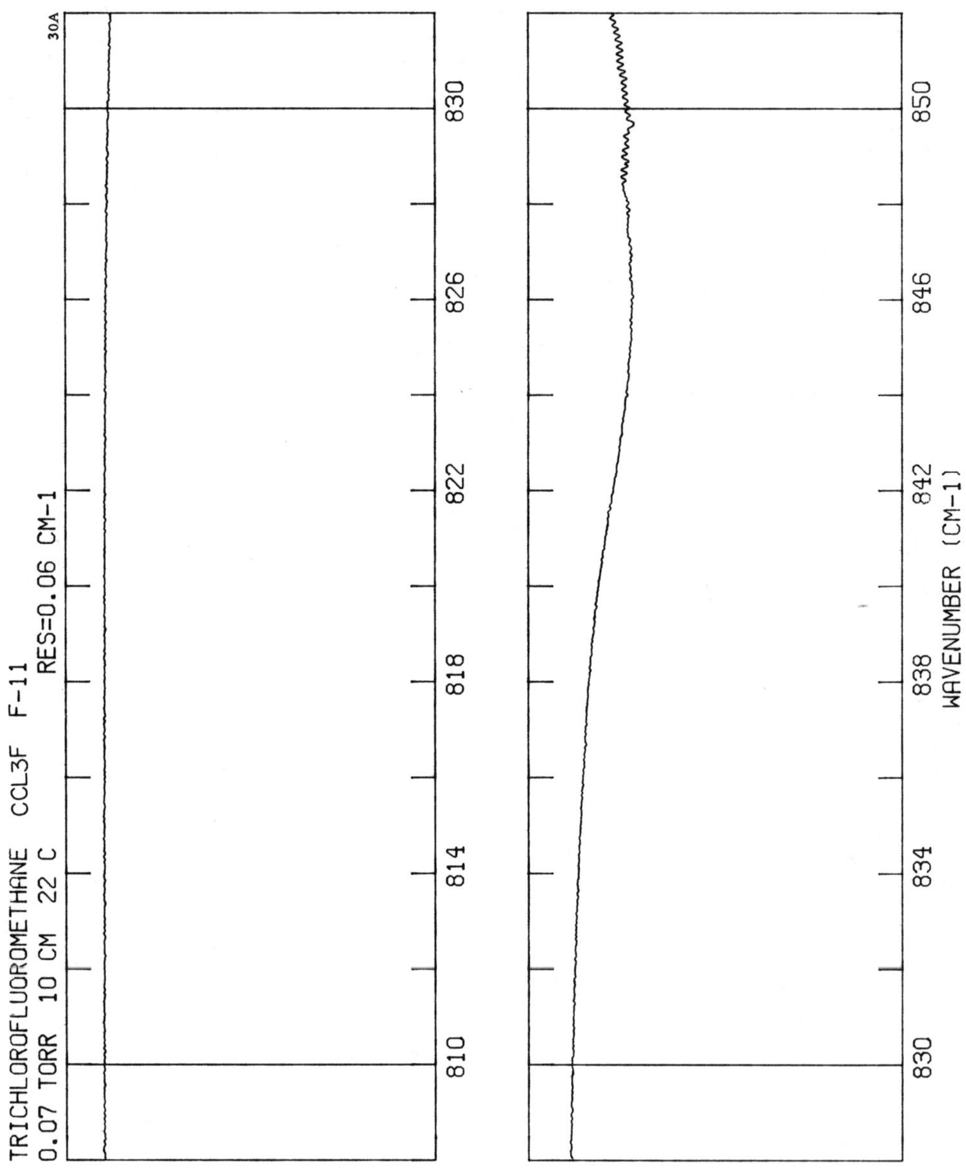
TRICHLOROFLUOROMETHANE CCL3F F-11
0.07 TORR 10 CM 22 C RES=0.06 CM-1
30A
810
814
818
822
826
830
830
834
838
842
846
850
WAVENUMBER (CM-1)

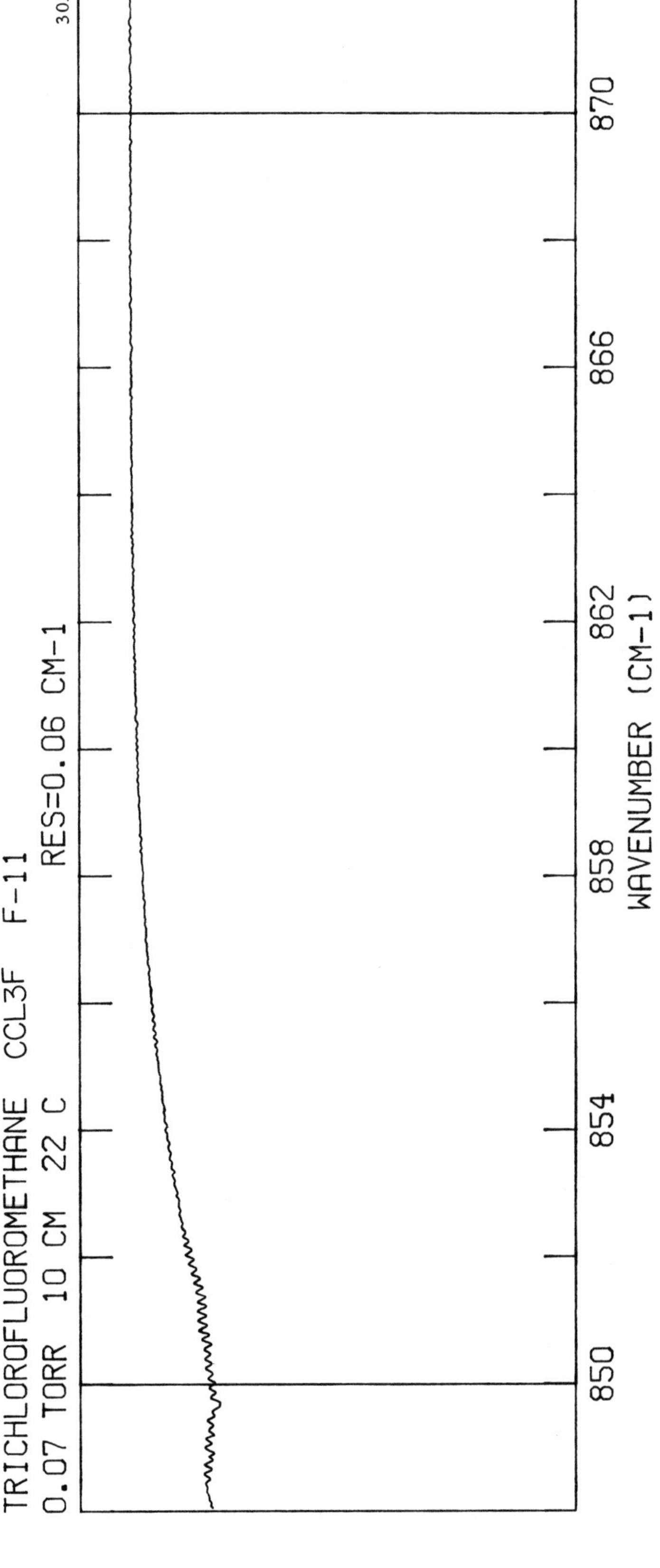

TRICHLOROFLUOROMETHANE CCL3F F-11
0.07 TORR 10 CM 22 C RES=0.06 CM-1
30A
850
854
858
862
866
870
WAVENUMBER (CM-1)

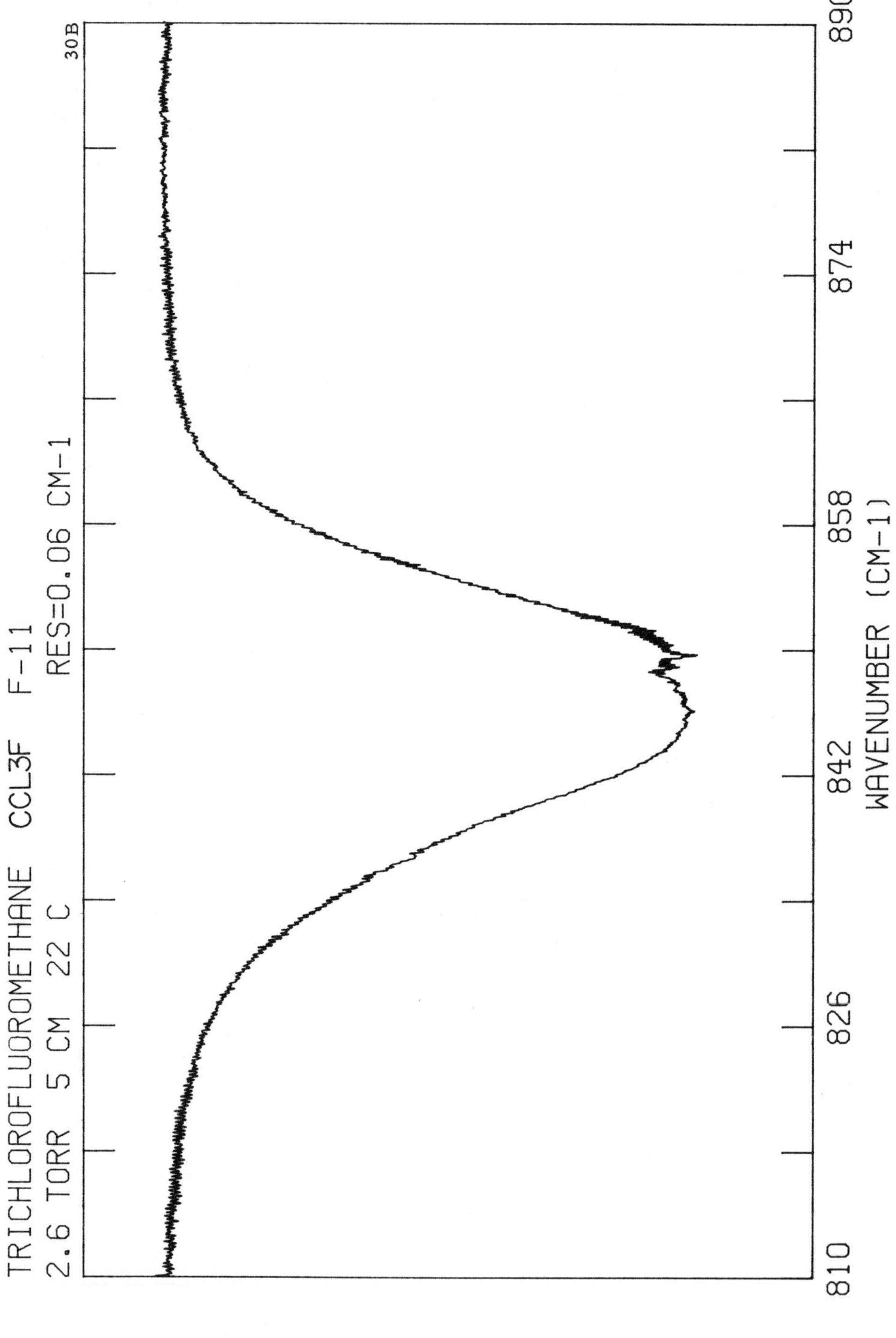
TRICHLOROFLUOROMETHANE CCL3F F-11
2.6 TORR 5 CM 22 C RES=0.06 CM-1
30B
810
826
842
858
874
890
WAVENUMBER (CM-1)

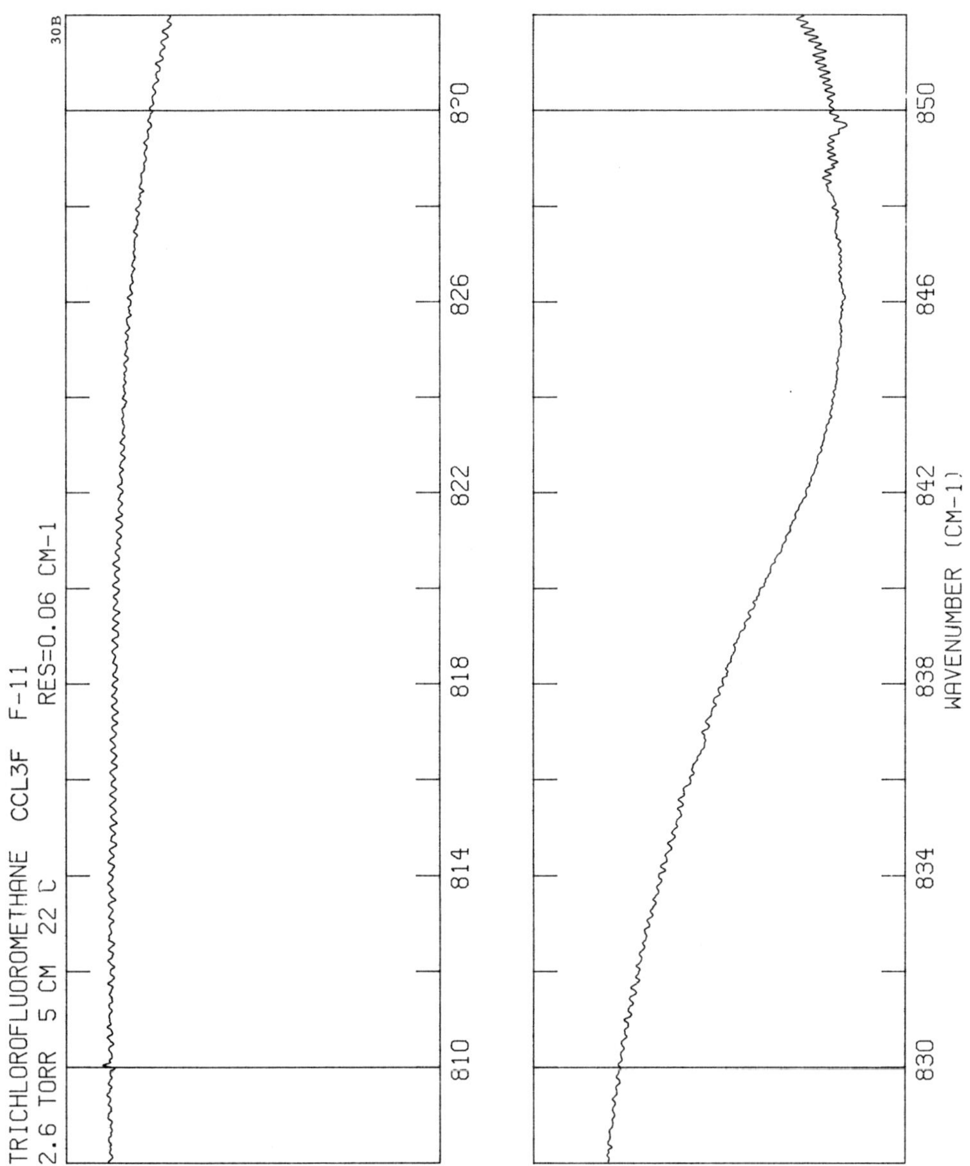
TRICHLOROFLUOROMETHANE CCL3F F-11
2.6 TORR 5 CM 22 C RES=0.06 CM-1
30B
810
814
818
822
826
830
830
834
838
842
846
850
WAVENUMBER (CM-1)

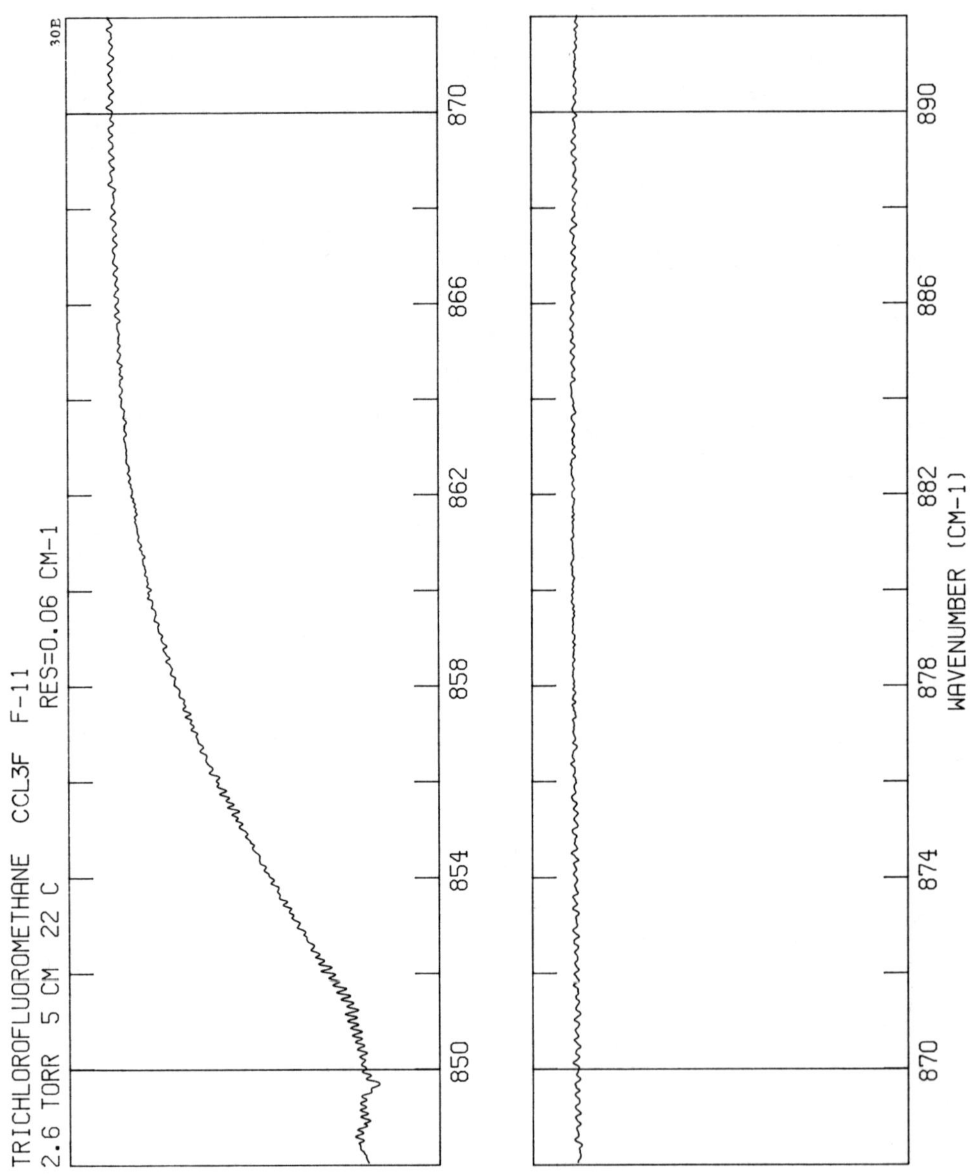
TRICHLOROFLUOROMETHANE CCL3F F-11
2.6 TORR 5 CM 22 C RES=0.06 CM-1
850
854
858
862
866
870
870
874
878
882
886
890
WAVENUMBER (CM-1)
30E

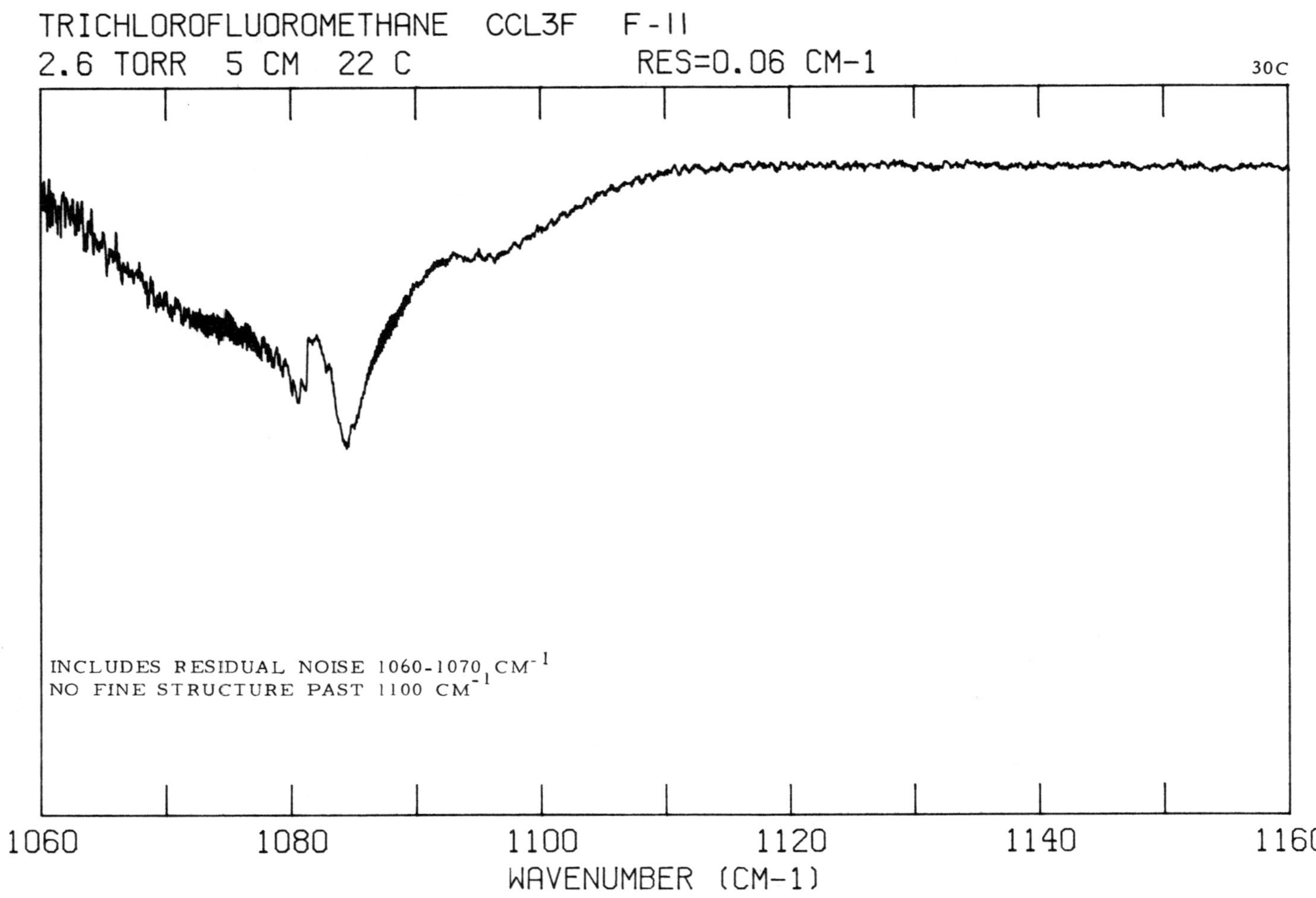
TRICHLOROFLUOROMETHANE CCL3F F-11
2.6 TORR 5 CM 22 C RES=0.06 CM-1
30C
INCLUDES RESIDUAL NOISE 1060-1070 CM-1
NO FINE STRUCTURE PAST 1100 CM-1
1060
1080
1100
1120
1140
1160
WAVENUMBER (CM-1)

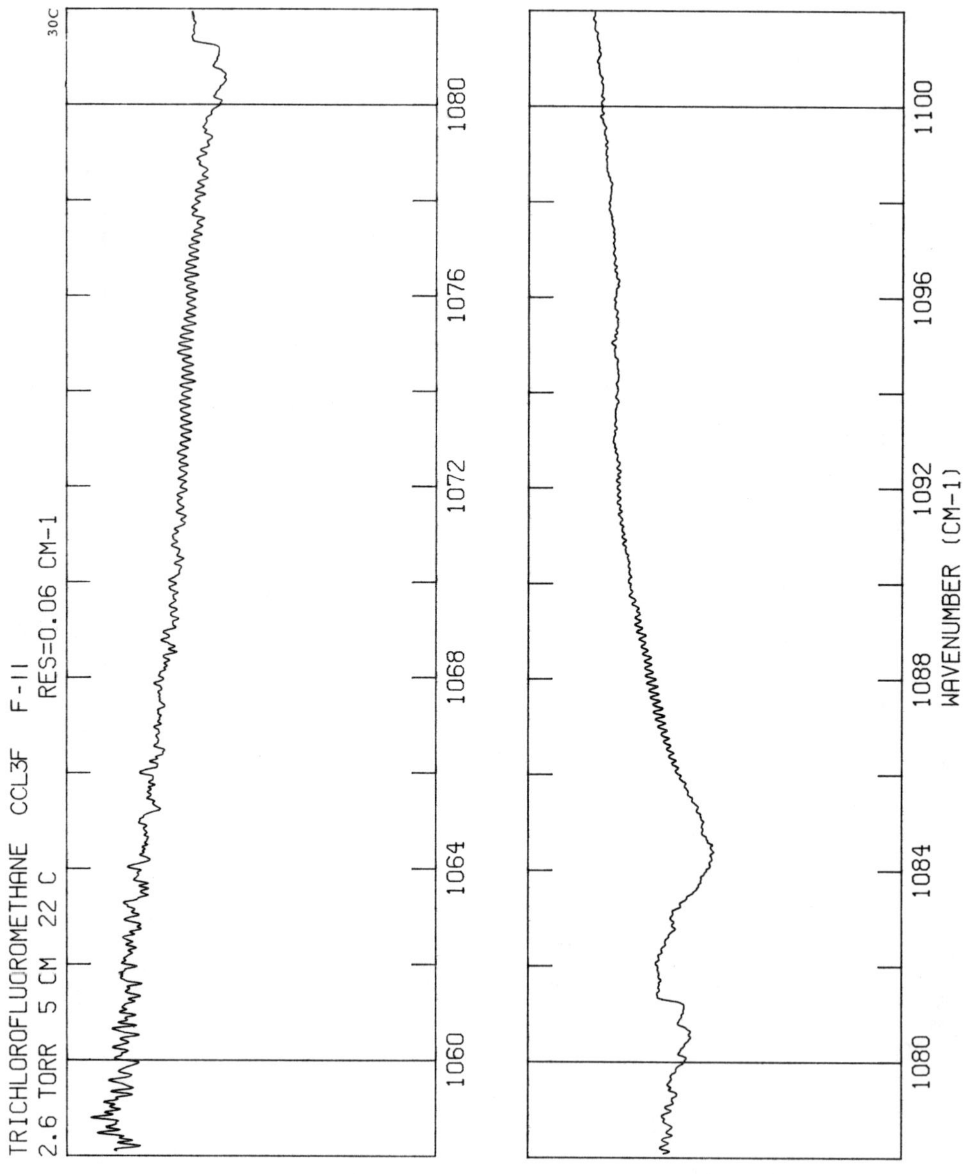
TRICHLOROFLUOROMETHANE CCL3F F-11
2.6 TORR 5 CM 22 C RES=0.06 CM-1
30C
1060
1064
1068
1072
1076
1080
1080
1084
1088
1092
1096
1100
WAVENUMBER (CM-1)

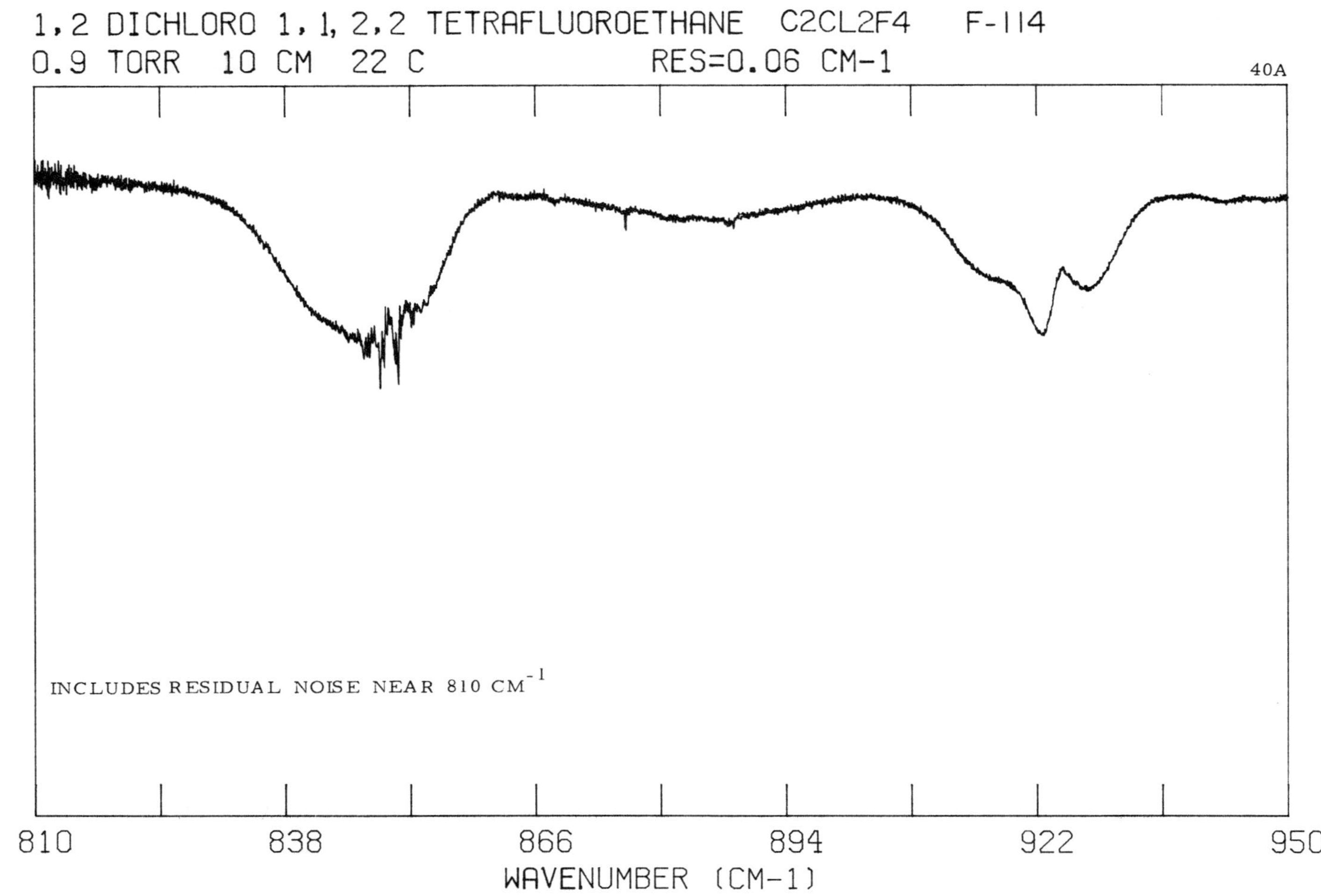
1,2 DICHLORO 1,1,2,2 TETRAFLUOROETHANE C2CL2F4 F-114
0.9 TORR 10 CM 22 C RES=0.06 CM-1
40A
INCLUDES RESIDUAL NOISE NEAR 810 CM-1
810
838
866
894
922
950
WAVENUMBER (CM-1)

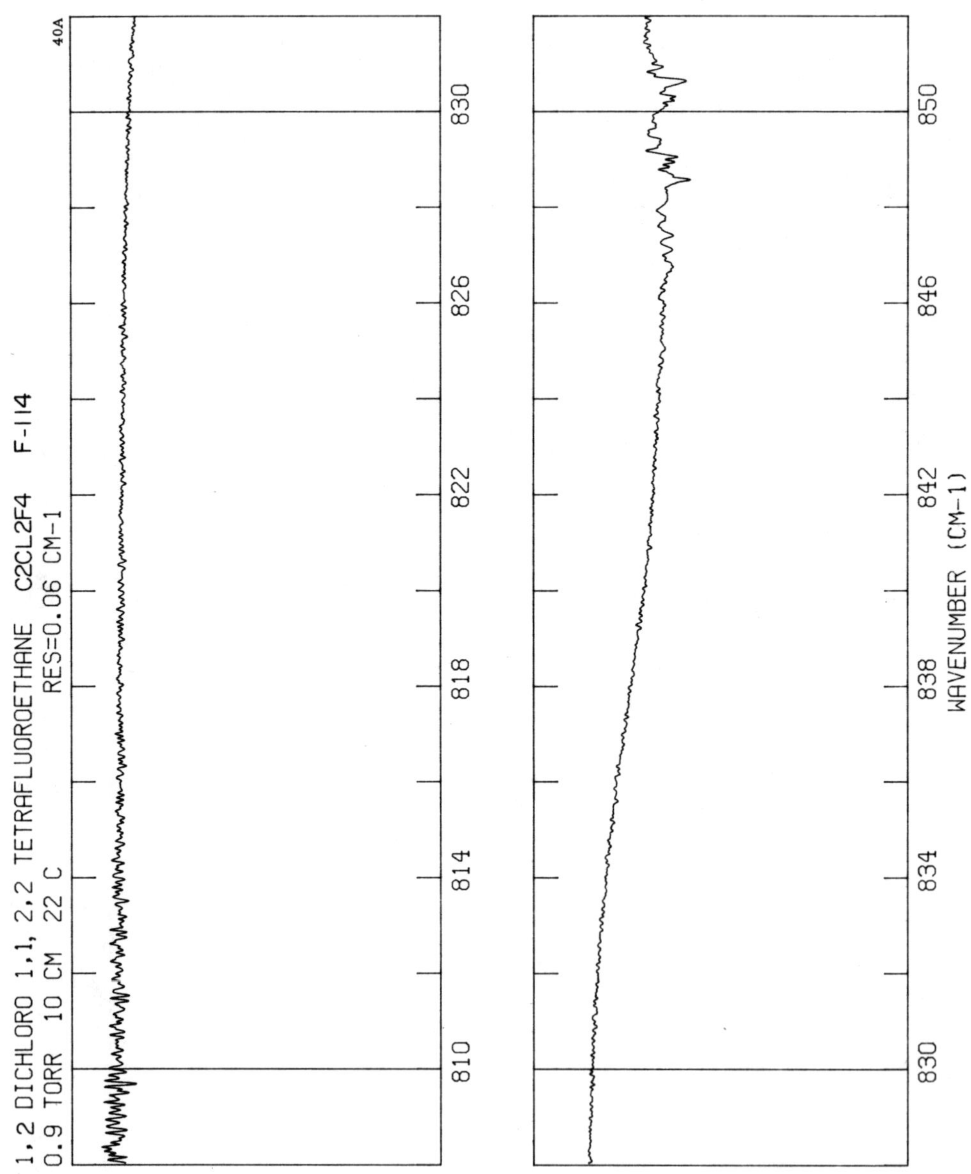
1,2 DICHLORO 1,1, 2,2 TETRAFLUOROETHANE C2CL2F4 F-114
0.9 TORR 10 CM 22 C RES=0.06 CM-1
40A
810
814
818
822
826
830
830
834
838
842
846
850
WAVENUMBER (CM-1)

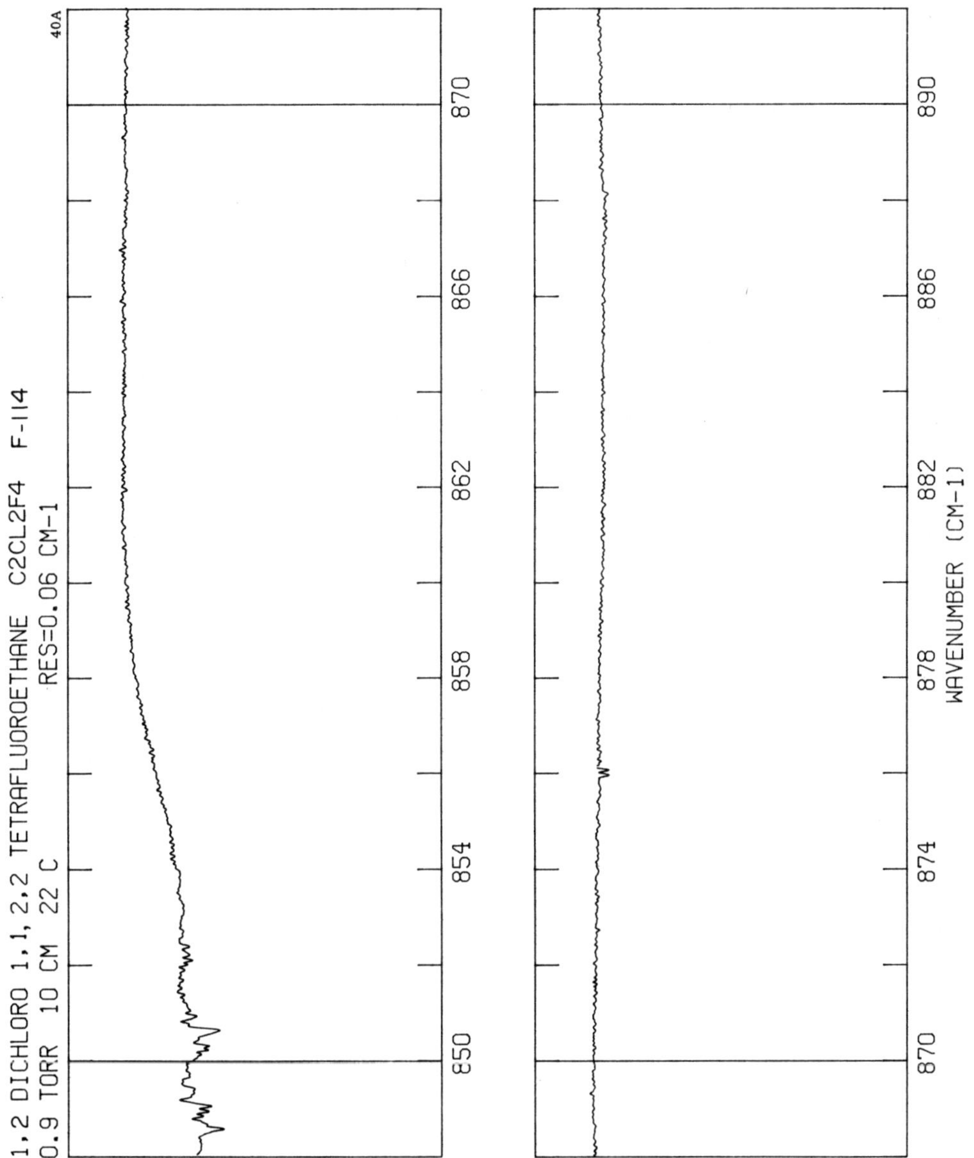
1,2 DICHLORO 1,1,2,2 TETRAFLUOROETHANE C2CL2F4 F-114
0.9 TORR 10 CM 22 C
RES=0.06 CM-1
40A
850
854
858
862
866
870
870
874
878
882
886
890
WAVENUMBER (CM-1)

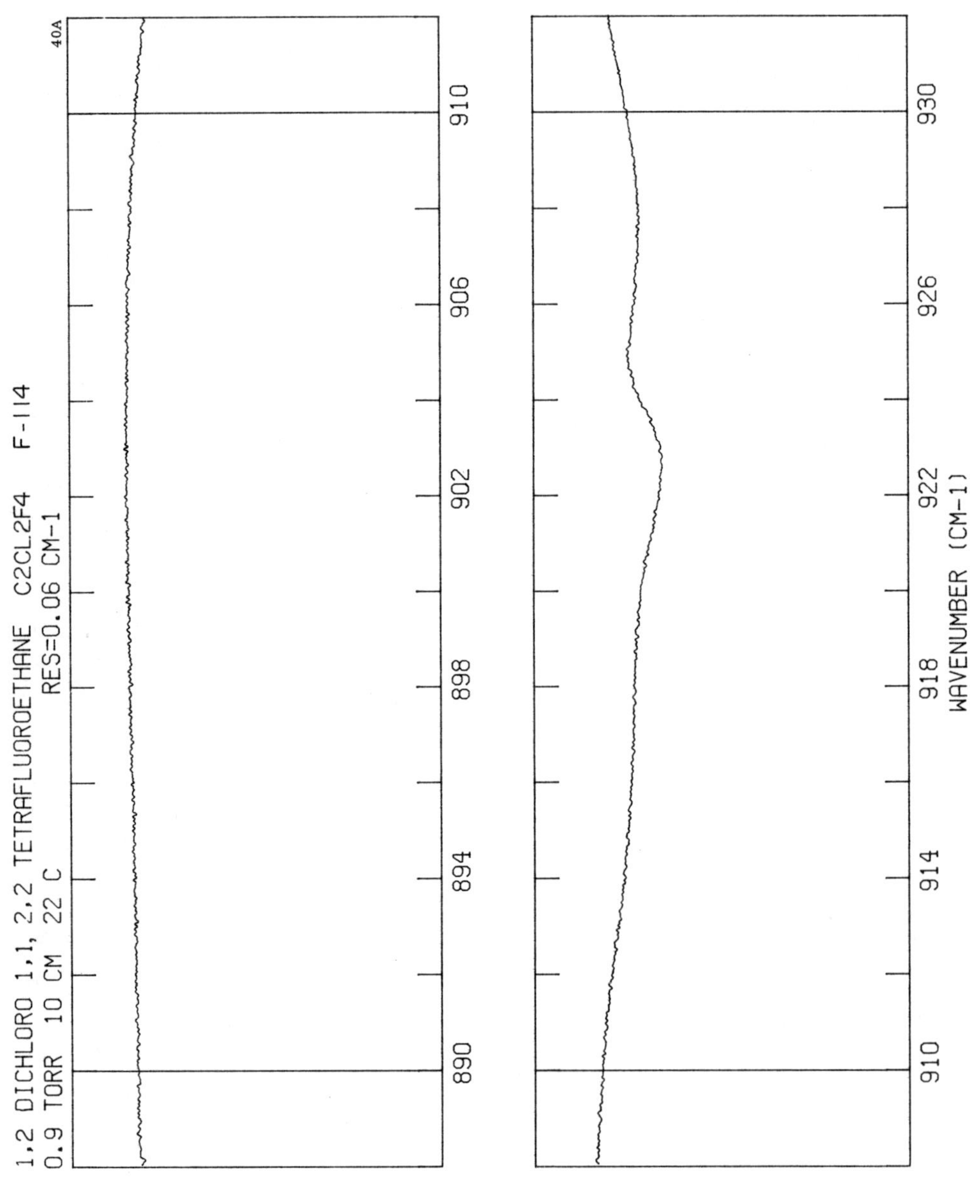

1,2 DICHLORO 1,1, 2,2 TETRAFLUOROETHANE C2CL2F4 F-114
0.9 TORR 10 CM 22 C RES=0.06 CM-1
40A
890
894
898
902
906
910
910
914
918
922
926
930
WAVENUMBER (CM-1)

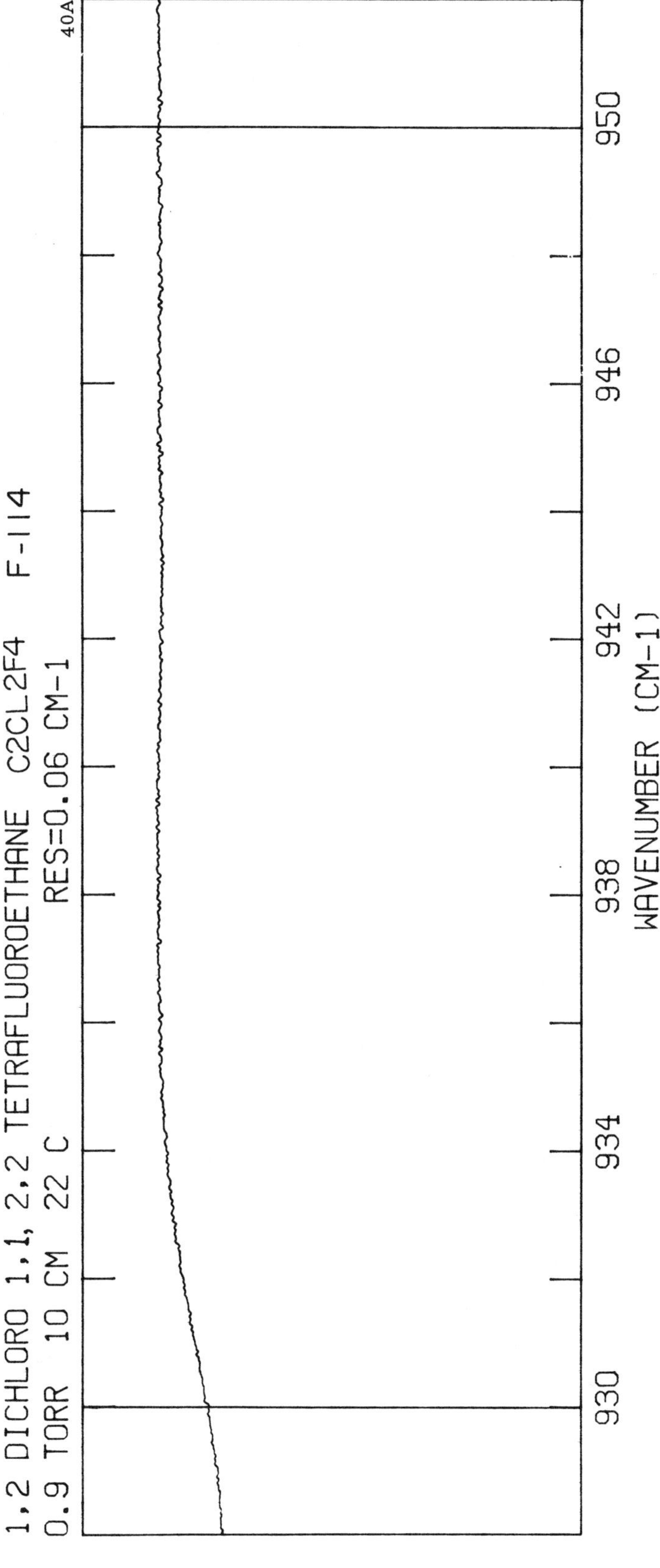
1.2 DICHLORO 1,1, 2,2 TETRAFLUOROETHANE C2CL2F4 F-114
0.9 TORR 10 CM 22 C RES=0.06 CM-1
40A
930
934
938
942
946
950
WAVENUMBER (CM-1)

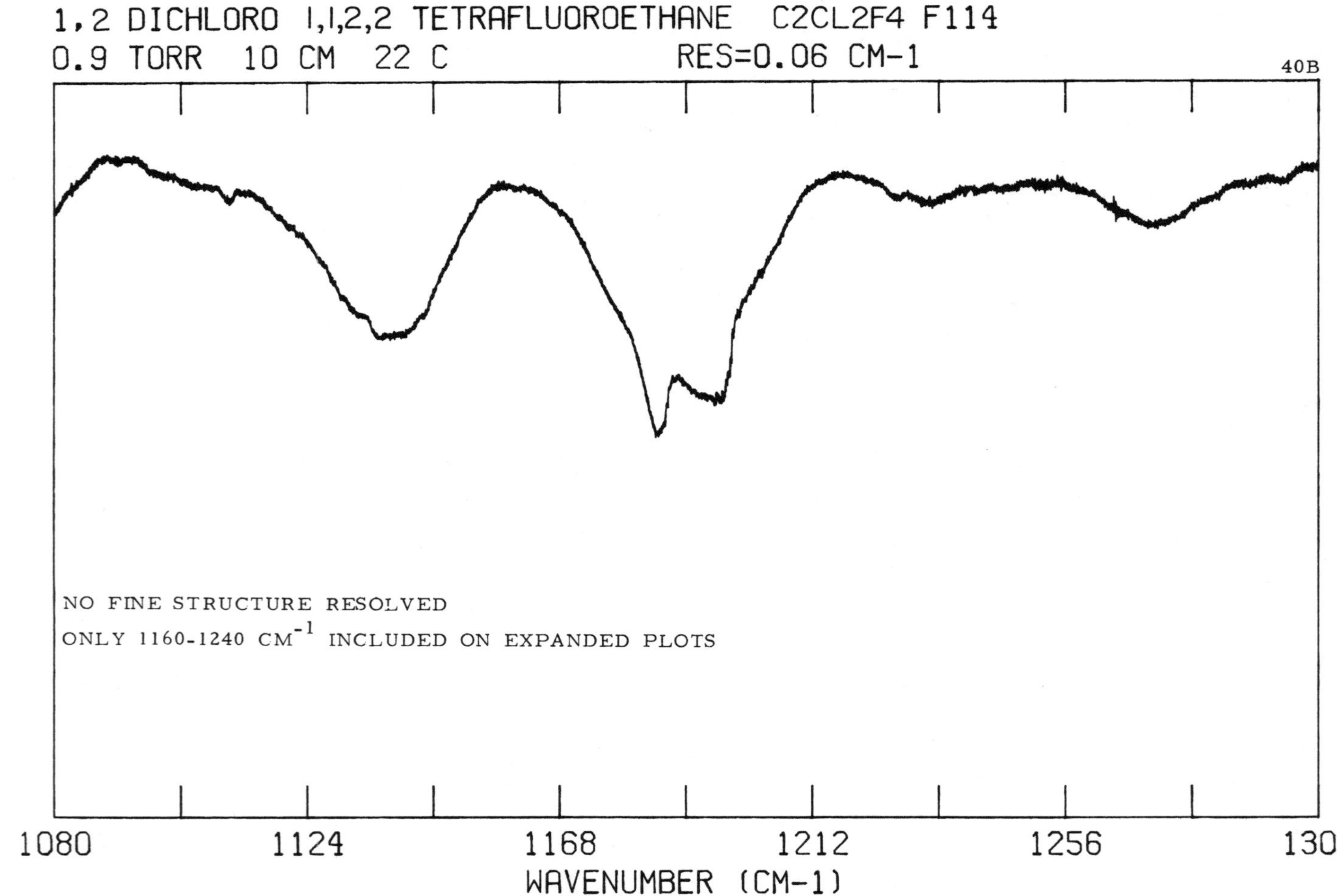
1,2 DICHLORO 1,1,2,2 TETRAFLUOROETHANE C2CL2F4 F114
0.9 TORR 10 CM 22 C RES=0.06 CM-1
40B
NO FINE STRUCTURE RESOLVED
ONLY 1160-1240 CM-1 INCLUDED ON EXPANDED PLOTS
1080
1124
1168
1212
1256
1300
WAVENUMBER (CM-1)

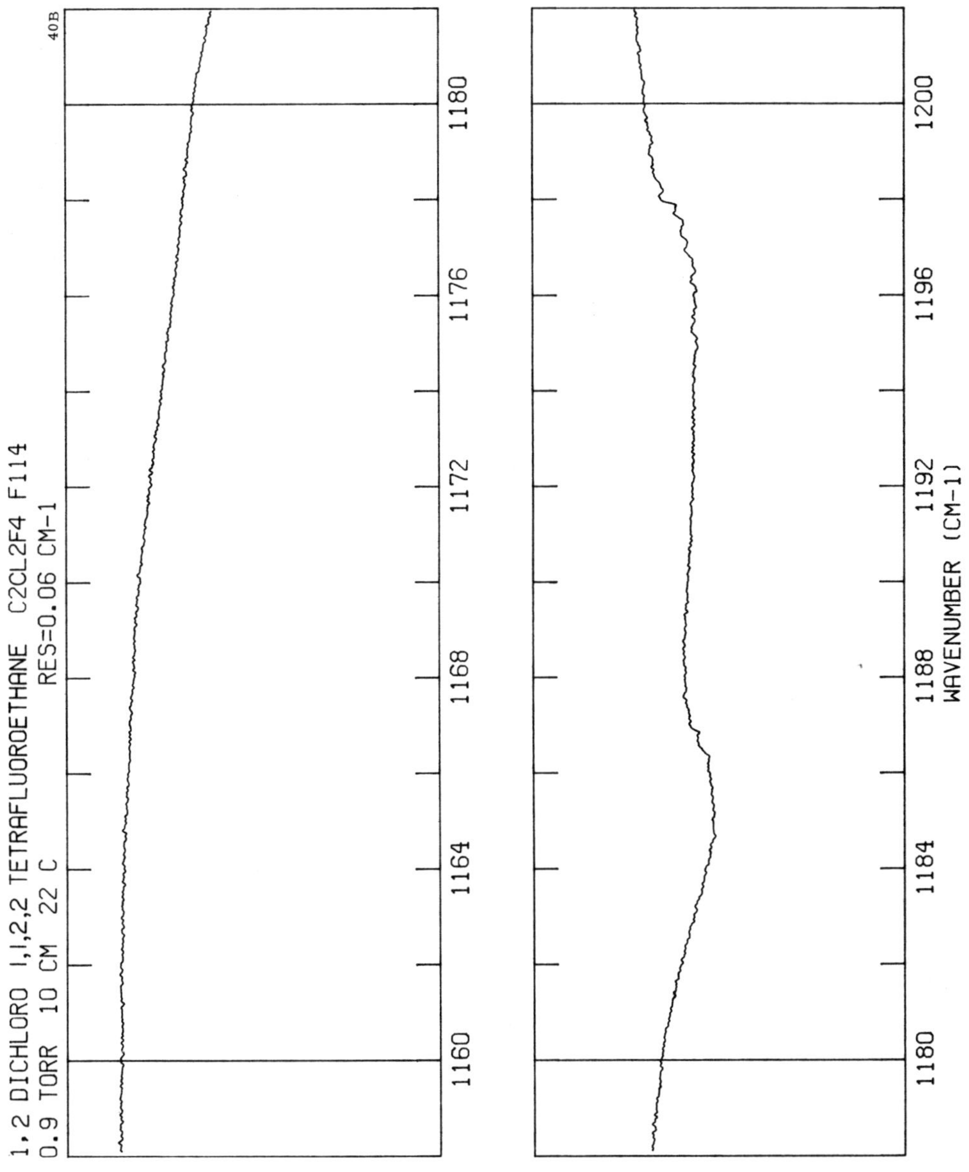
1,2 DICHLORO 1,1,2,2 TETRAFLUOROETHANE C2CL2F4 F114
0.9 TORR 10 CM 22 C RES=0.06 CM-1
40B
1160
1164
1168
1172
1176
1180
1180
1184
1188
1192
1196
1200
WAVENUMBER (CM-1)

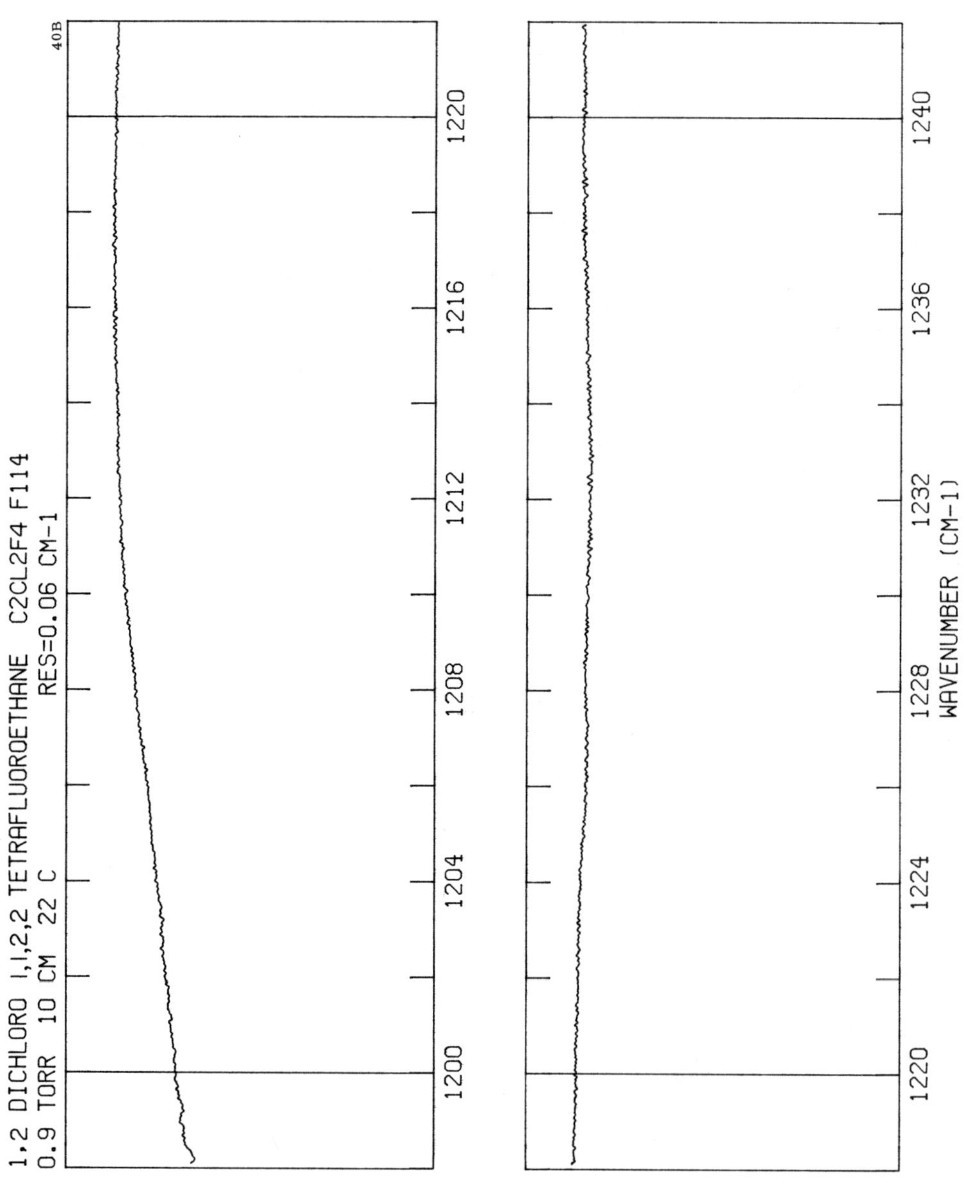
1,2 DICHLORO 1,1,2,2 TETRAFLUOROETHANE C2CL2F4 F114
0.9 TORR 10 CM 22 C RES=0.06 CM-1
40B
1200
1204
1208
1212
1216
1220
1220
1224
1228
1232
1236
1240
WAVENUMBER (CM-1)

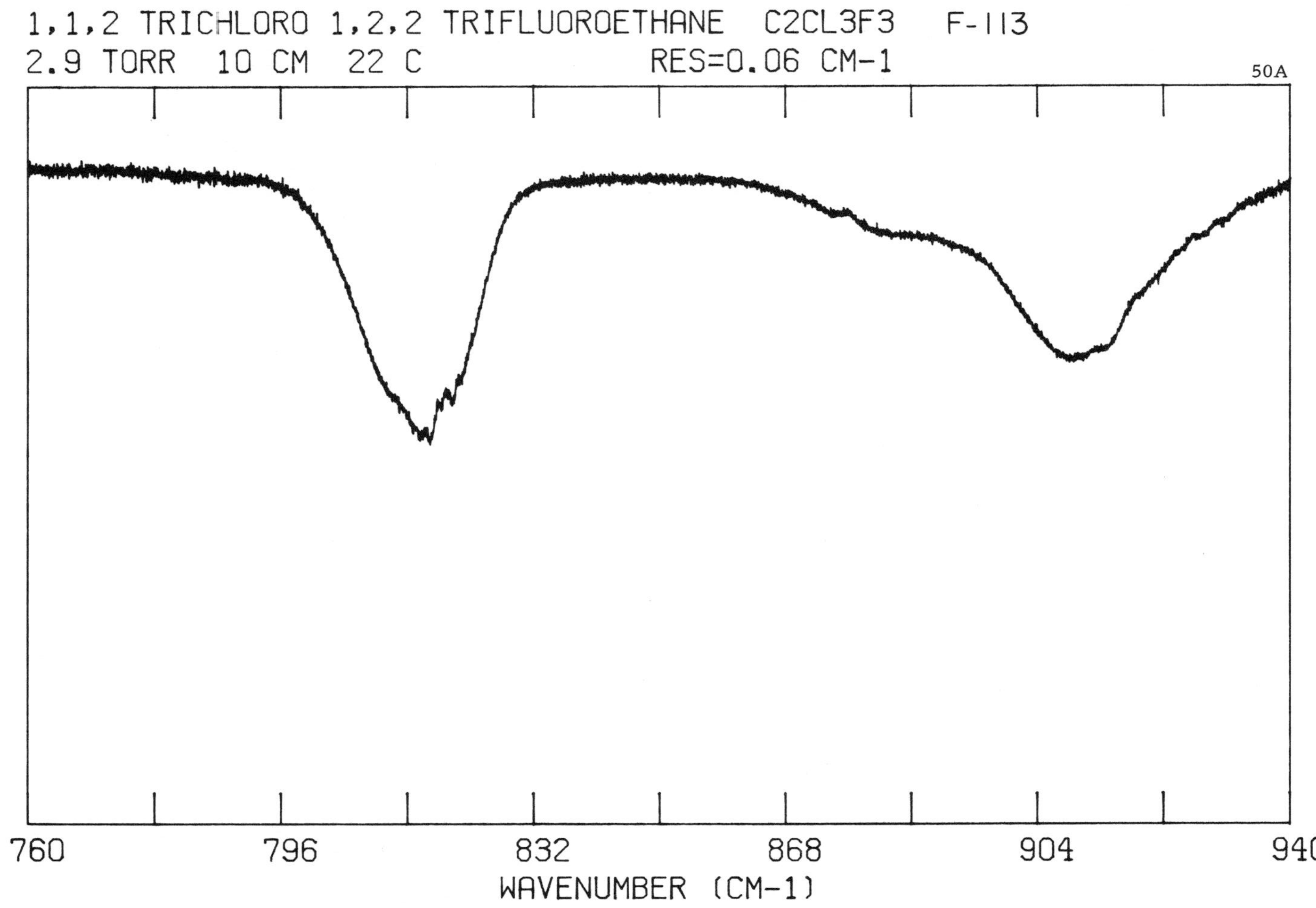

1,1,2 TRICHLORO 1,2,2 TRIFLUOROETHANE C2CL3F3 F-113
2.9 TORR 10 CM 22 C RES=0.06 CM-1
50A
760
796
832
868
904
940
WAVENUMBER (CM-1)

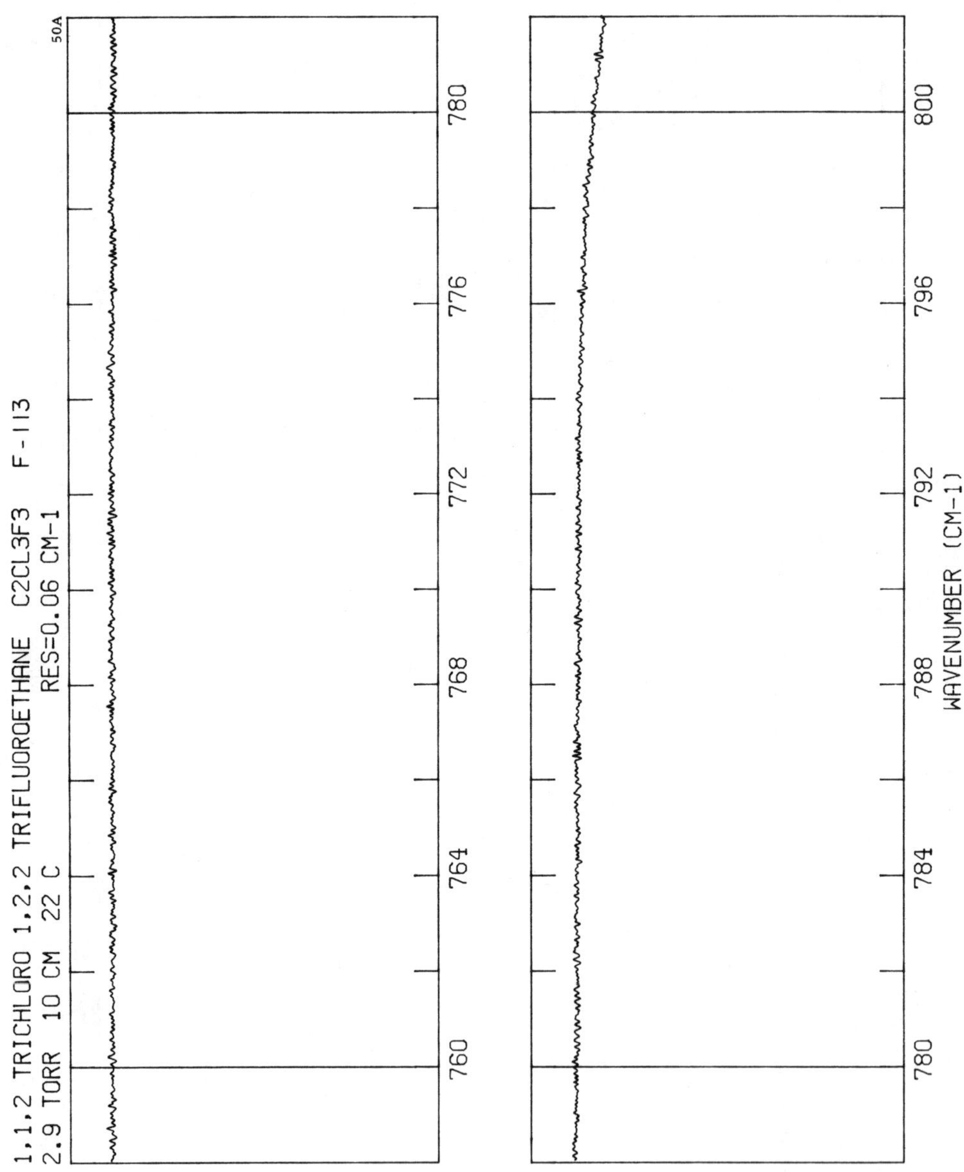
1,1,2 TRICHLORO 1,2,2 TRIFLUOROETHANE C2CL3F3 F-113
2.9 TORR 10 CM 22 C RES=0.06 CM-1
50A
760
764
768
772
776
780
780
784
788
792
796
800
WAVENUMBER (CM-1)

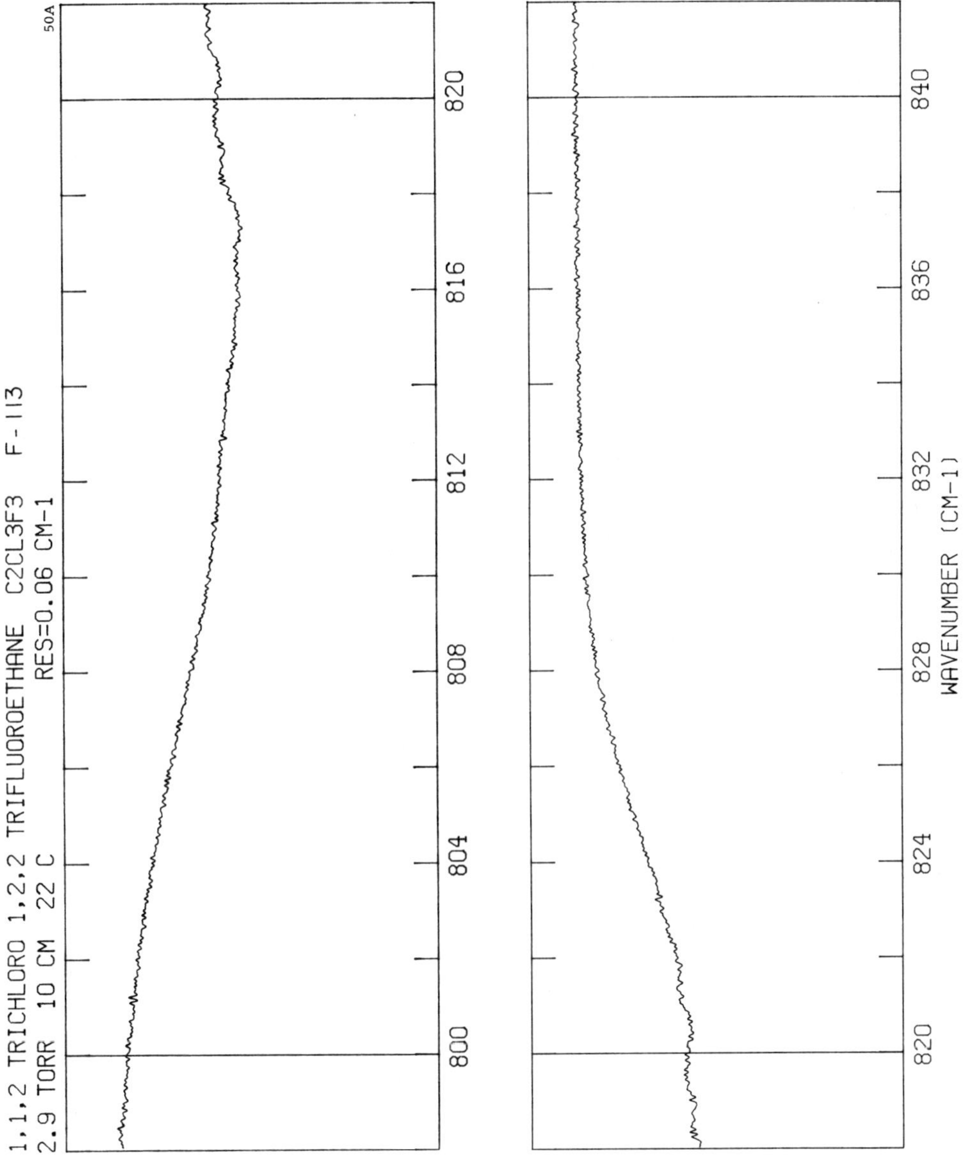

1,1,2 TRICHLORO 1,2,2 TRIFLUOROETHANE C2CL3F3 F-113
2.9 TORR 10 CM 22 C RES=0.06 CM-1
50A
800
804
808
812
816
820
820
824
828
832
836
840
WAVENUMBER (CM-1)

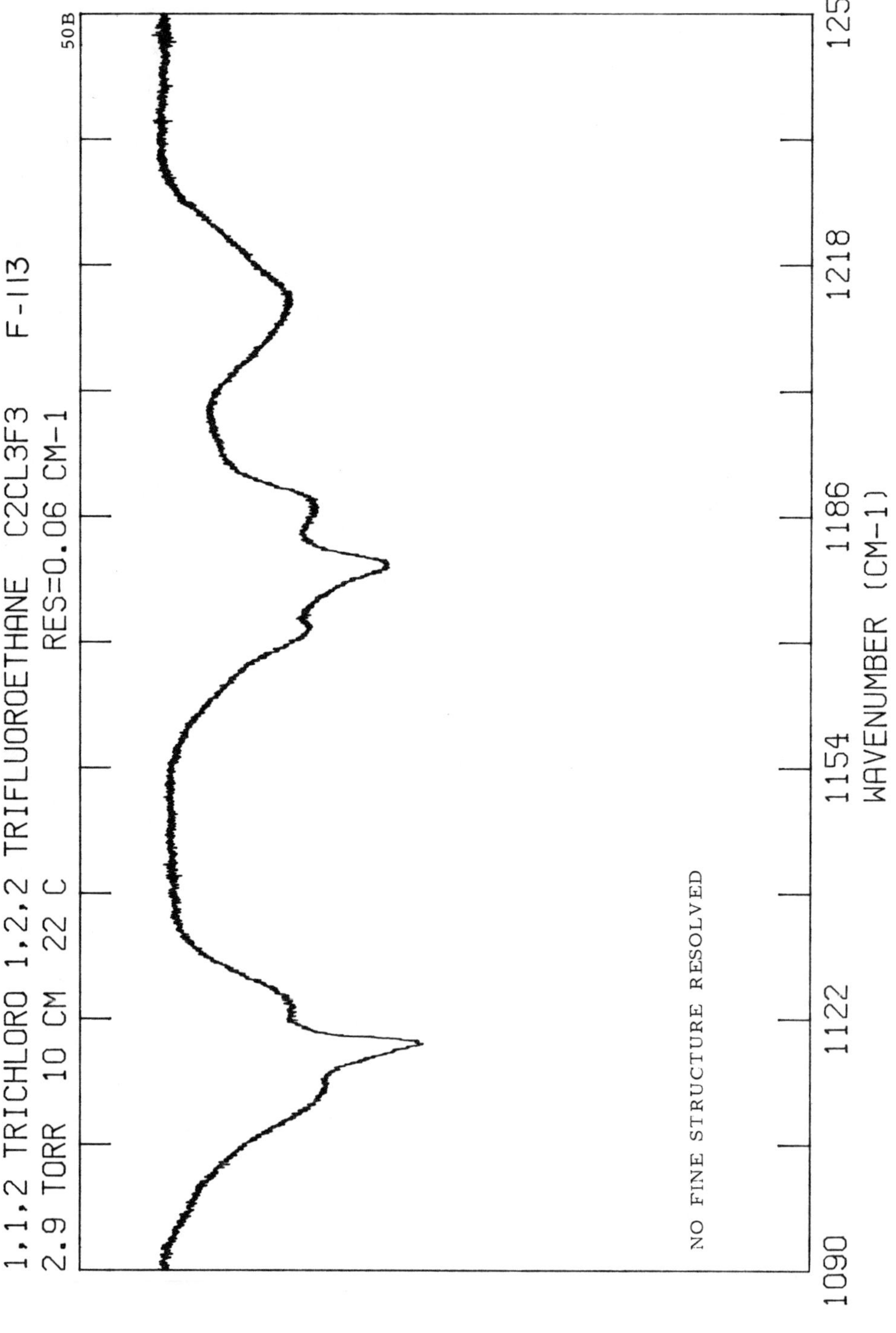
1,1,2 TRICHLORO 1,2,2 TRIFLUOROETHANE C2CL3F3 F-113
2.9 TORR 10 CM 22 C RES=0.06 CM-1
50B
NO FINE STRUCTURE RESOLVED
1090
1122
1154
1186
1218
1250
WAVENUMBER (CM-1)

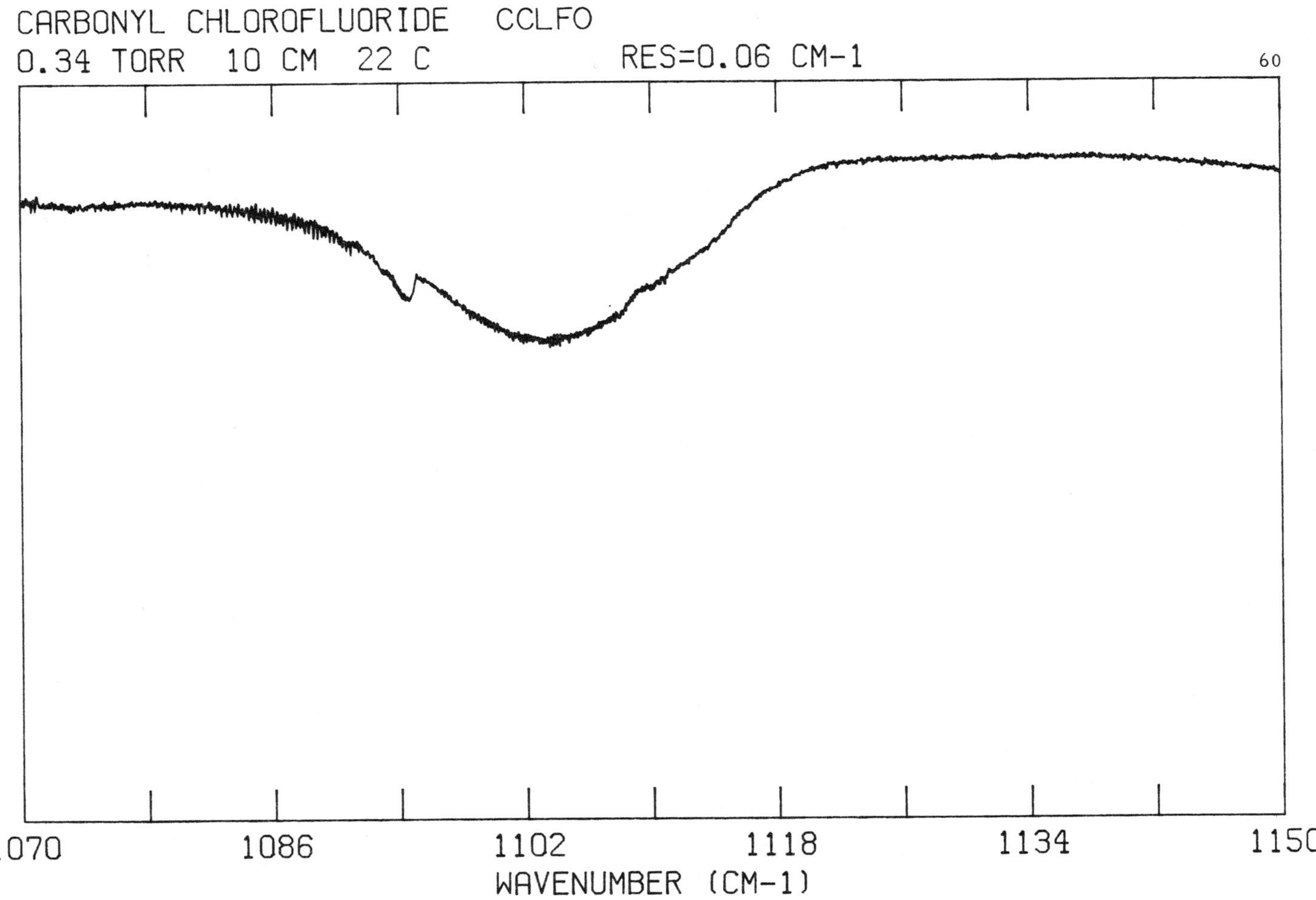

CARBONYL CHLOROFLUORIDE CCLFO
0.34 TORR 10 CM 22 C
RES=0.06 CM-1
60
1070
1086
1102
1118
1134
1150
WAVENUMBER (CM-1)

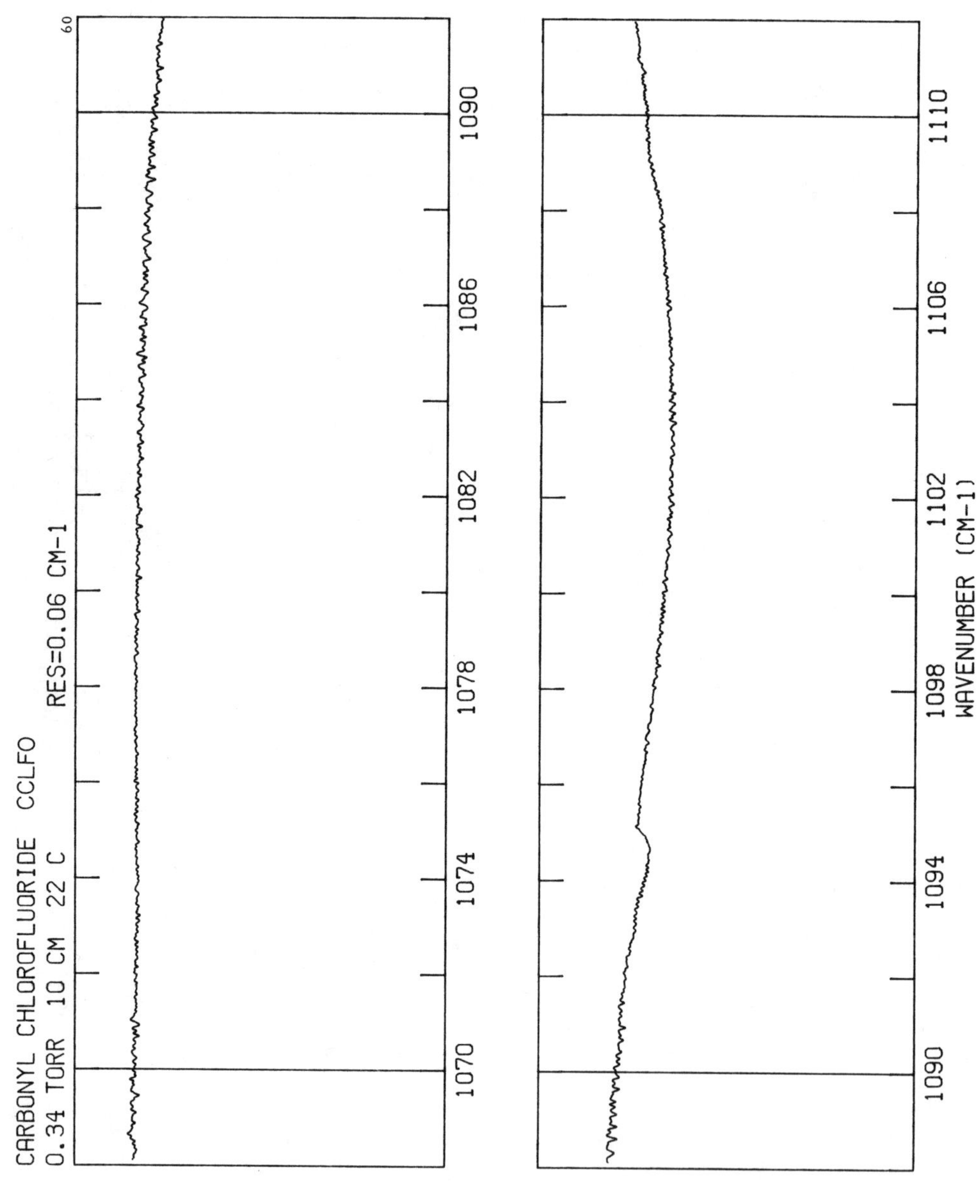

CARBONYL CHLOROFLUORIDE CCLFO
0.34 TORR 10 CM 22 C
RES=0.06 CM-1
60
1070
1074
1078
1082
1086
1090
1090
1094
1098
1102
1106
1110
WAVENUMBER (CM-1)

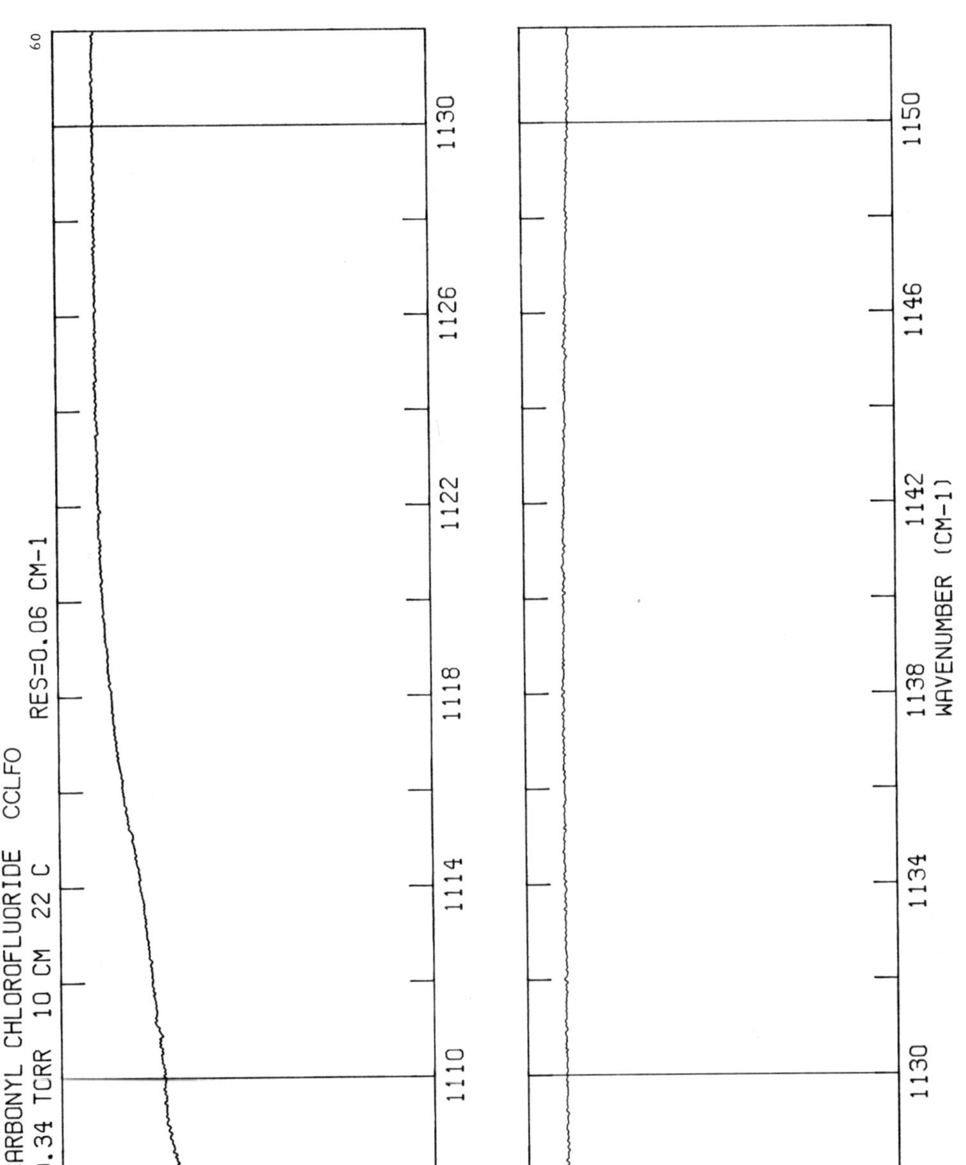
CARBONYL CHLOROFLUORIDE CCLFO
0.34 TORR 10 CM 22 C
RES=0.06 CM-1
60
1110
1114
1118
1122
1126
1130
1134
1138
1142
1146
1150
WAVENUMBER (CM-1)

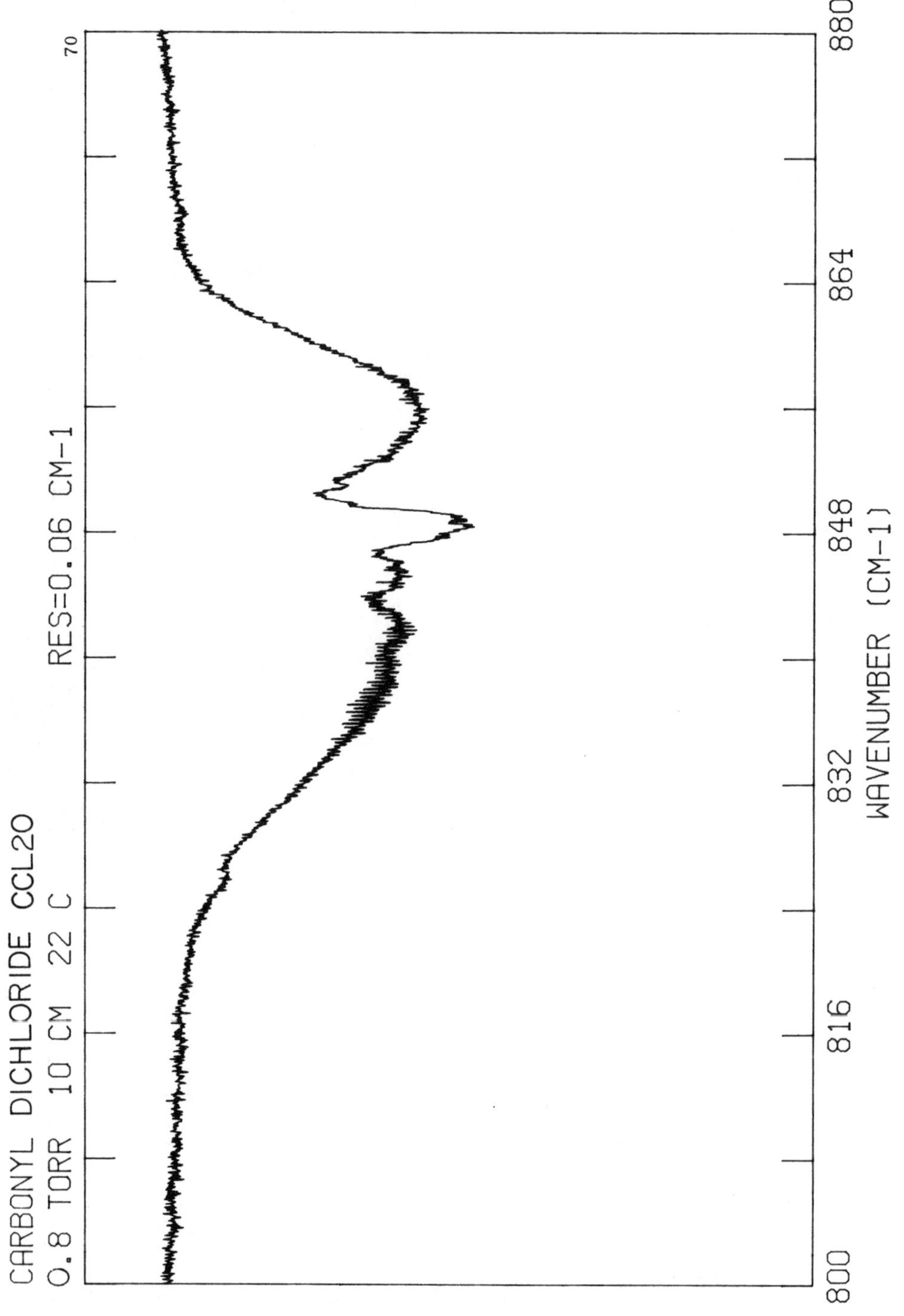
CARBONYL DICHLORIDE CCL2O
0.8 TORR 10 CM 22 C
RES=0.06 CM-1
70
800
816
832
848
864
880
WAVENUMBER (CM-1)

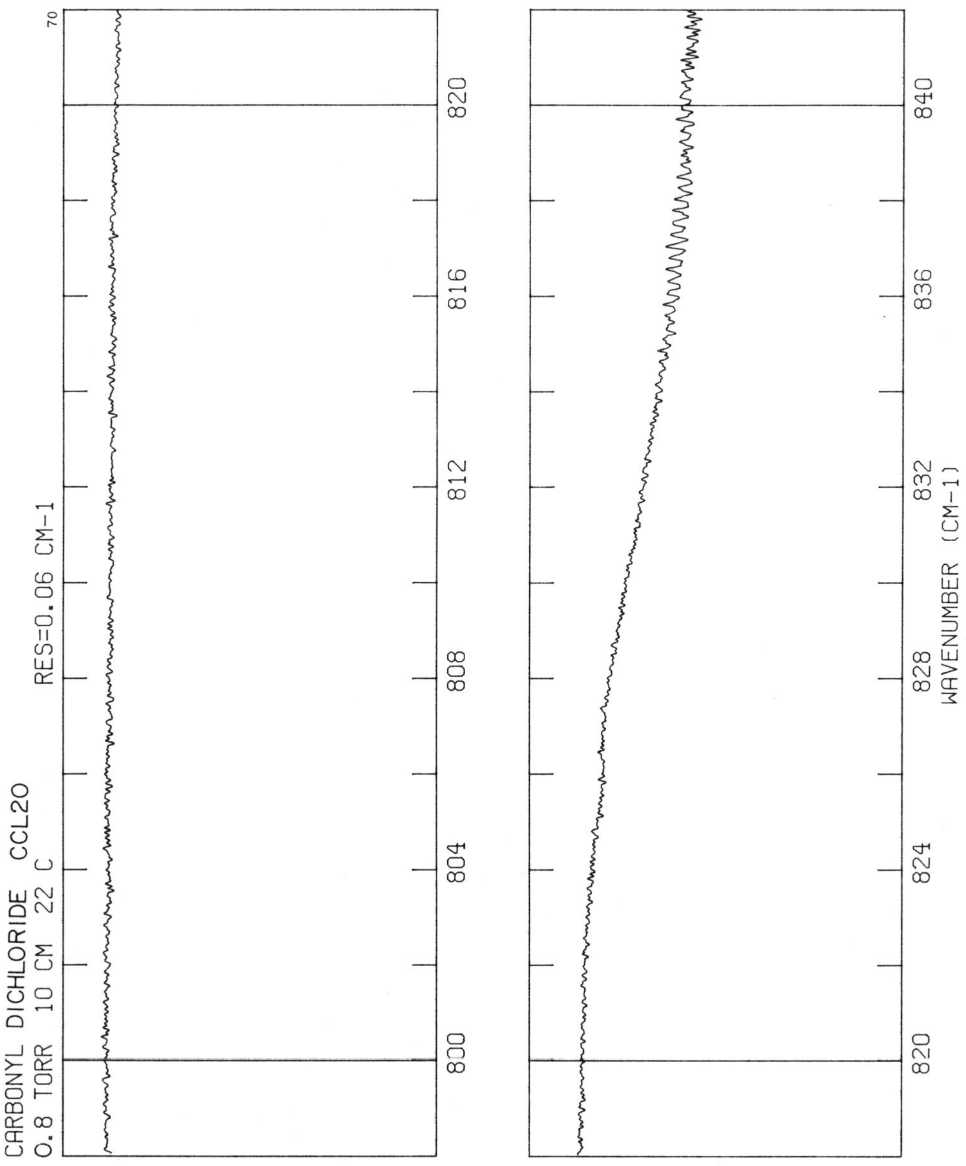
CARBONYL DICHLORIDE CCL2O
0.8 TORR 10 CM 22 C
RES=0.06 CM-1
70
800
804
808
812
816
820
820
824
828
832
836
840
WAVENUMBER (CM-1)

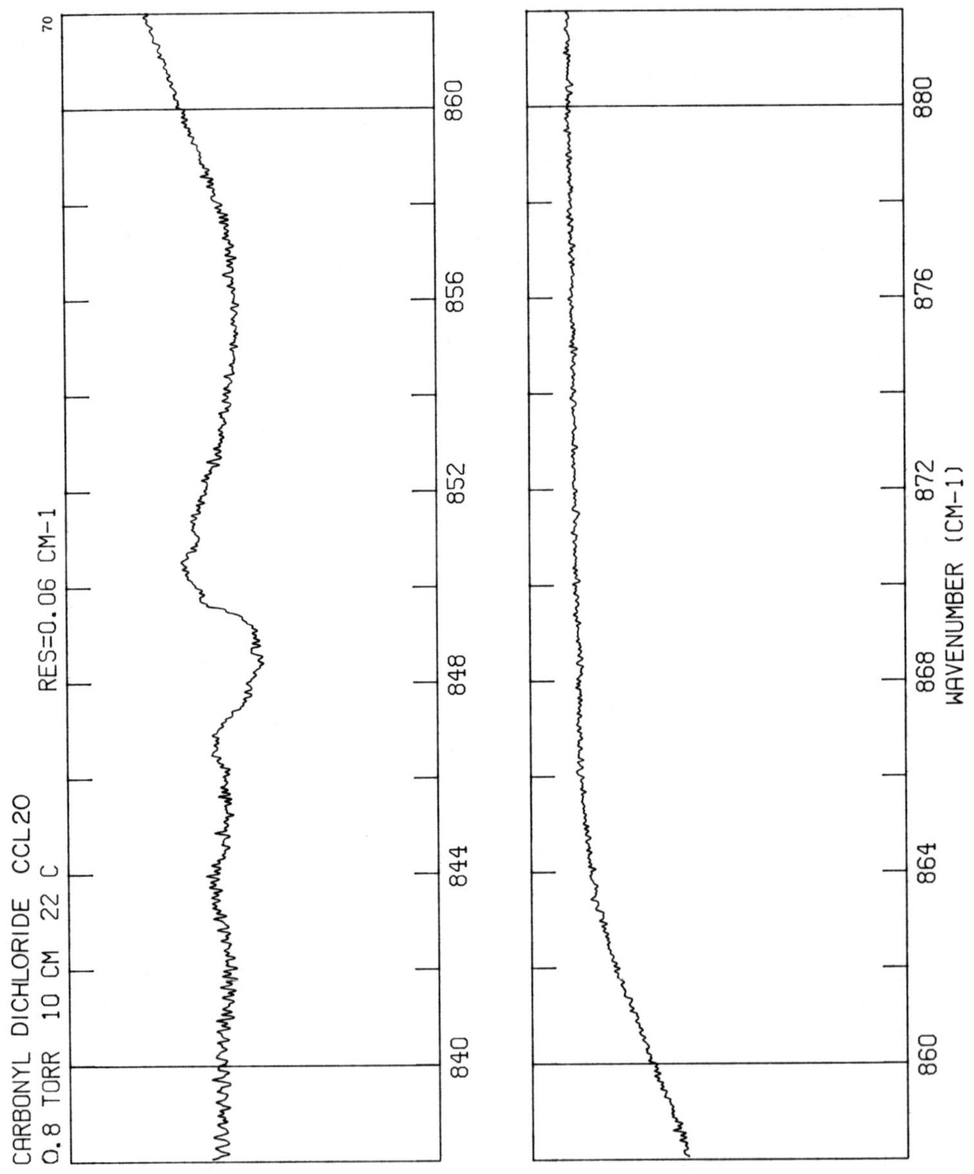
CARBONYL DICHLORIDE CCL2O
0.8 TORR 10 CM 22 C
RES=0.06 CM-1
70
840
844
848
852
856
860
860
864
868
872
876
880
WAVENUMBER (CM-1)

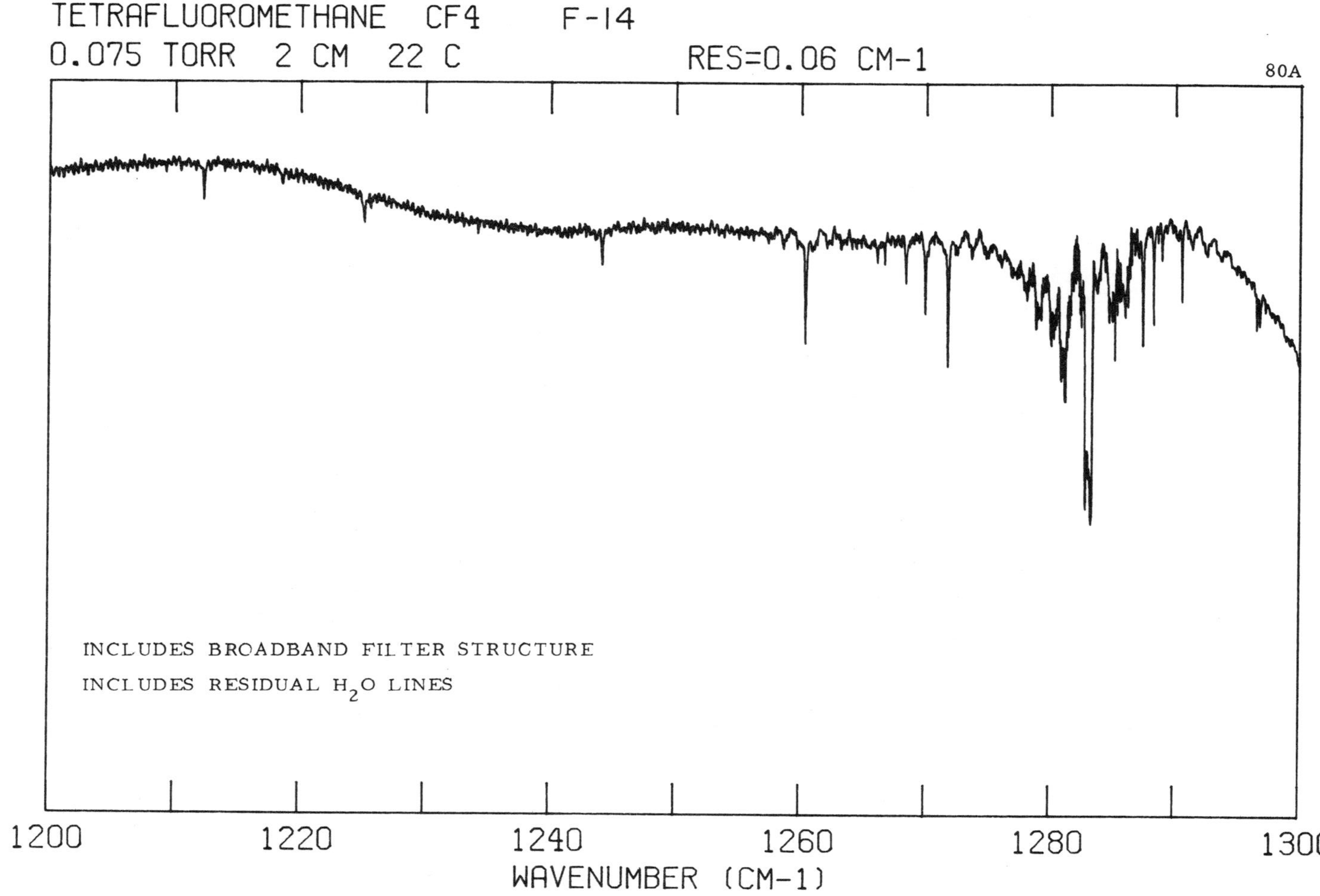
TETRAFLUOROMETHANE CF4 F-14
0.075 TORR 2 CM 22 C RES=0.06 CM-1
80A
INCLUDES BROADBAND FILTER STRUCTURE
INCLUDES RESIDUAL H_2O LINES
1200
1220
1240
1260
1280
1300
WAVENUMBER (CM-1)

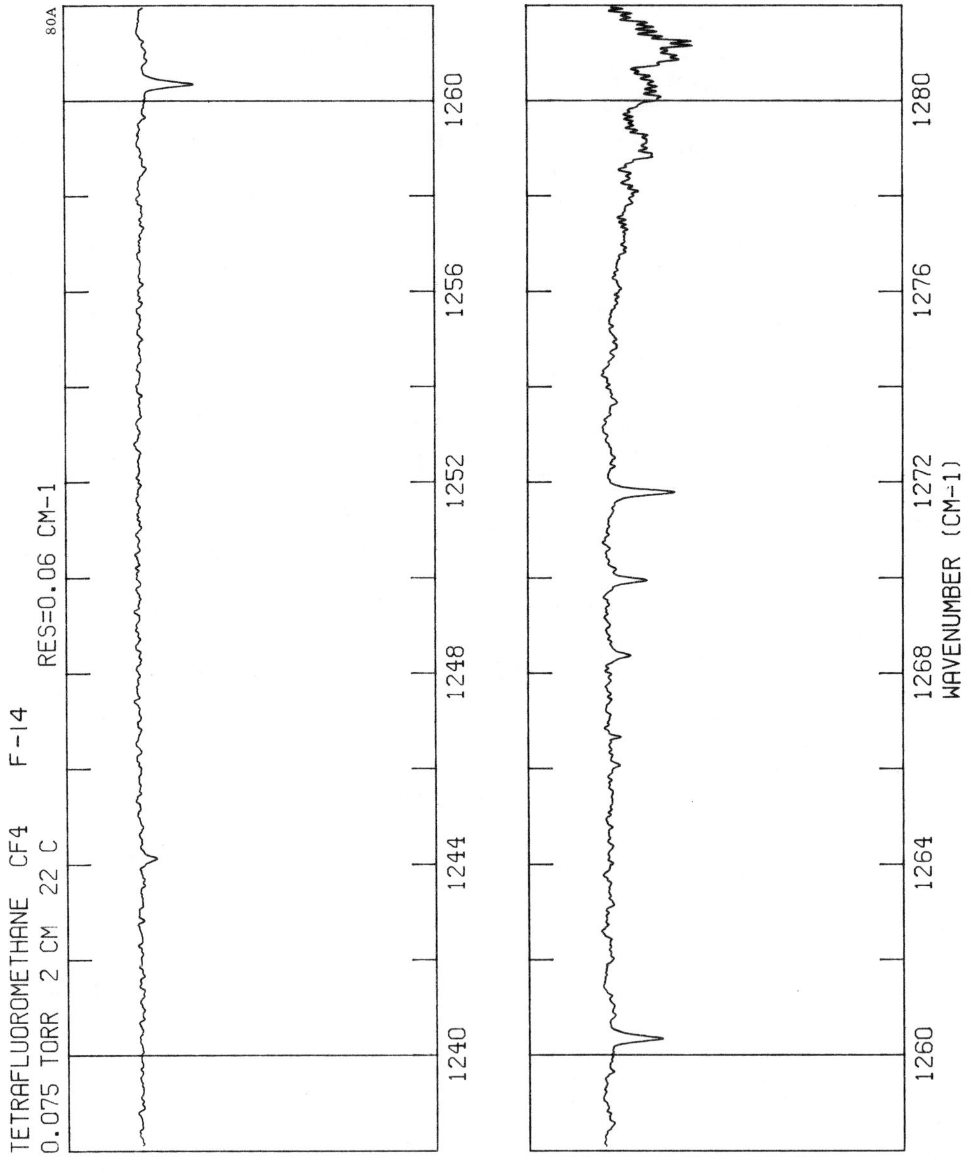
TETRAFLUOROMETHANE CF4 F-14
0.075 TORR 2 CM 22 C
RES=0.06 CM-1
80A
1240
1244
1248
1252
1256
1260
1260
1264
1268
1272
1276
1280
WAVENUMBER (CM-1)

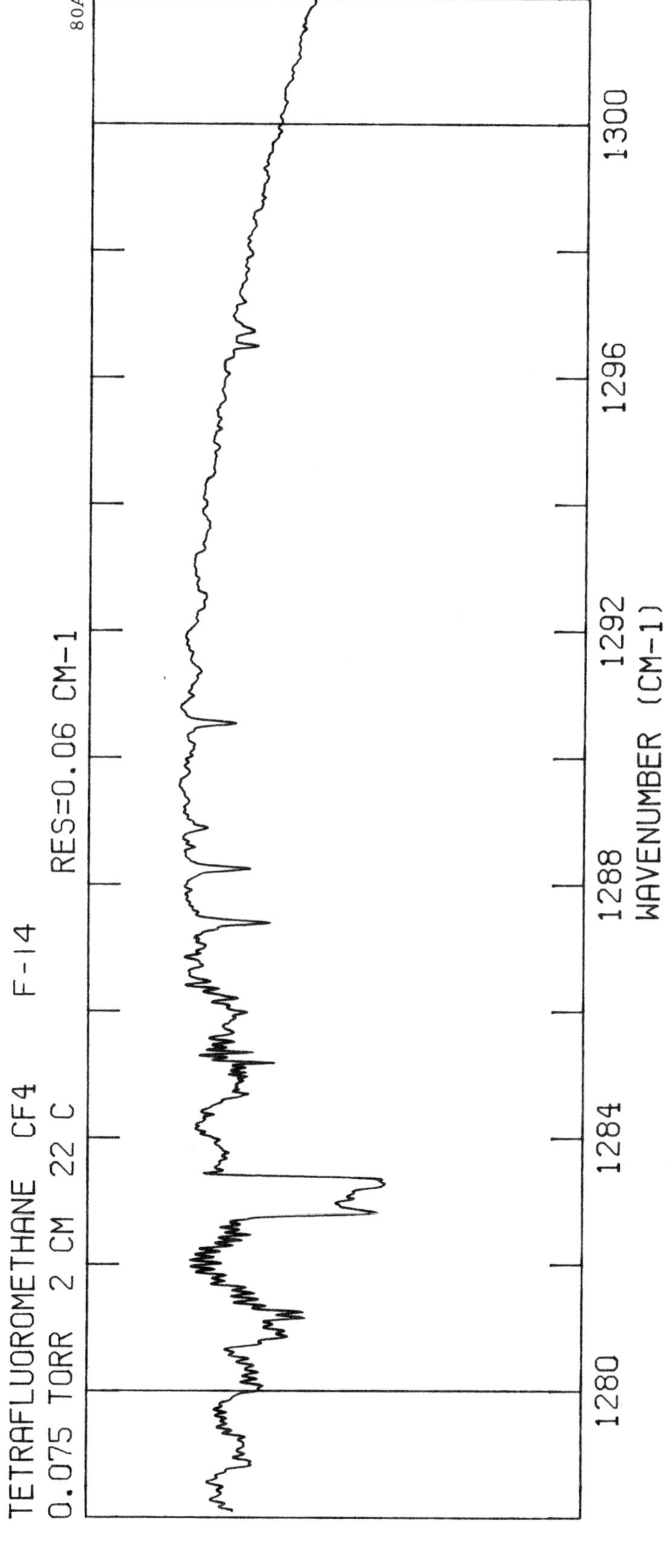
TETRAFLUOROMETHANE CF4 F-14
0.075 TORR 2 CM 22 C RES=0.06 CM-1
80A
1280
1284
1288
1292
1296
1300
WAVENUMBER (CM-1)

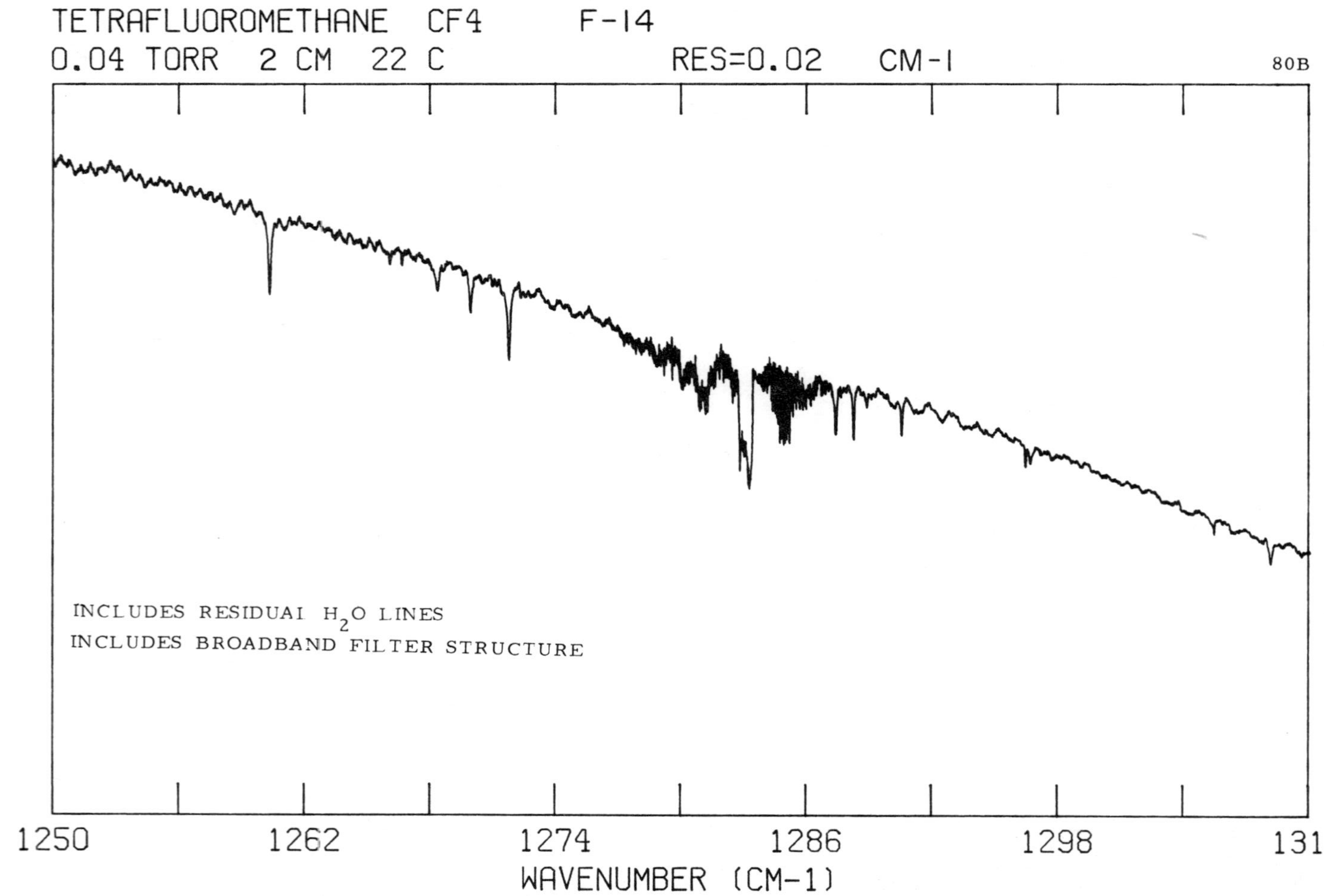

TETRAFLUOROMETHANE CF4 F-14
0.04 TORR 2 CM 22 C RES=0.02 CM-1
80B
INCLUDES RESIDUAI H_2O LINES
INCLUDES BROADBAND FILTER STRUCTURE
1250
1262
1274
1286
1298
1310
WAVENUMBER (CM-1)

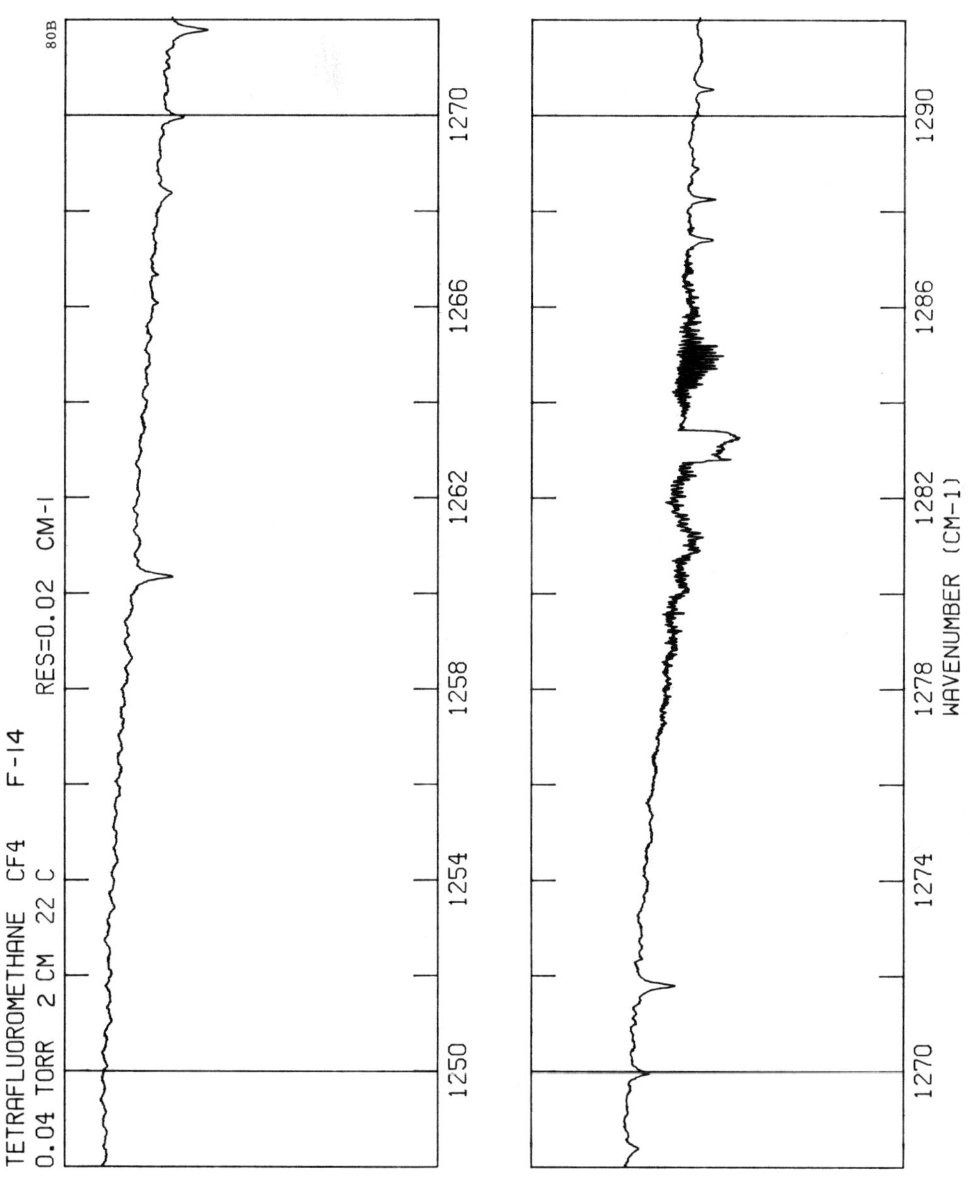
TETRAFLUOROMETHANE CF4 F-14
0.04 TORR 2 CM 22 C
RES=0.02 CM-1
80B
1250
1254
1258
1262
1266
1270
1270
1274
1278
1282
1286
1290
WAVENUMBER (CM-1)

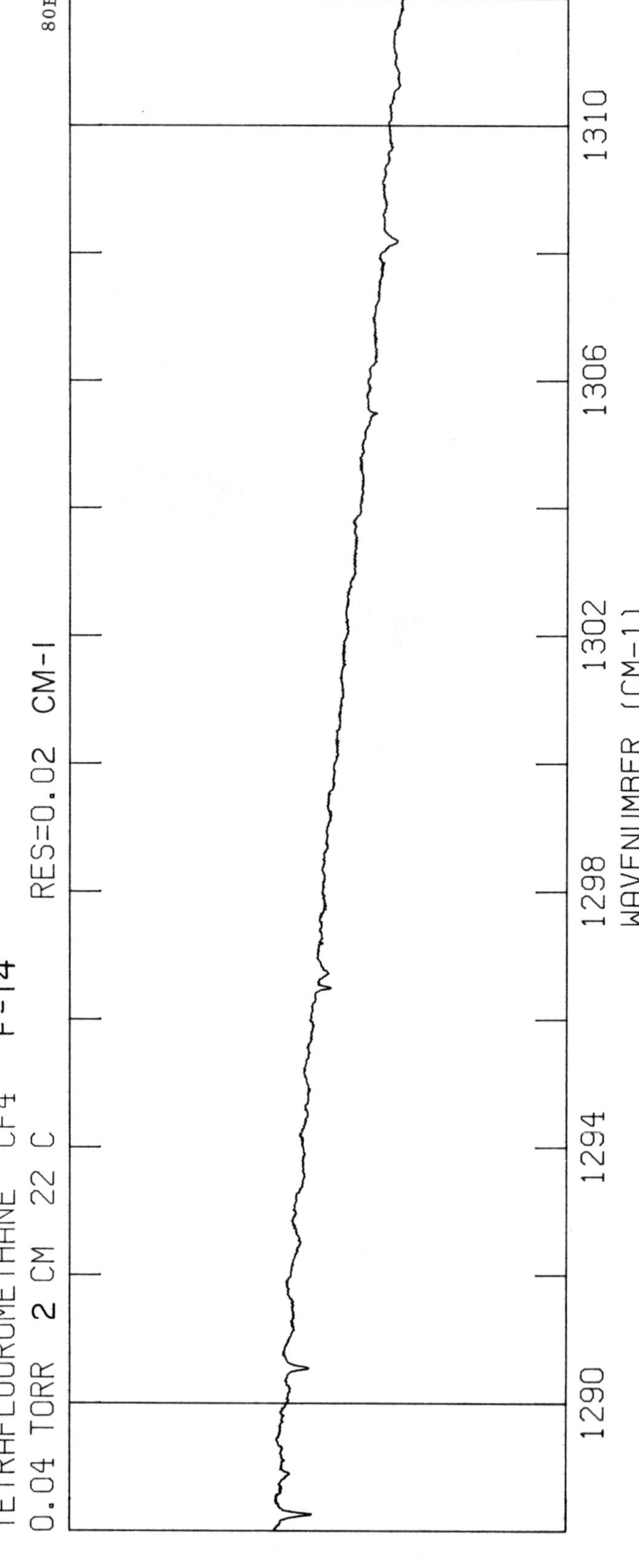
TETRAFLUOROMETHANE CF4 F-14
0.04 TORR 2 CM 22 C RES=0.02 CM-1
80B
1290
1294
1298
1302
1306
1310
WAVENUMBER (CM-1)

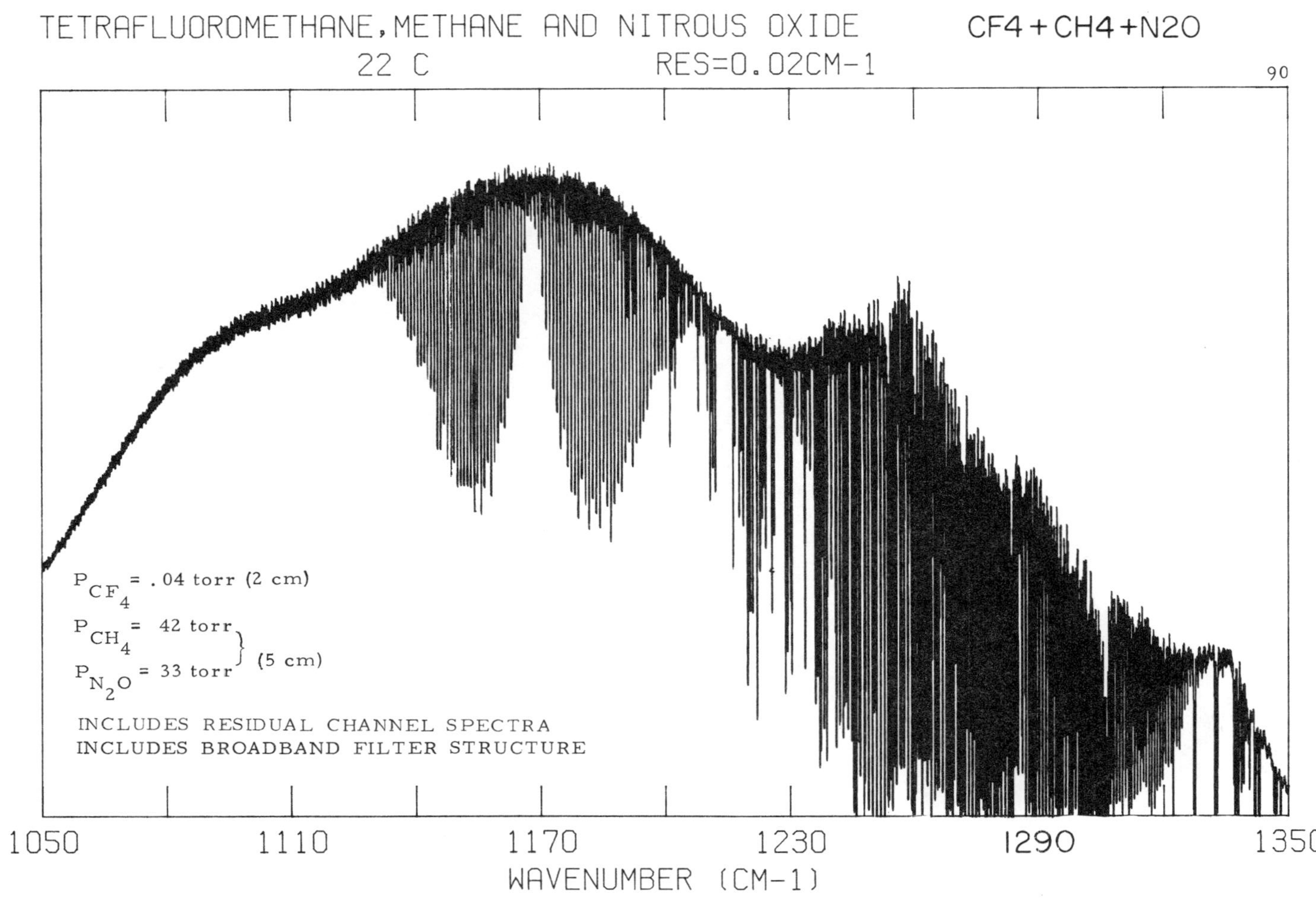

TETRAFLUOROMETHANE, METHANE AND NITROUS OXIDE
CF4+CH4+N2O
22 C
RES=0.02CM-1
90
P_{CF_4} = .04 torr (2 cm)
P_{CH_4} = 42 torr
P_{N_2O} = 33 torr
(5 cm)
INCLUDES RESIDUAL CHANNEL SPECTRA
INCLUDES BROADBAND FILTER STRUCTURE
1050
1110
1170
1230
1290
1350
WAVENUMBER (CM-1)

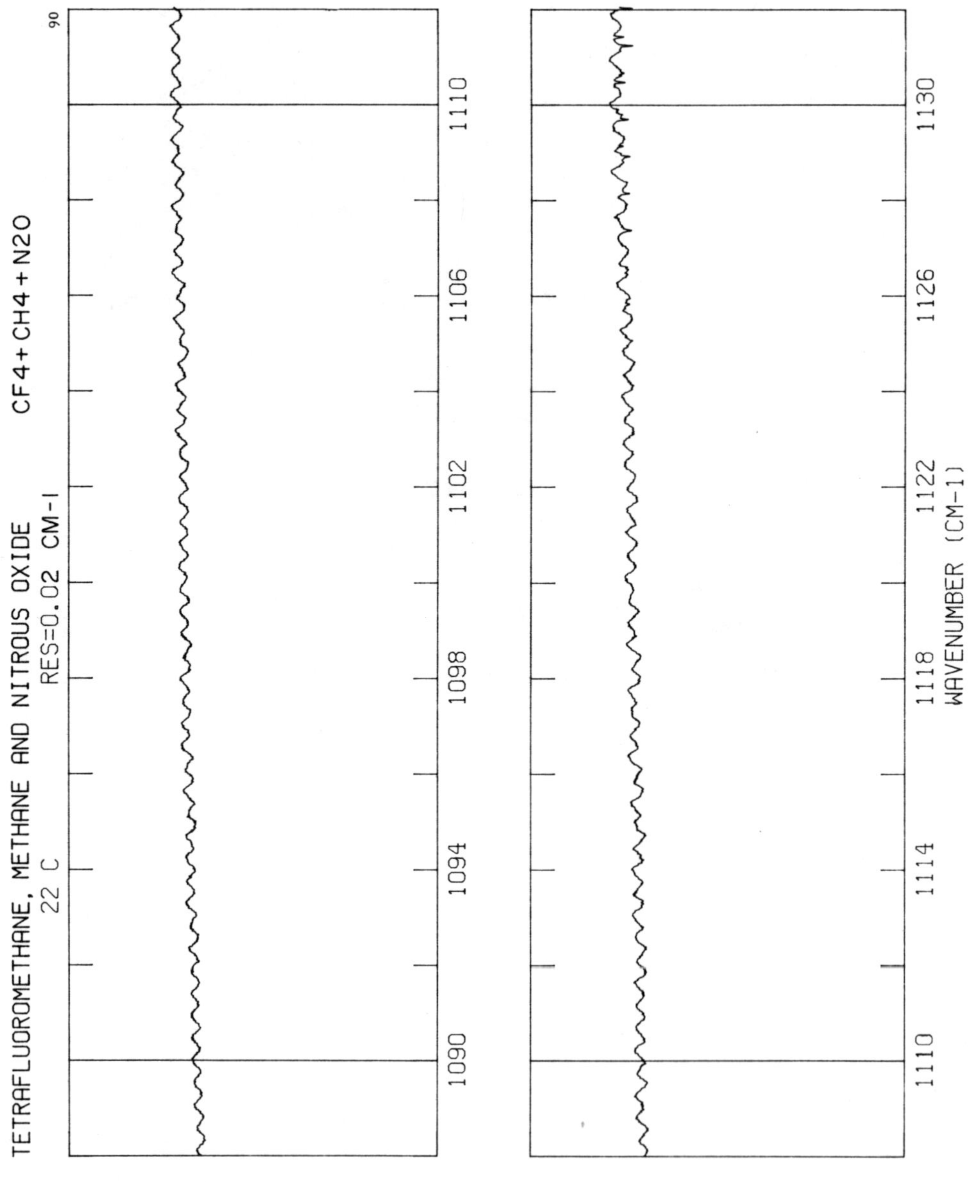
TETRAFLUOROMETHANE, METHANE AND NITROUS OXIDE
CF4+CH4+N2O
22 C
RES=0.02 CM-1
90
1090
1094
1098
1102
1106
1110
1110
1114
1118
1122
1126
1130
WAVENUMBER (CM-1)

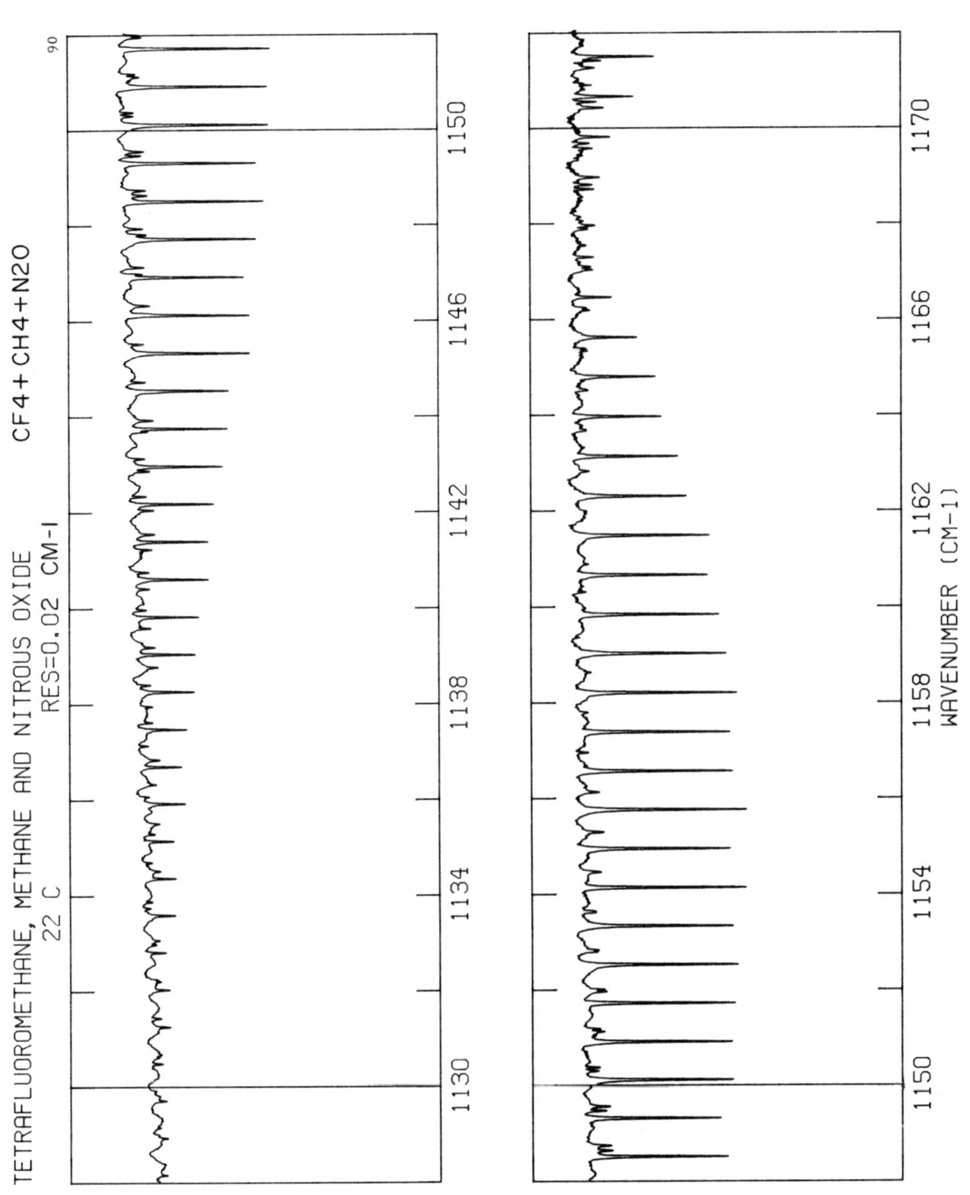
TETRAFLUOROMETHANE, METHANE AND NITROUS OXIDE
CF4+CH4+N2O
22 C
RES=0.02 CM-I
90
1130
1134
1138
1142
1146
1150
1154
1158
1162
1166
1170
WAVENUMBER (CM-1)

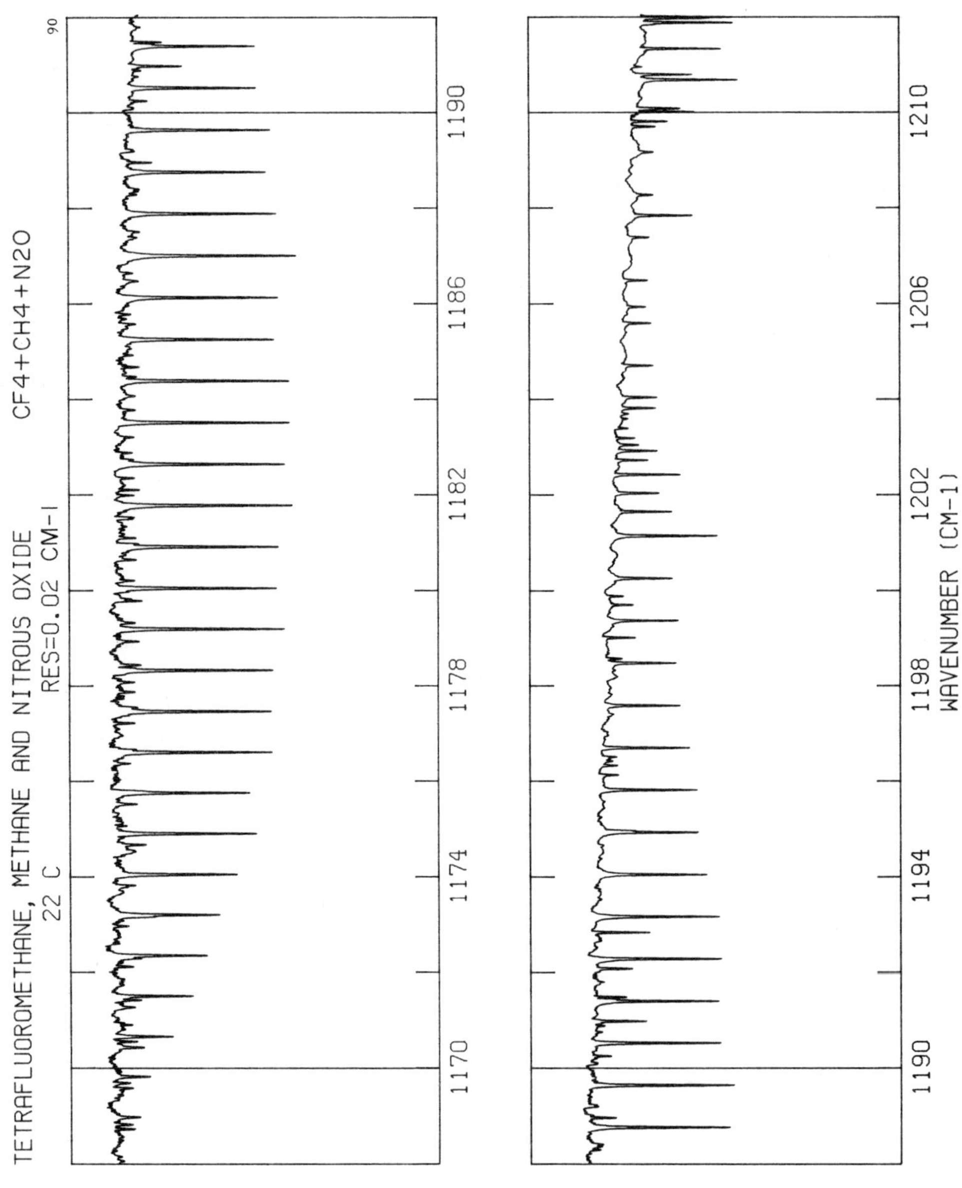
TETRAFLUOROMETHANE, METHANE AND NITROUS OXIDE
CF4+CH4+N2O
22 C
RES=0.02 CM-1
90
1170
1174
1178
1182
1186
1190
1190
1194
1198
1202
1206
1210
WAVENUMBER (CM-1)

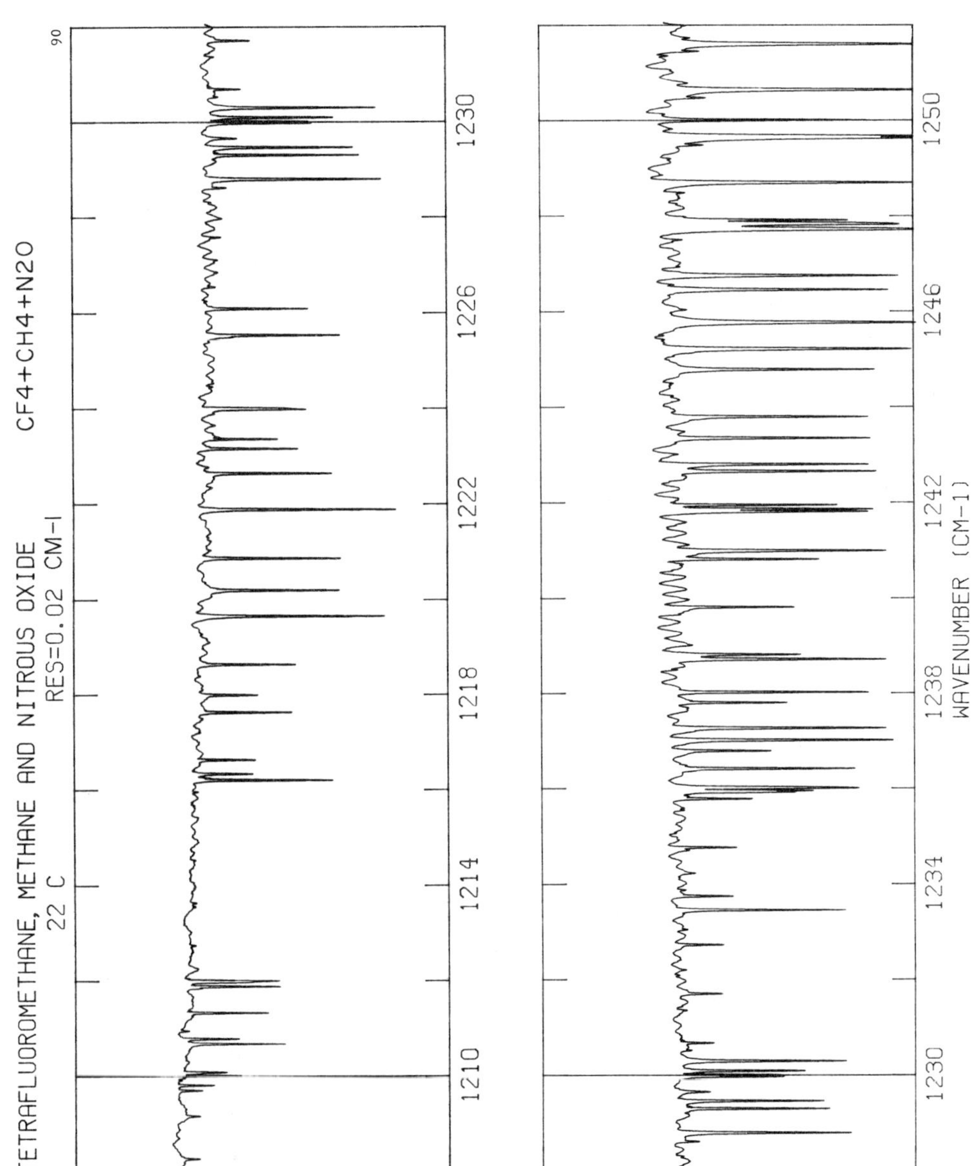

TETRAFLUOROMETHANE, METHANE AND NITROUS OXIDE
CF4+CH4+N2O
22 C
RES=0.02 CM-1
90
1210
1214
1218
1222
1226
1230
1230
1234
1238
1242
1246
1250
WAVENUMBER (CM-1)

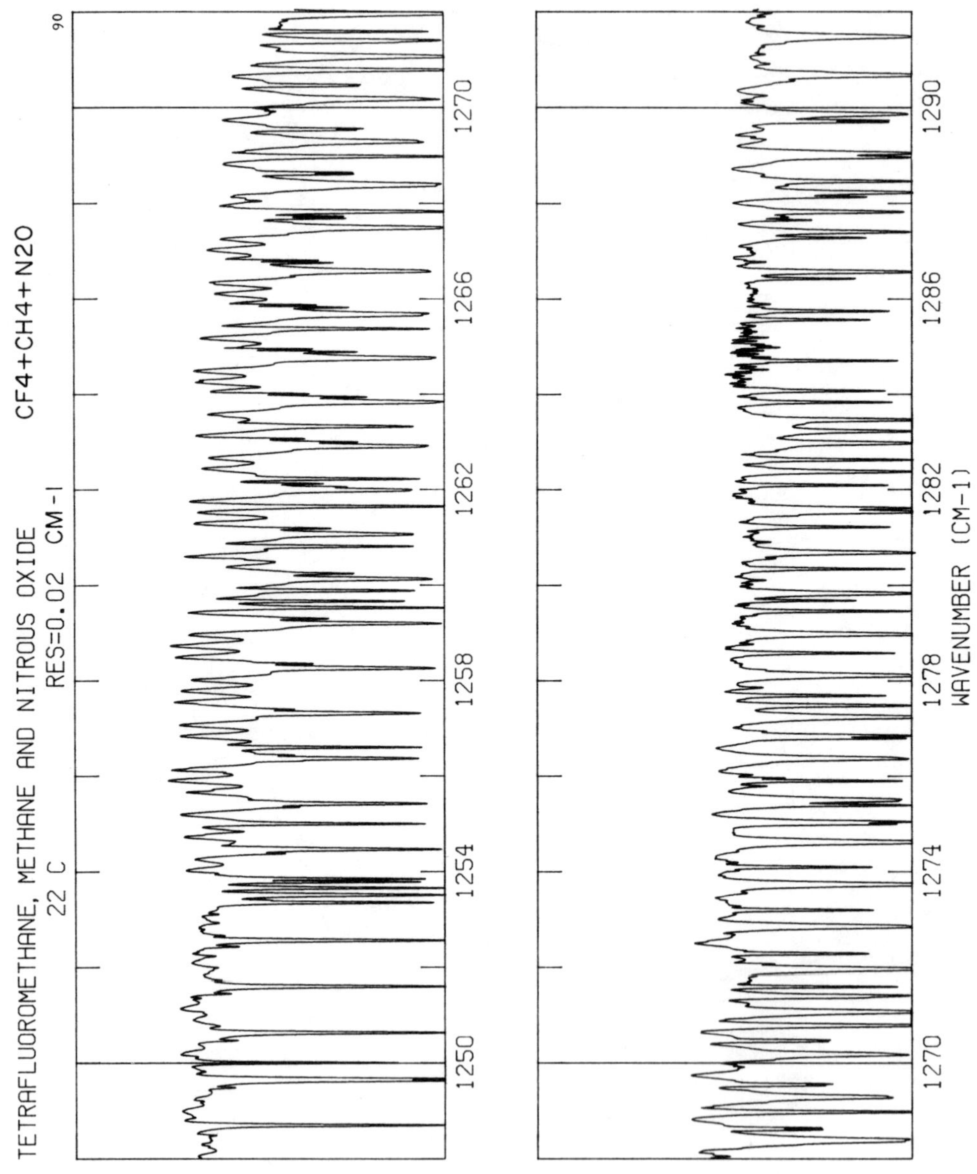
TETRAFLUOROMETHANE, METHANE AND NITROUS OXIDE
CF4+CH4+N2O
22 C
RES=0.02 CM-1
90
1250
1254
1258
1262
1266
1270
1274
1278
1282
1286
1290
WAVENUMBER (CM-1)

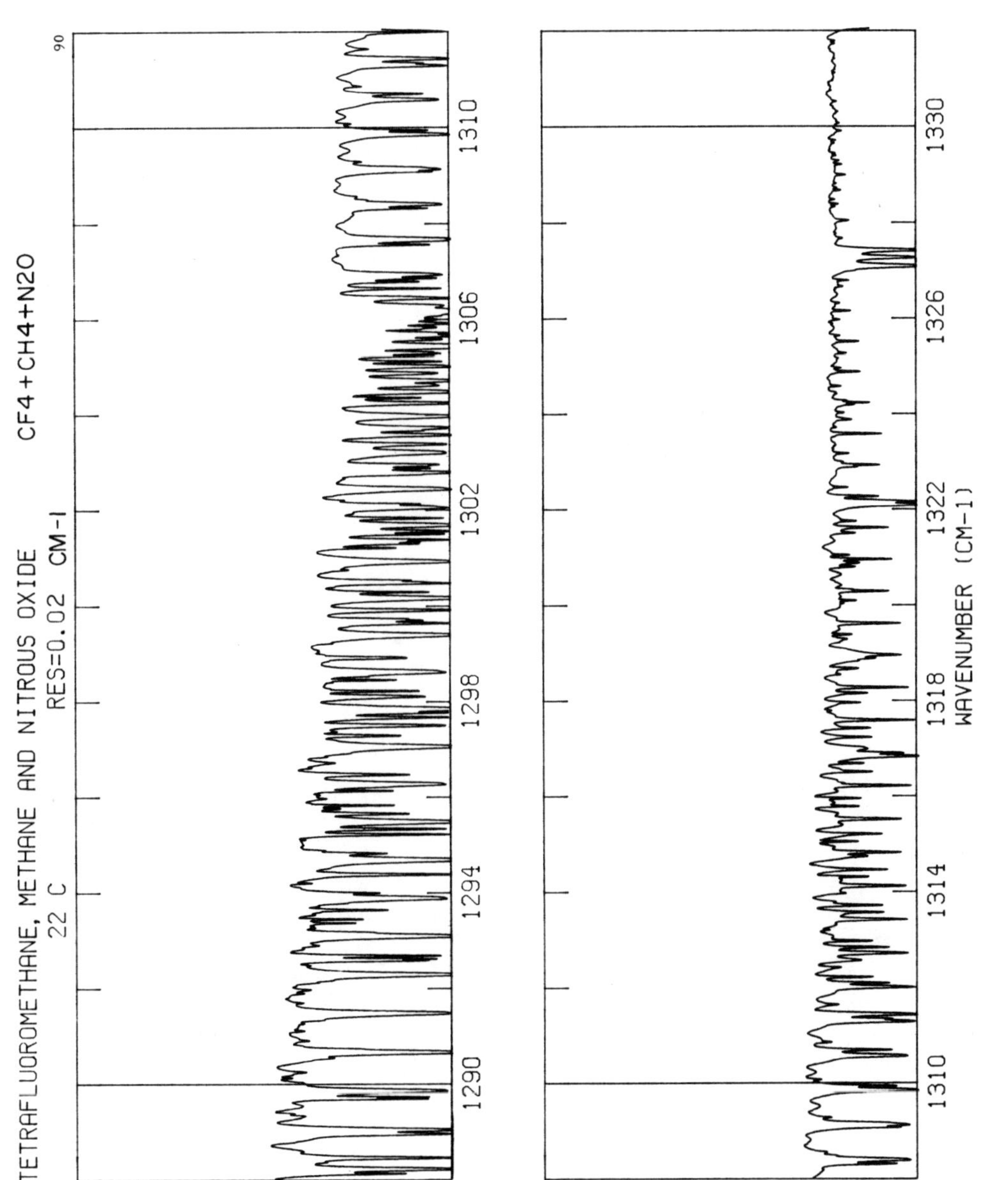

TETRAFLUOROMETHANE, METHANE AND NITROUS OXIDE
22 C
RES=0.02 CM-1
CF4+CH4+N2O
90
1290
1294
1298
1302
1306
1310
1314
1318
1322
1326
1330
WAVENUMBER (CM-1)

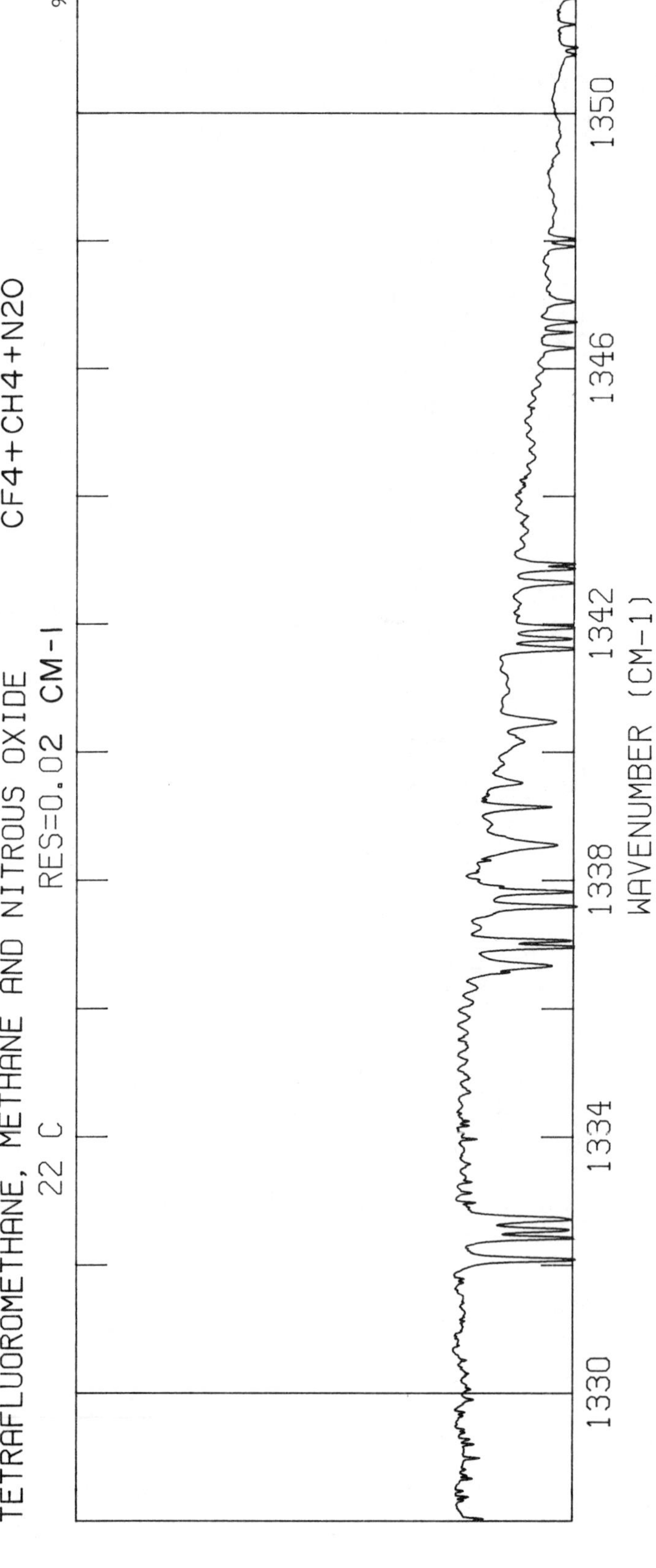
TETRAFLUOROMETHANE, METHANE AND NITROUS OXIDE
CF4+CH4+N2O
22 C
RES=0.02 CM-1
90
1330
1334
1338
1342
1346
1350
WAVENUMBER (CM-1)

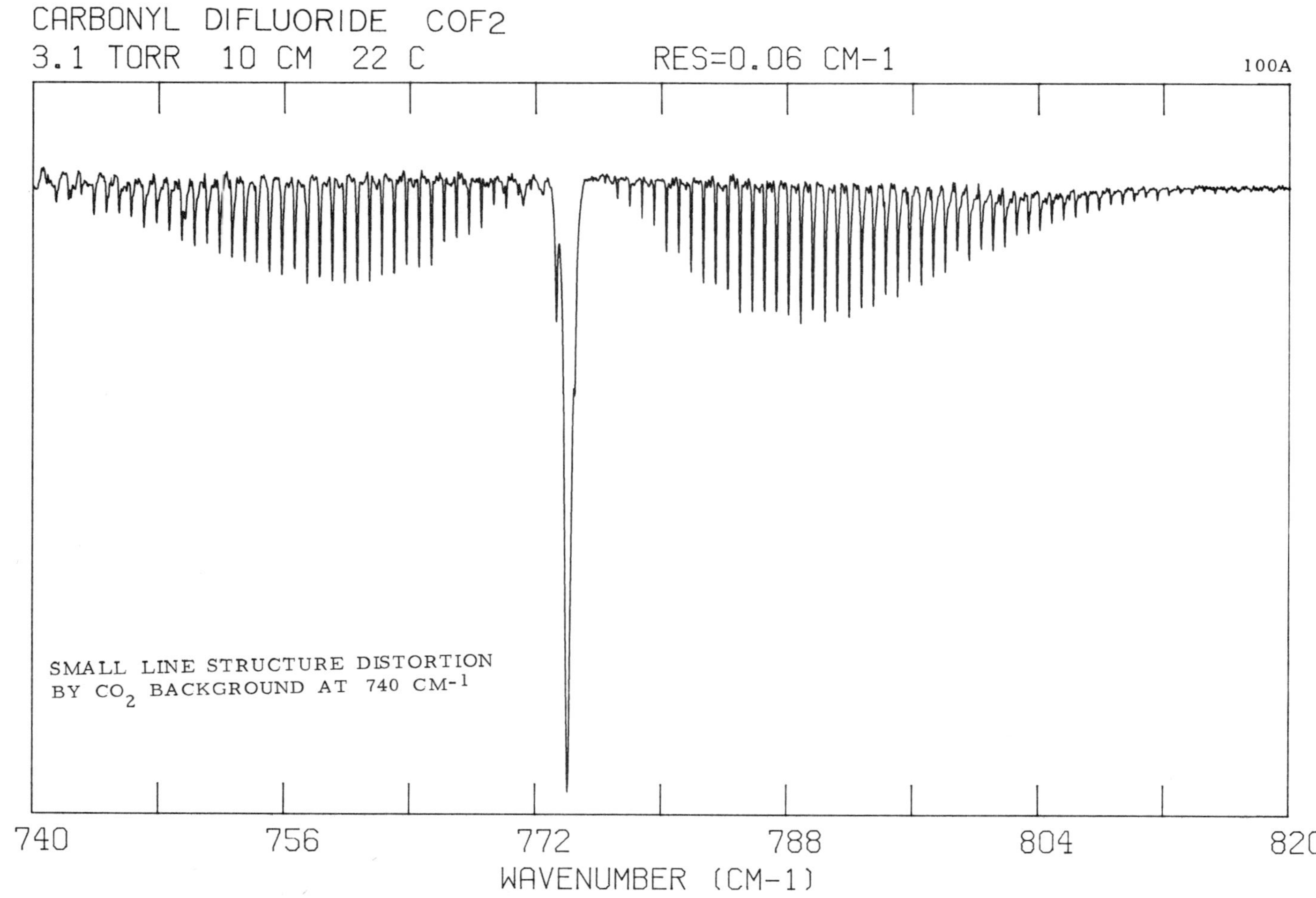
CARBONYL DIFLUORIDE COF2
3.1 TORR 10 CM 22 C
RES=0.06 CM-1
100A
SMALL LINE STRUCTURE DISTORTION
BY CO2 BACKGROUND AT 740 CM-1
740
756
772
788
804
820
WAVENUMBER (CM-1)

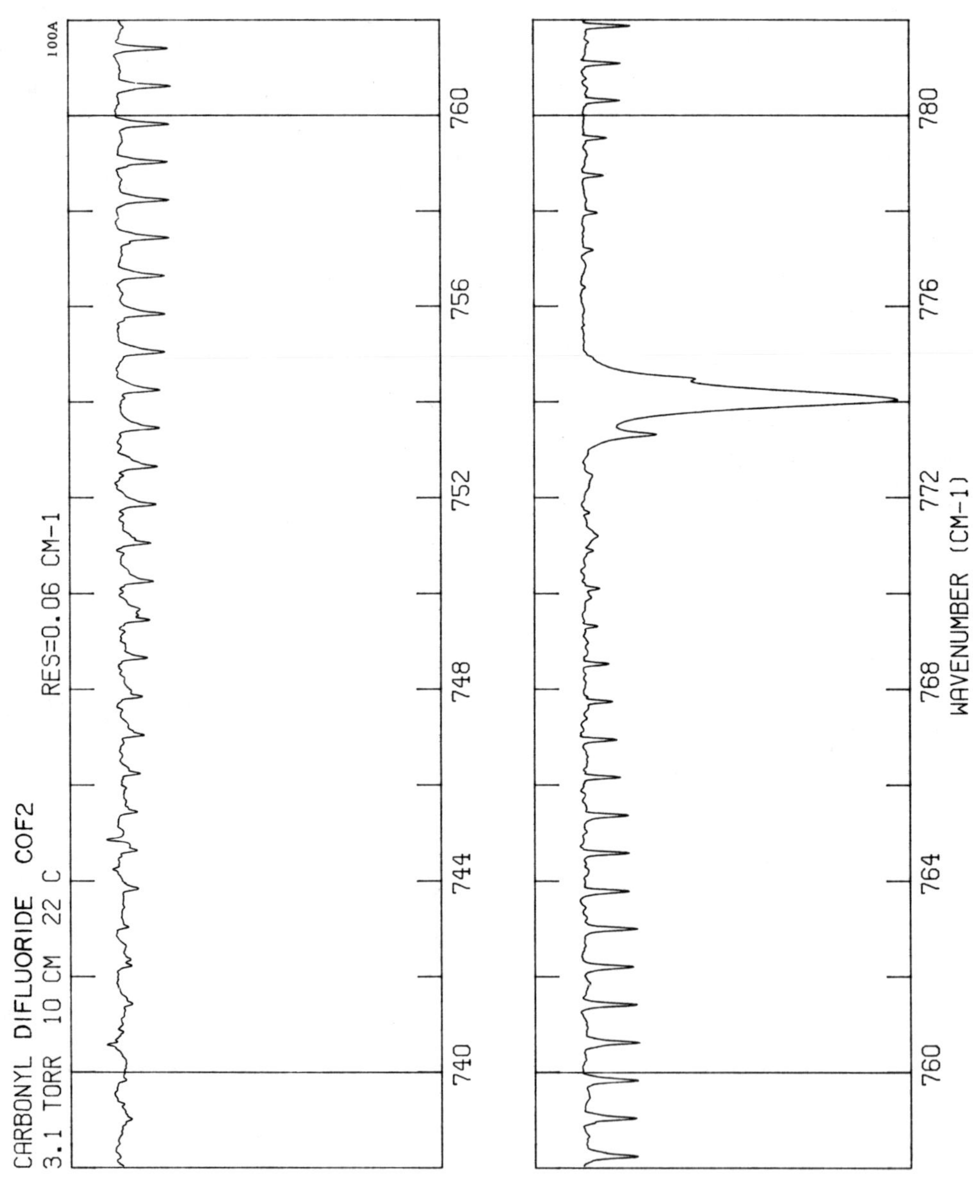
CARBONYL DIFLUORIDE COF2
3.1 TORR 10 CM 22 C
RES=0.06 CM-1
100A
740
744
748
752
756
760
760
764
768
772
776
780
WAVENUMBER (CM-1)

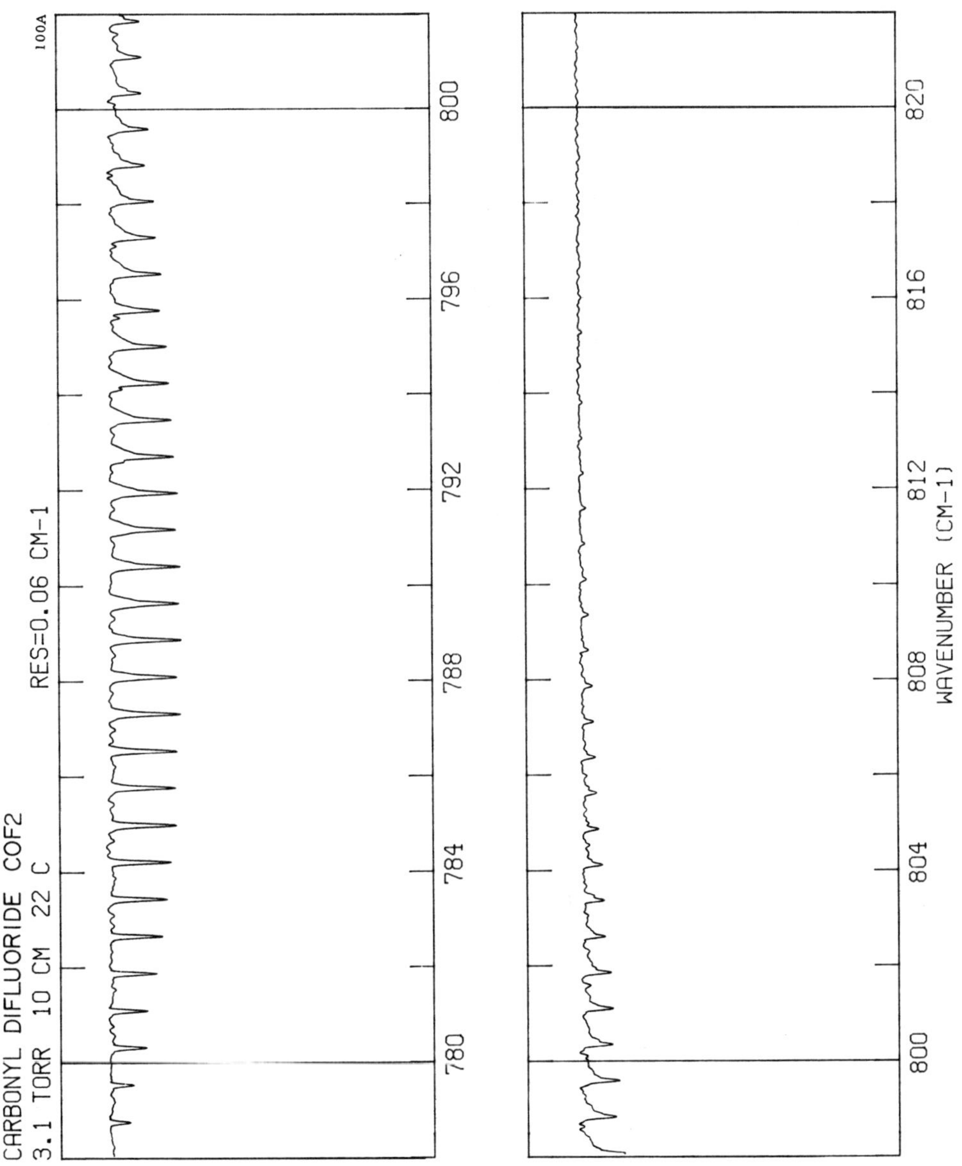
CARBONYL DIFLUORIDE COF2
3.1 TORR 10 CM 22 C
RES=0.06 CM-1
100A
780
784
788
792
796
800
800
804
808
812
816
820
WAVENUMBER (CM-1)

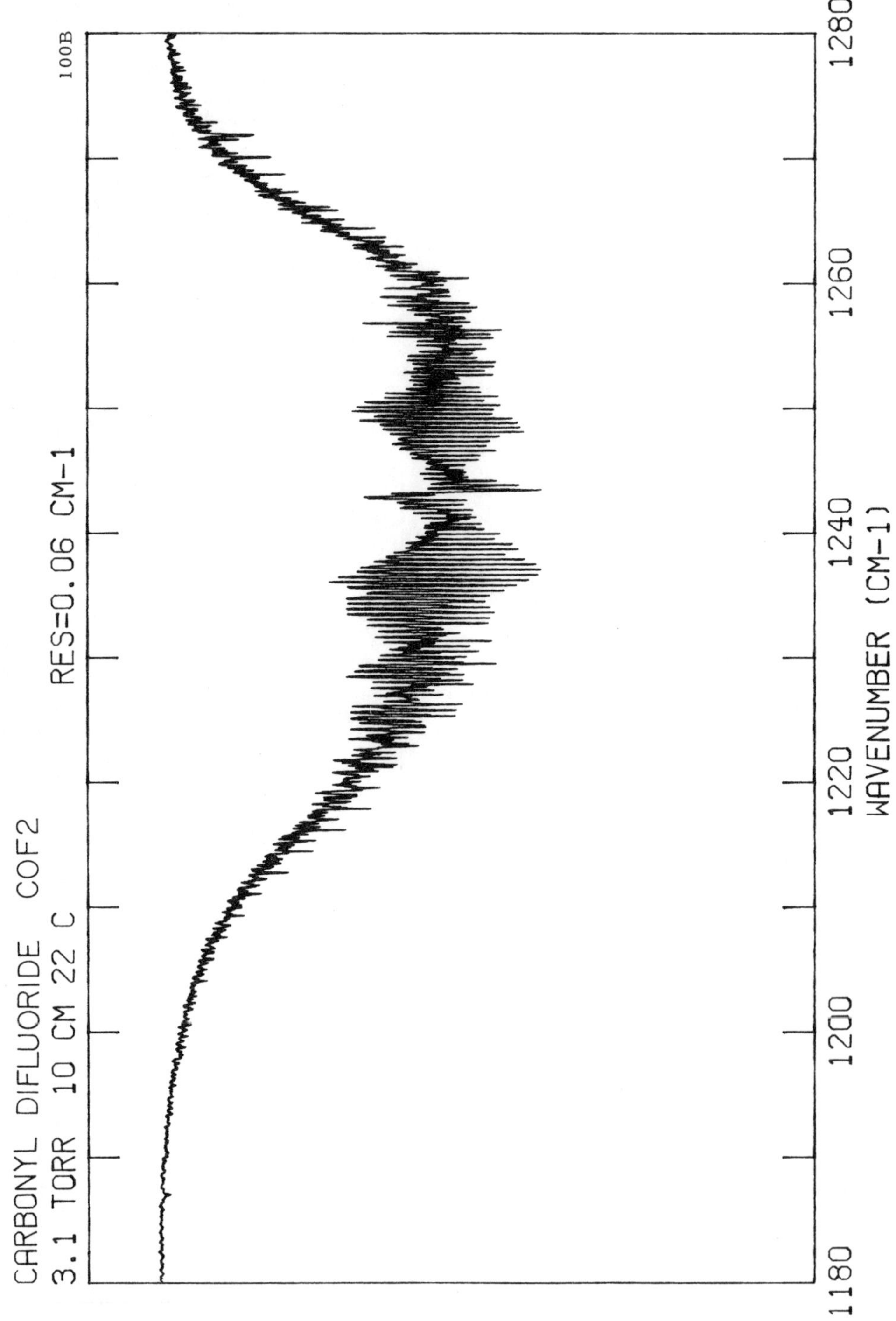
CARBONYL DIFLUORIDE COF2
3.1 TORR 10 CM 22 C
RES=0.06 CM-1
100B
1180
1200
1220
1240
1260
1280
WAVENUMBER (CM-1)

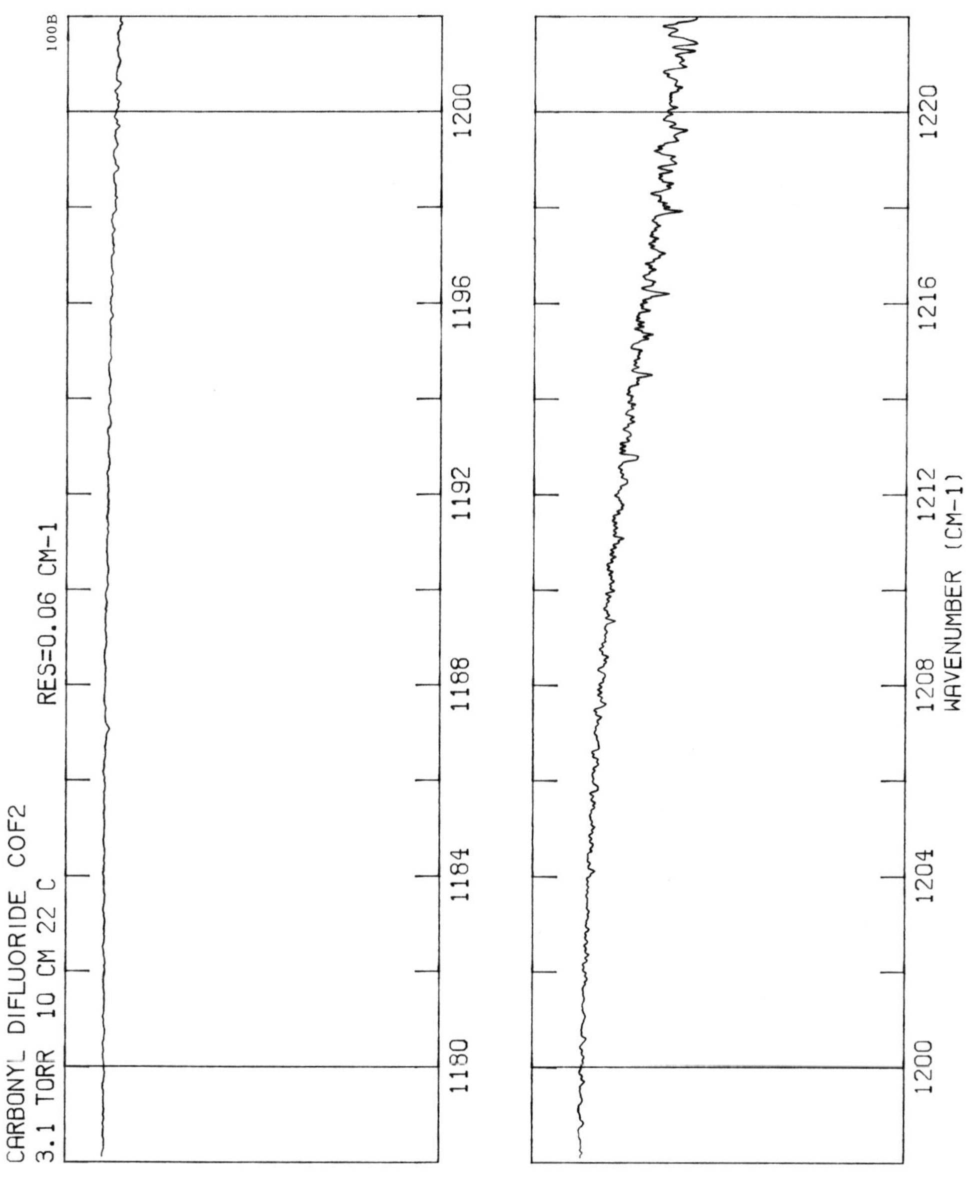
CARBONYL DIFLUORIDE COF2
3.1 TORR 10 CM 22 C
RES=0.06 CM-1
100B
1180
1184
1188
1192
1196
1200
1200
1204
1208
1212
1216
1220
WAVENUMBER (CM-1)

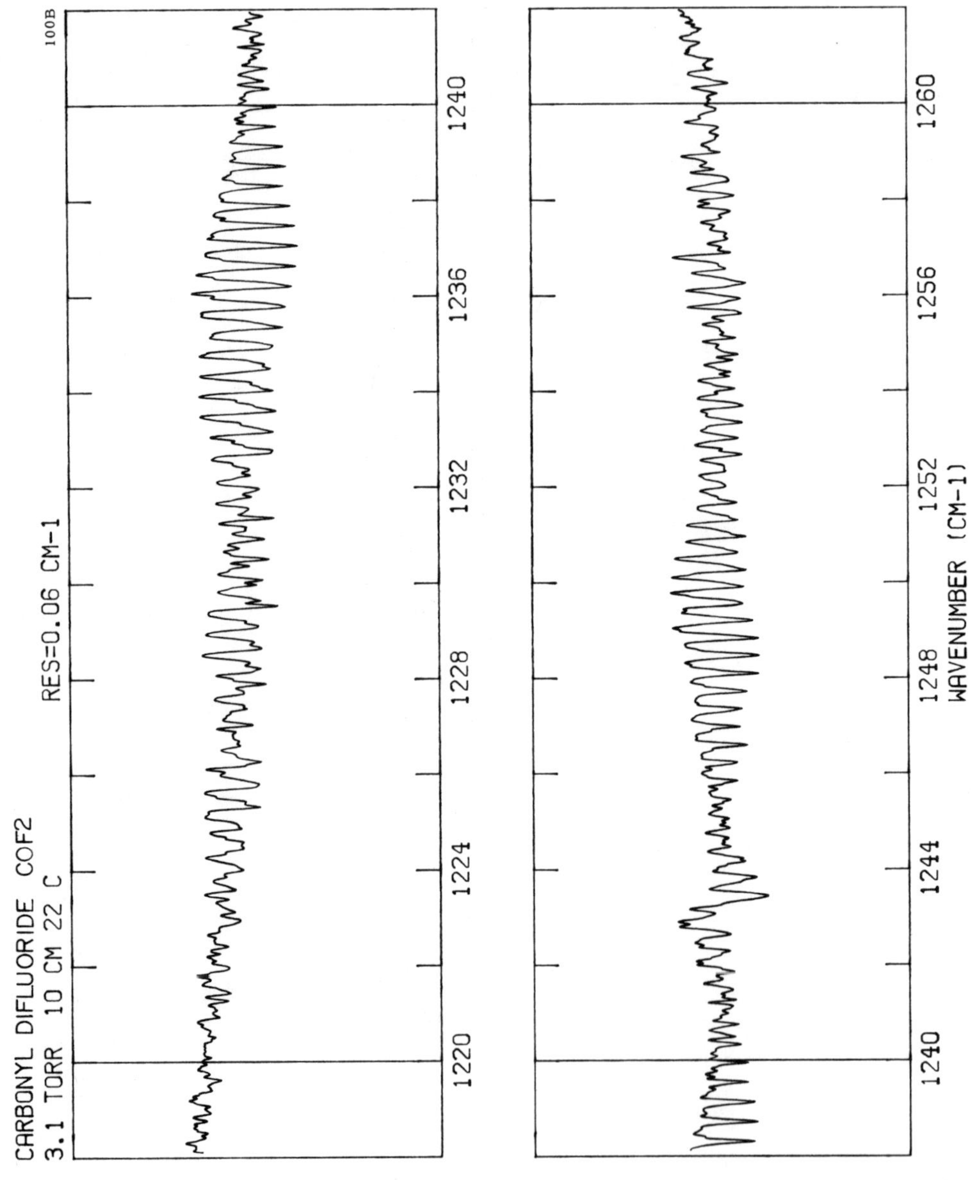
CARBONYL DIFLUORIDE COF2
3.1 TORR 10 CM 22 C
RES=0.06 CM-1
100B
1220
1224
1228
1232
1236
1240
1240
1244
1248
1252
1256
1260
WAVENUMBER (CM-1)

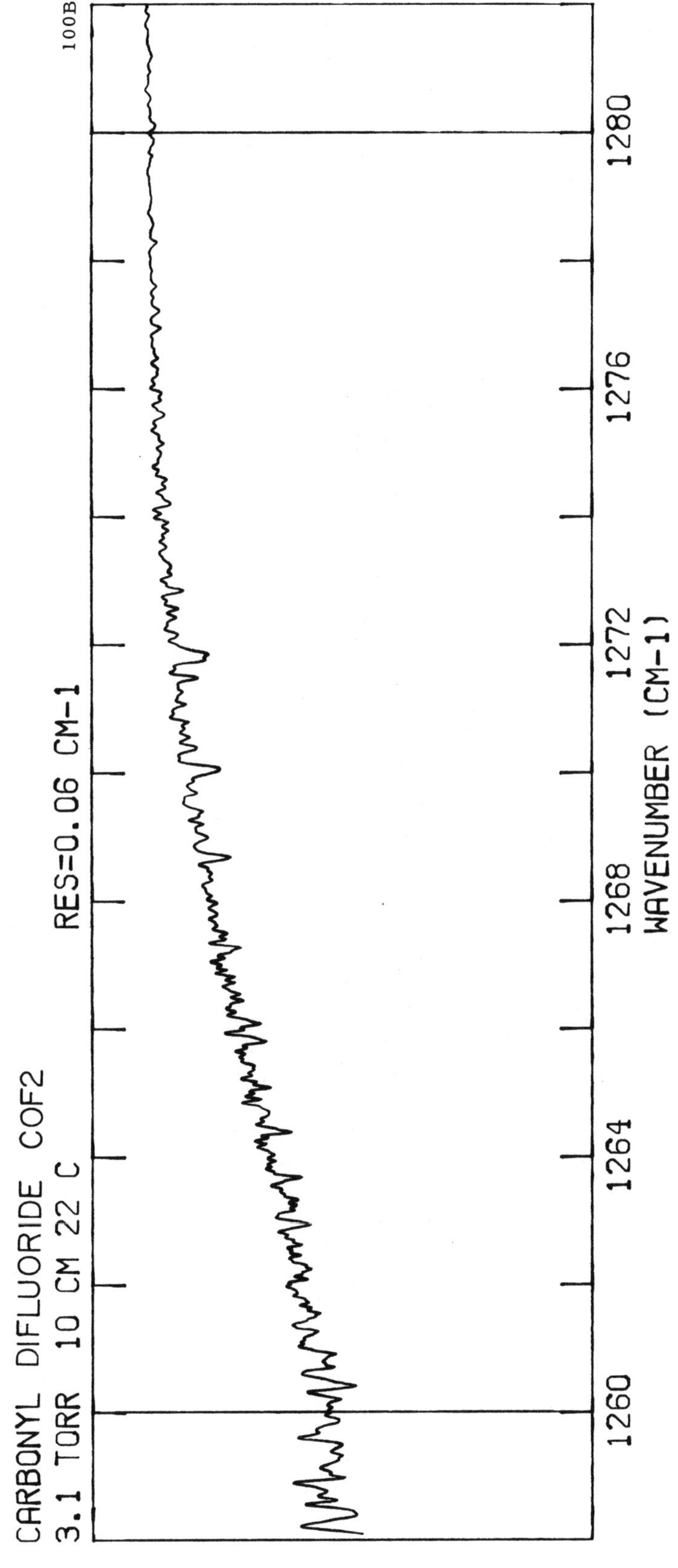
CARBONYL DIFLUORIDE COF2
3.1 TORR 10 CM 22 C
RES=0.06 CM-1
100B
1260
1264
1268
1272
1276
1280
WAVENUMBER (CM-1)

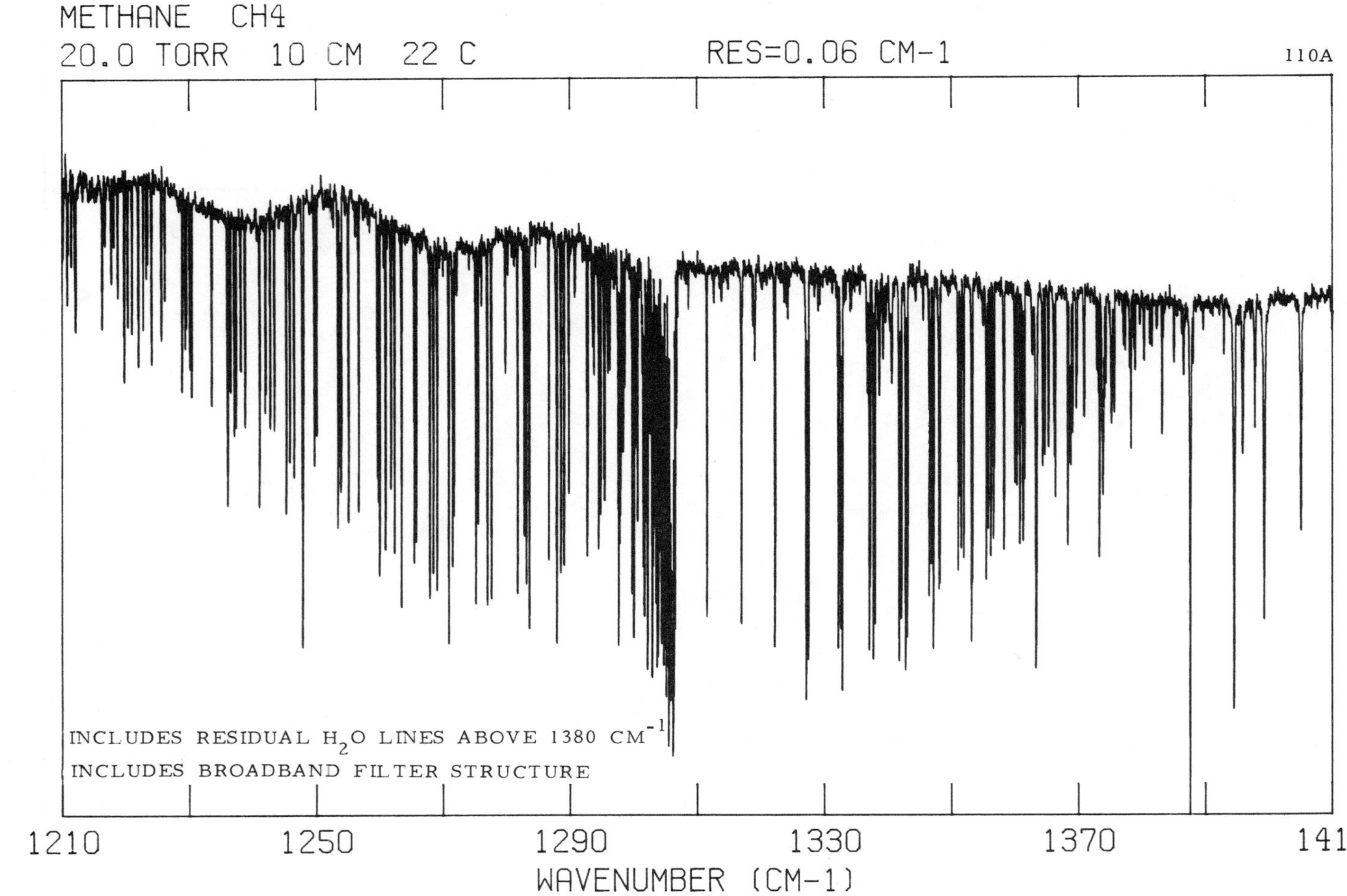
METHANE CH4
20.0 TORR 10 CM 22 C
RES=0.06 CM-1
110A
INCLUDES RESIDUAL H_2O LINES ABOVE 1380 CM^{-1}
INCLUDES BROADBAND FILTER STRUCTURE
1210
1250
1290
1330
1370
1410
WAVENUMBER (CM-1)

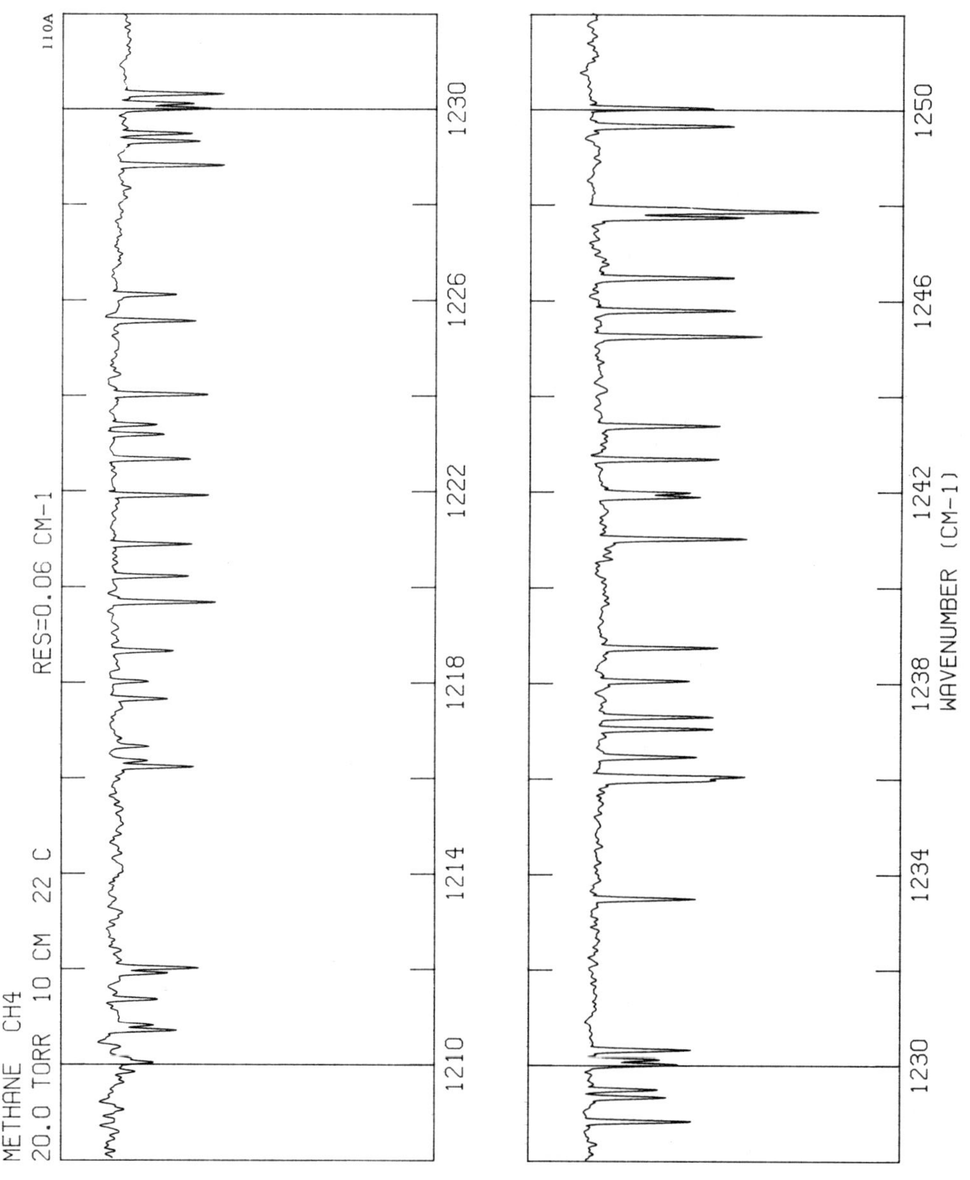
METHANE CH4
20.0 TORR 10 CM 22 C
RES=0.06 CM-1
110A
1210
1214
1218
1222
1226
1230
1234
1238
1242
1246
1250
WAVENUMBER (CM-1)

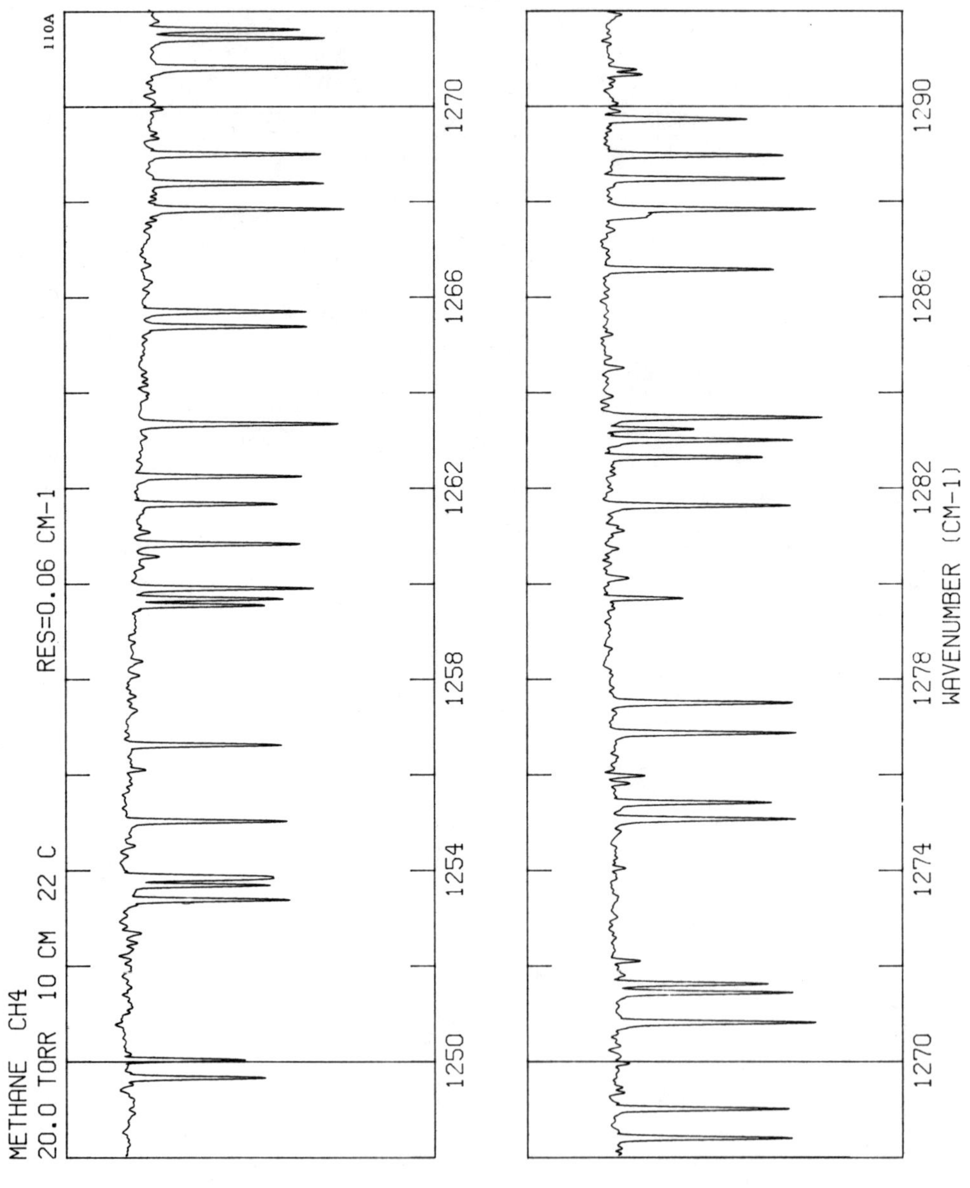
METHANE CH4
20.0 TORR 10 CM 22 C
RES=0.06 CM-1
110A
1250
1254
1258
1262
1266
1270
1274
1278
1282
1286
1290
WAVENUMBER (CM-1)

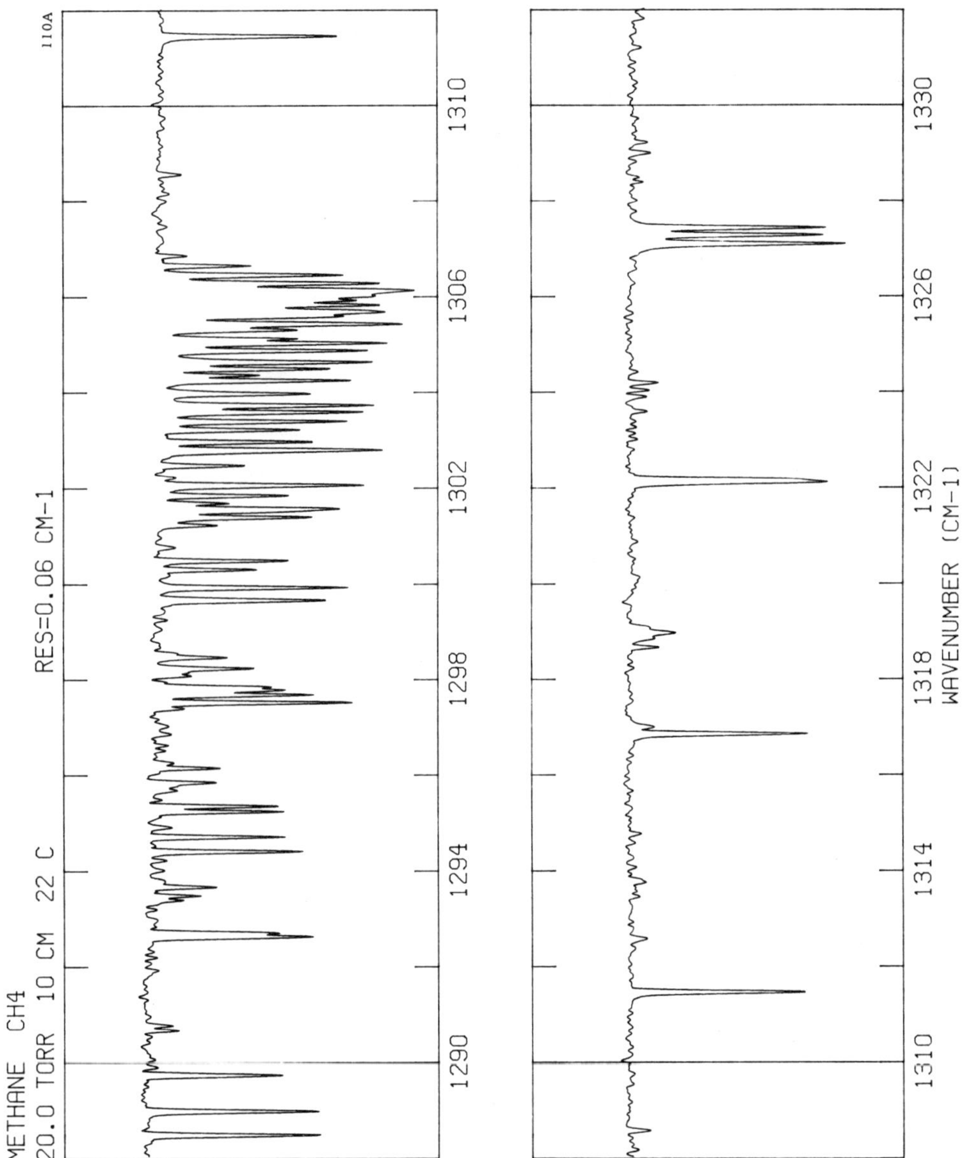
METHANE CH4
20.0 TORR 10 CM 22 C
RES=0.06 CM-1
110A
1290
1294
1298
1302
1306
1310
1310
1314
1318
1322
1326
1330
WAVENUMBER (CM-1)

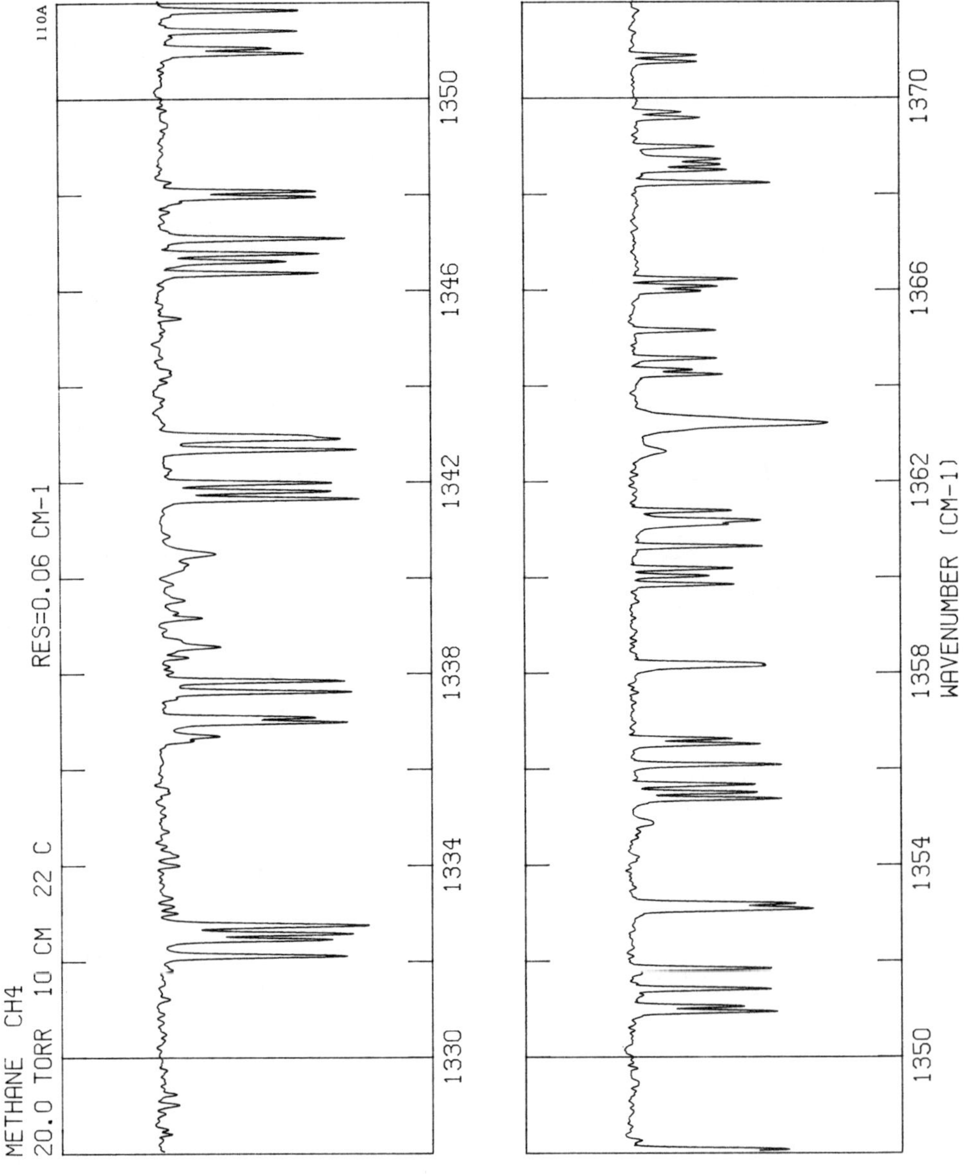
METHANE CH4
20.0 TORR 10 CM 22 C
RES=0.06 CM-1
110A
1330
1334
1338
1342
1346
1350
1350
1354
1358
1362
1366
1370
WAVENUMBER (CM-1)

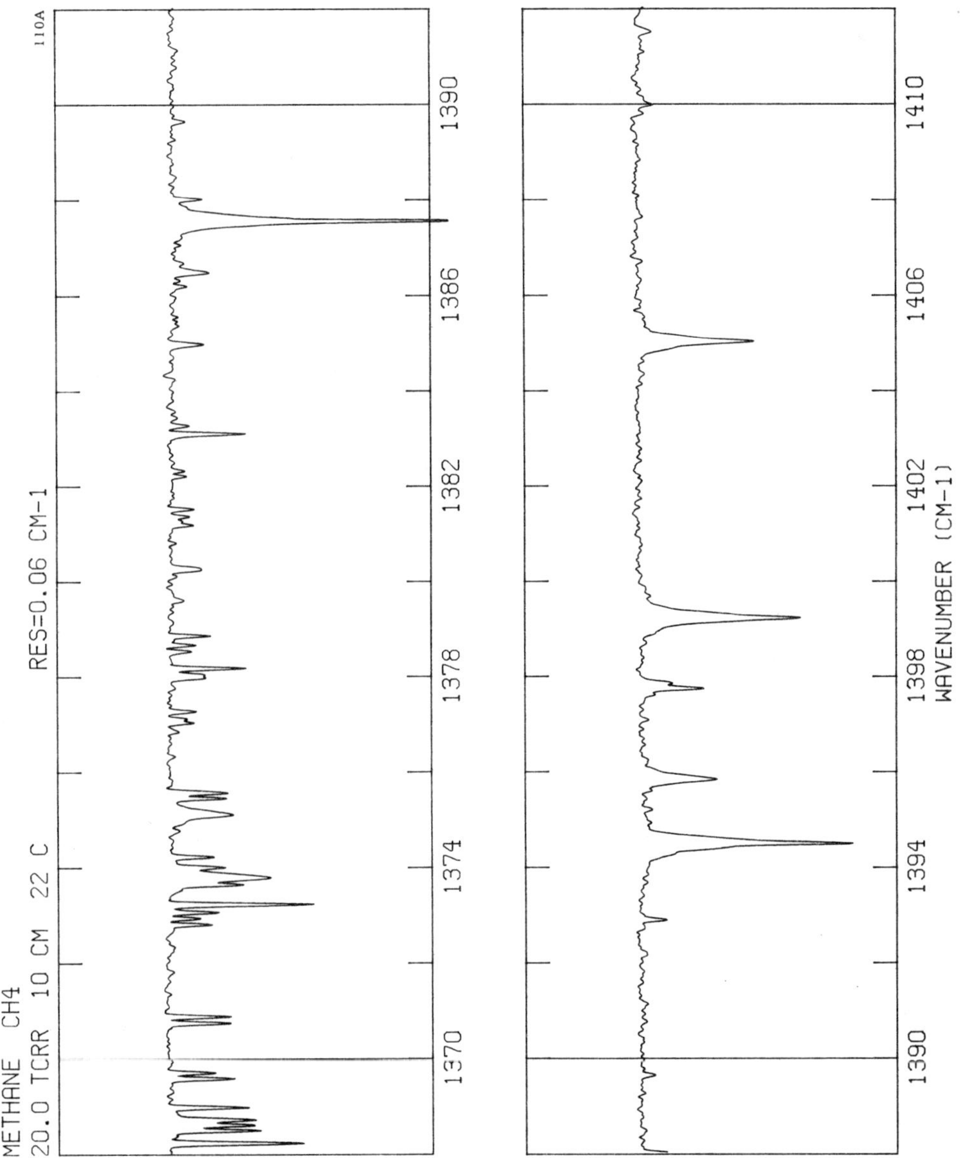
METHANE CH4
20.0 TORR 10 CM 22 C
RES=0.06 CM-1
110A
1370
1374
1378
1382
1386
1390
1390
1394
1398
1402
1406
1410
WAVENUMBER (CM-1)

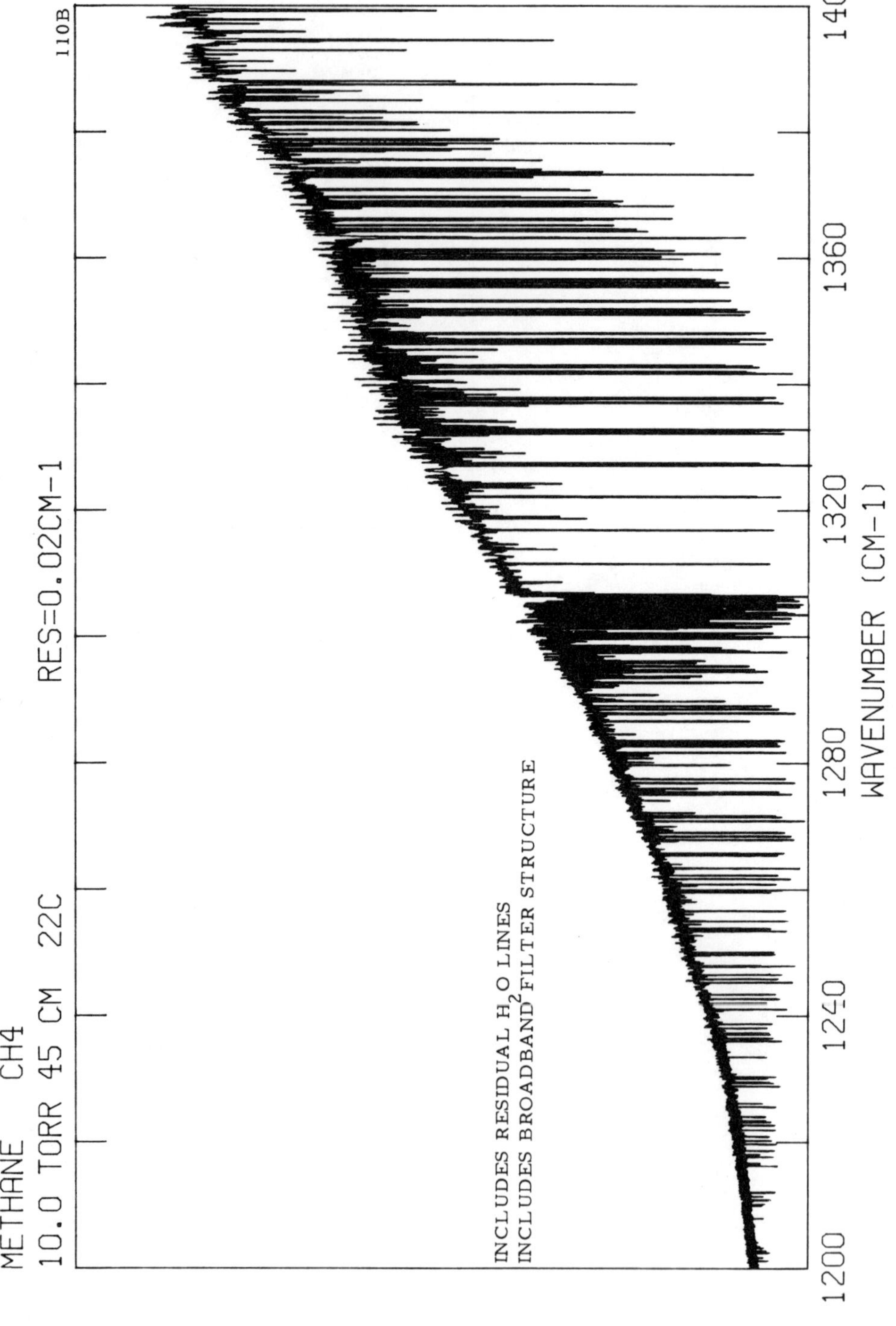

METHANE CH4
10.0 TORR 45 CM 22C
RES=0.02CM-1
110B
INCLUDES RESIDUAL H2O LINES
INCLUDES BROADBAND FILTER STRUCTURE
1200
1240
1280
1320
1360
1400
WAVENUMBER (CM-1)

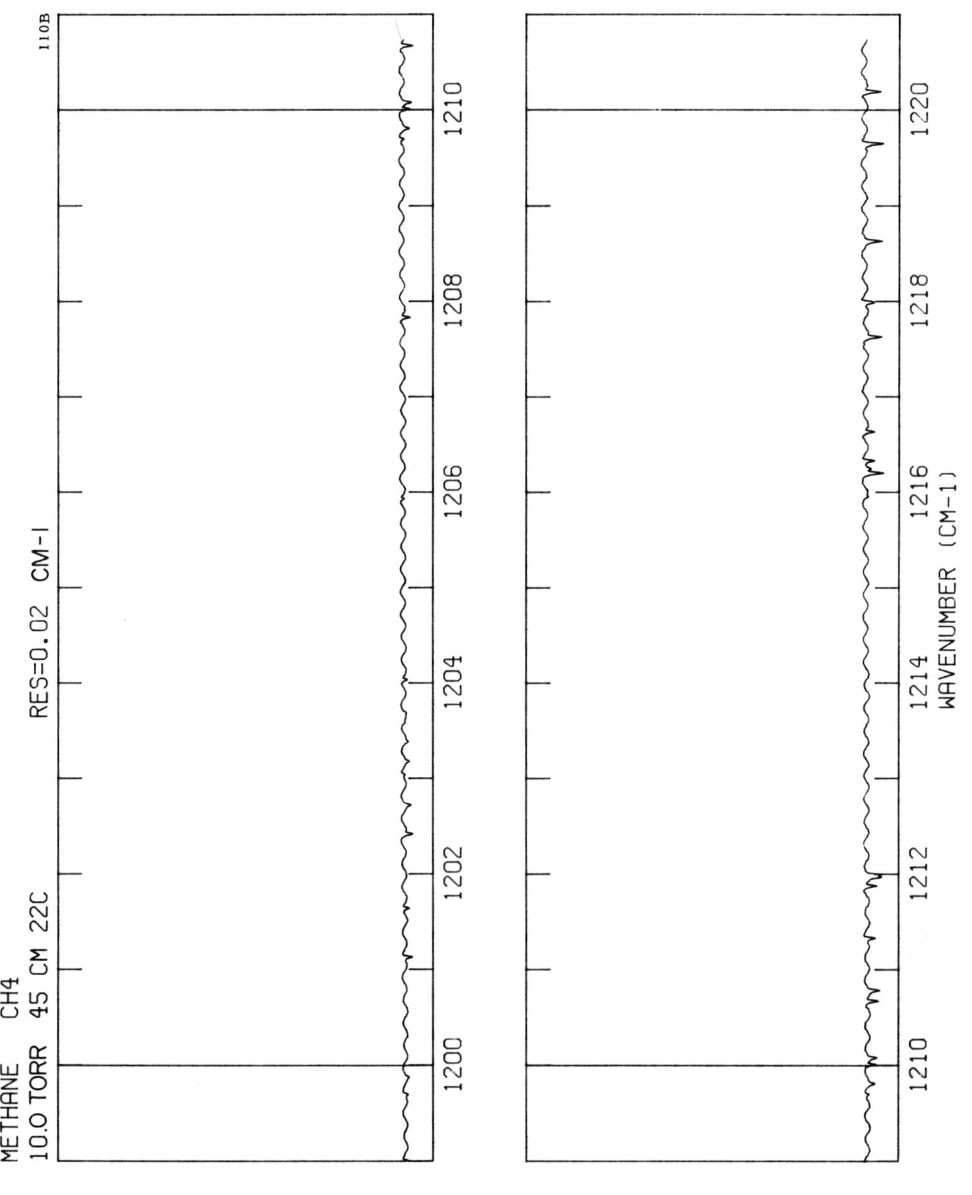
110B
METHANE CH4
10.0 TORR 45 CM 22C
RES=0.02 CM-1
1200
1202
1204
1206
1208
1210
1210
1212
1214
1216
1218
1220
WAVENUMBER (CM-1)

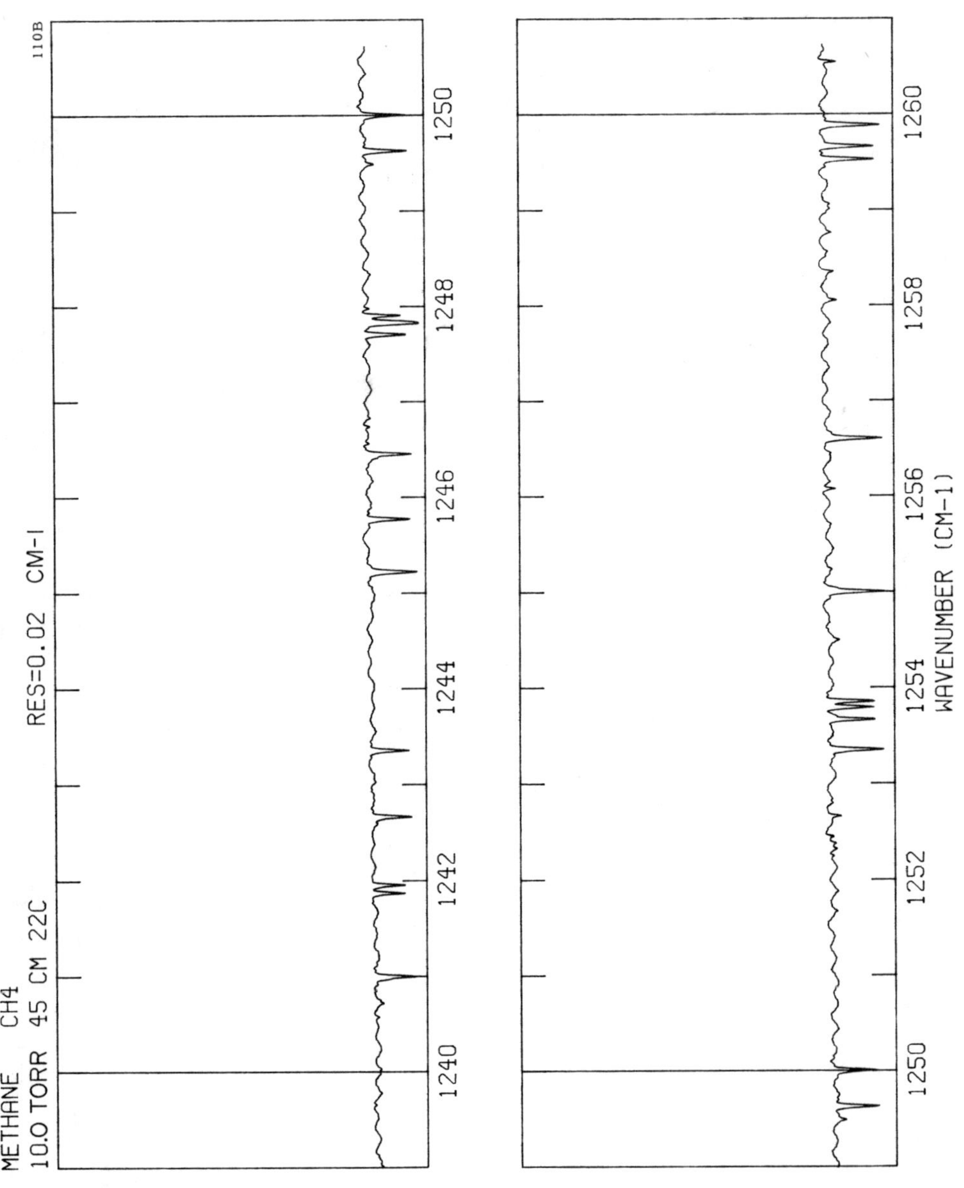
METHANE CH4
10.0 TORR 45 CM 22C
RES=0.02 CM-1
110B
1240
1242
1244
1246
1248
1250
1252
1254
1256
1258
1260
WAVENUMBER (CM-1)

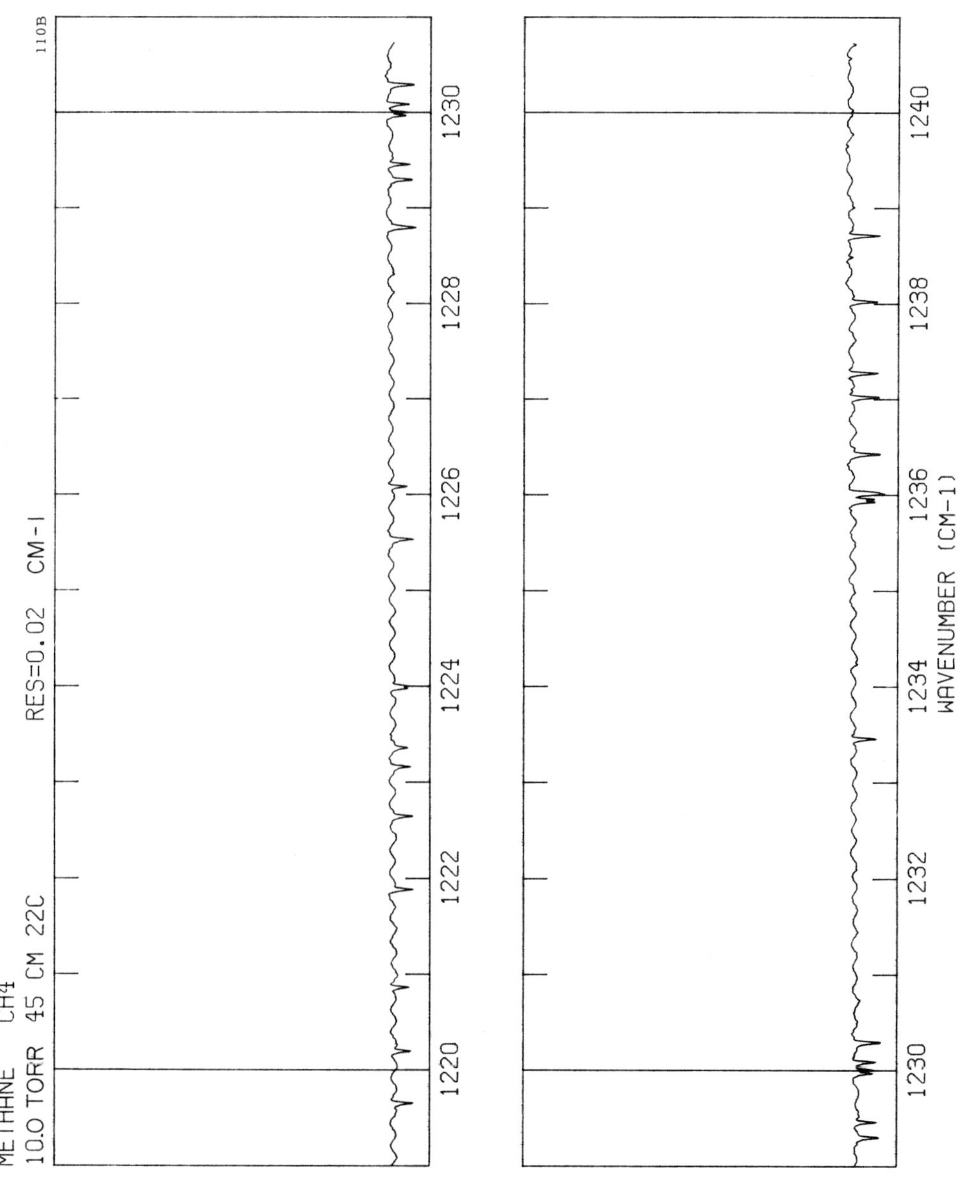
METHANE CH4
10.0 TORR 45 CM 22C
RES=0.02 CM-1
110B
1220
1222
1224
1226
1228
1230
1232
1234
1236
1238
1240
WAVENUMBER (CM-1)

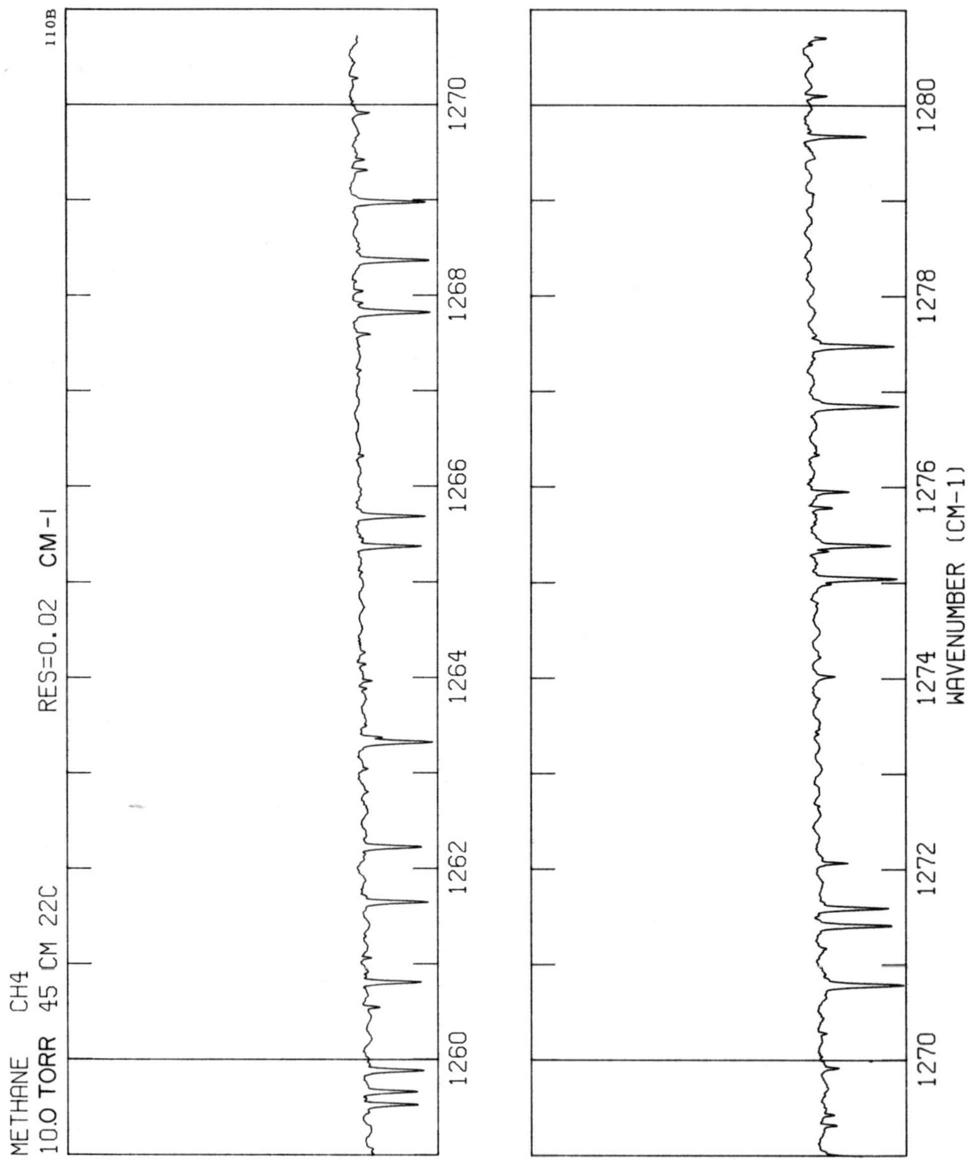
METHANE CH4
10.0 TORR 45 CM 22C
RES=0.02 CM-1
110B
1260
1262
1264
1266
1268
1270
1272
1274
1276
1278
1280
WAVENUMBER (CM-1)

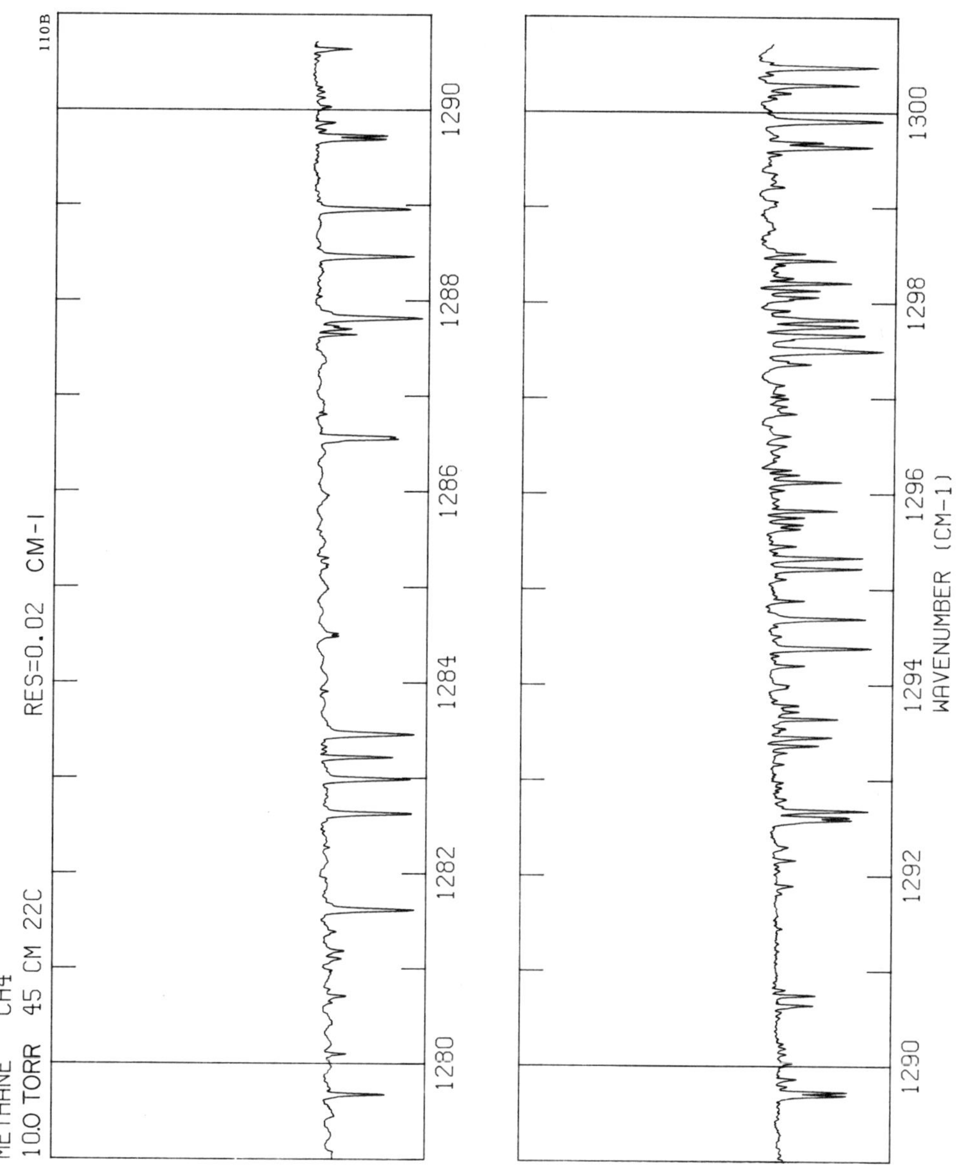
110B
METHANE CH4
10.0 TORR 45 CM 22C
RES=0.02 CM-1
1280
1282
1284
1286
1288
1290
1292
1294
1296
1298
1300
WAVENUMBER (CM-1)

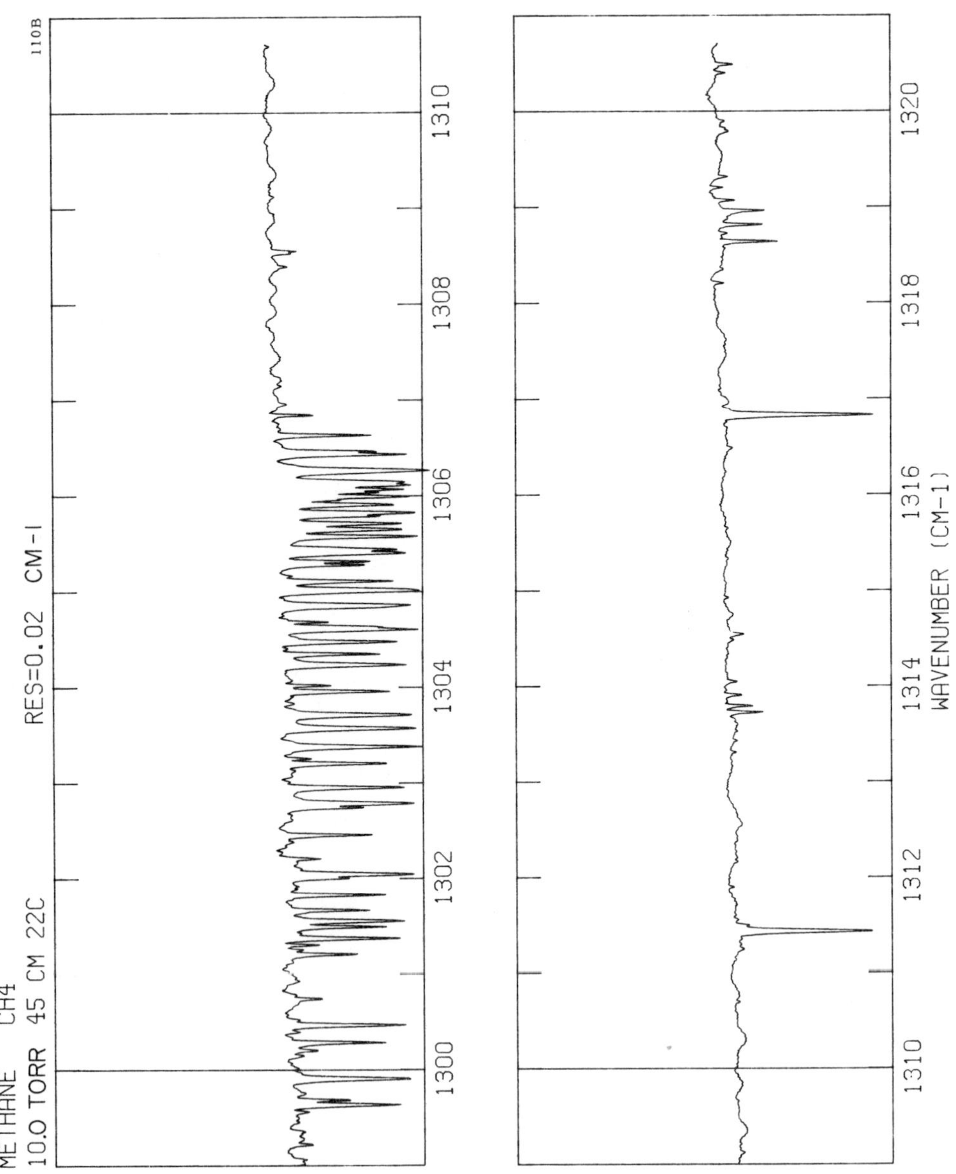
110B
METHANE CH4
10.0 TORR 45 CM 22C
RES=0.02 CM-1
1300
1302
1304
1306
1308
1310
1312
1314
1316
1318
1320
WAVENUMBER (CM-1)

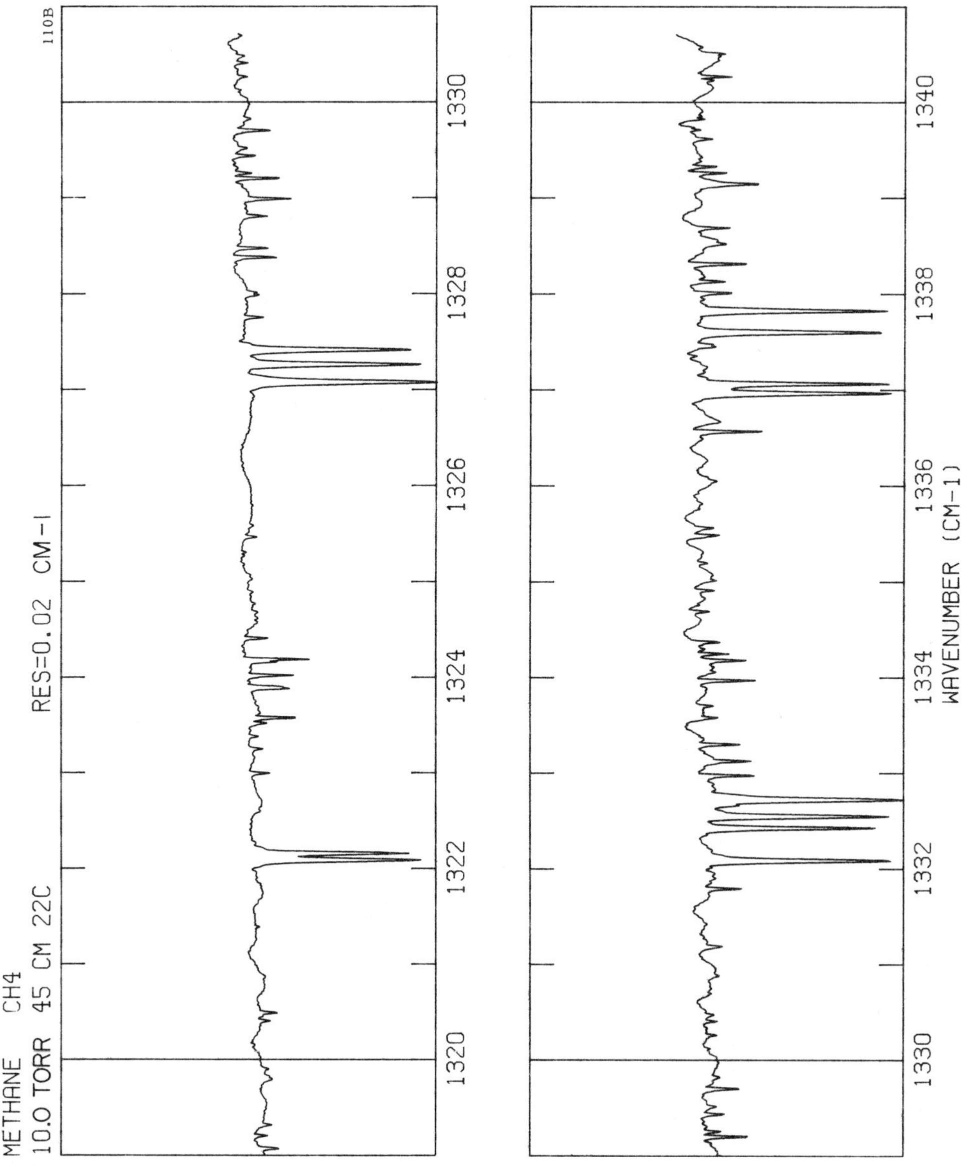
METHANE CH4
10.0 TORR 45 CM 22C
RES=0.02 CM-1
110B
1320
1322
1324
1326
1328
1330
1330
1332
1334
1336
1338
1340
WAVENUMBER (CM-1)

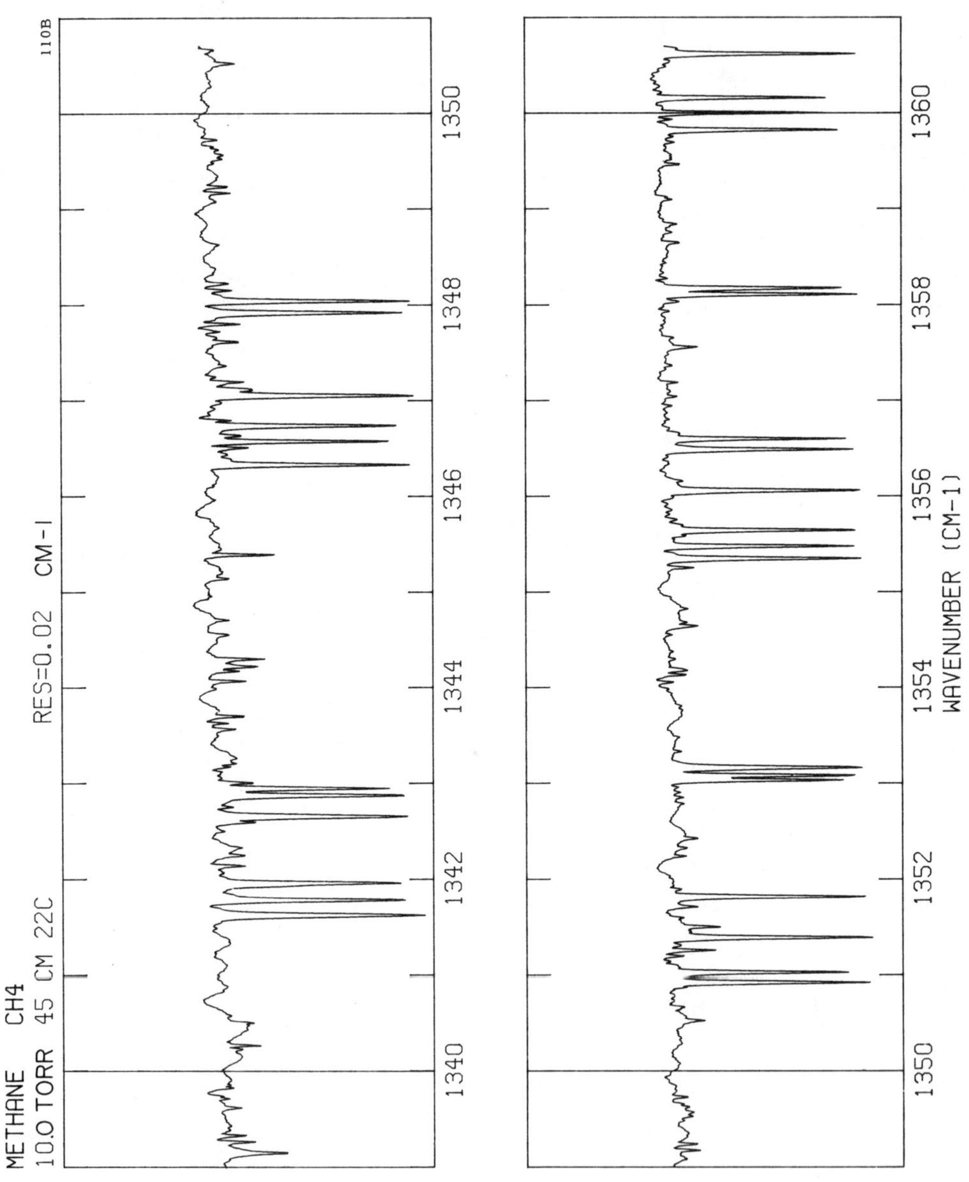
METHANE CH4
10.0 TORR 45 CM 22C
RES=0.02 CM-1
110B
1340
1342
1344
1346
1348
1350
1352
1354
1356
1358
1360
WAVENUMBER (CM-1)

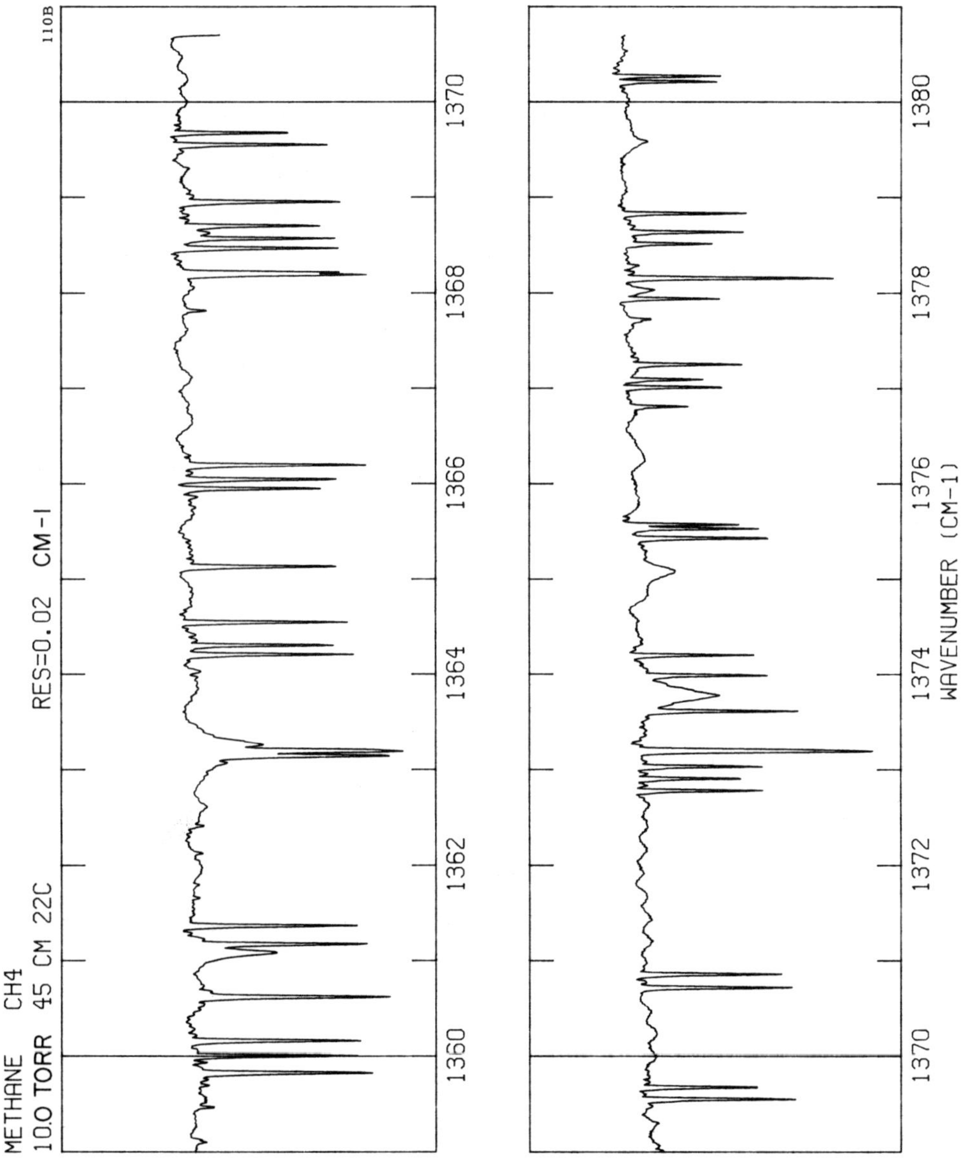
METHANE CH4
10.0 TORR 45 CM 22C
RES=0.02 CM-1
110B
1360
1362
1364
1366
1368
1370
1370
1372
1374
1376
1378
1380
WAVENUMBER (CM-1)

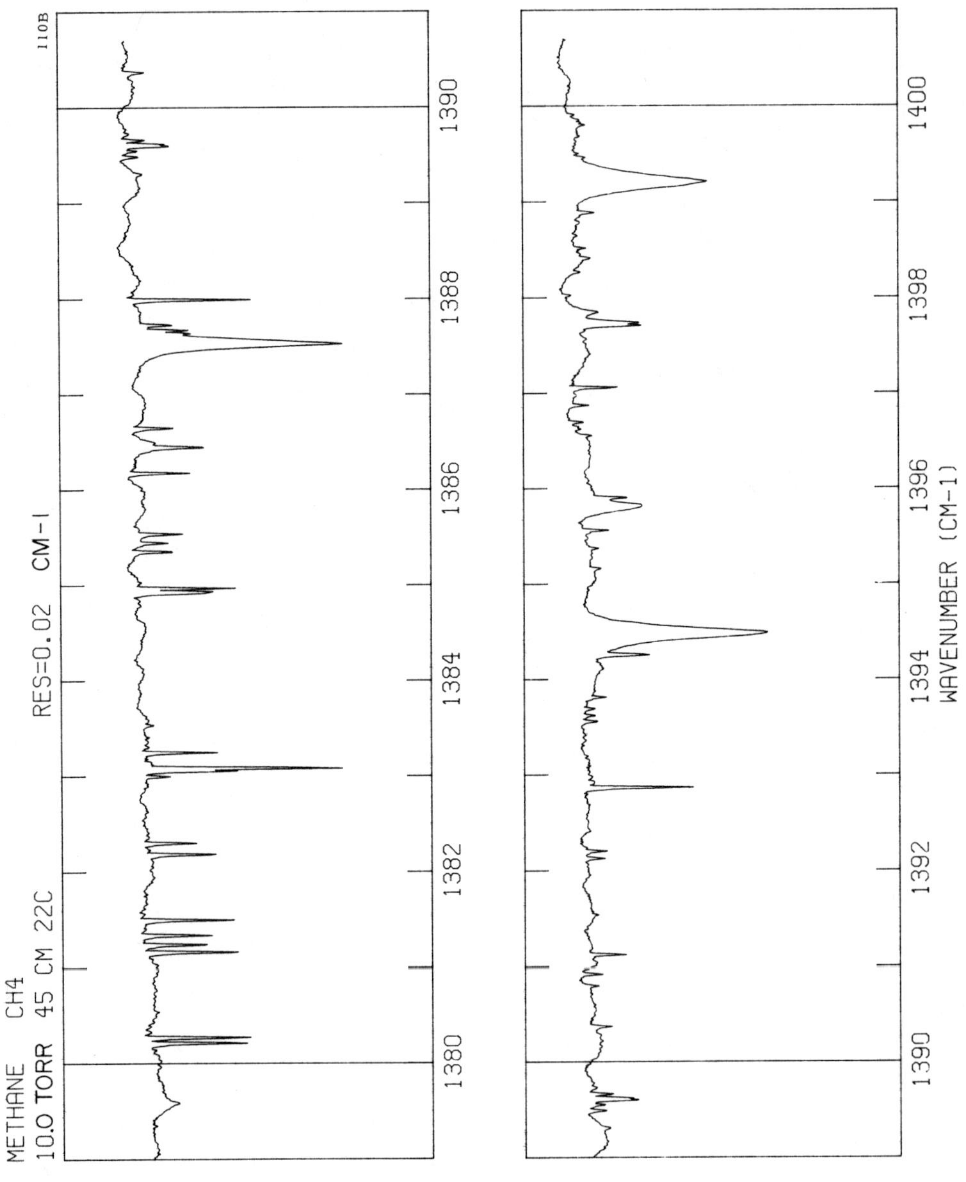
METHANE CH4
10.0 TORR 45 CM 22C
RES=0.02 CM-1
110B
1380
1382
1384
1386
1388
1390
1392
1394
1396
1398
1400
WAVENUMBER (CM-1)

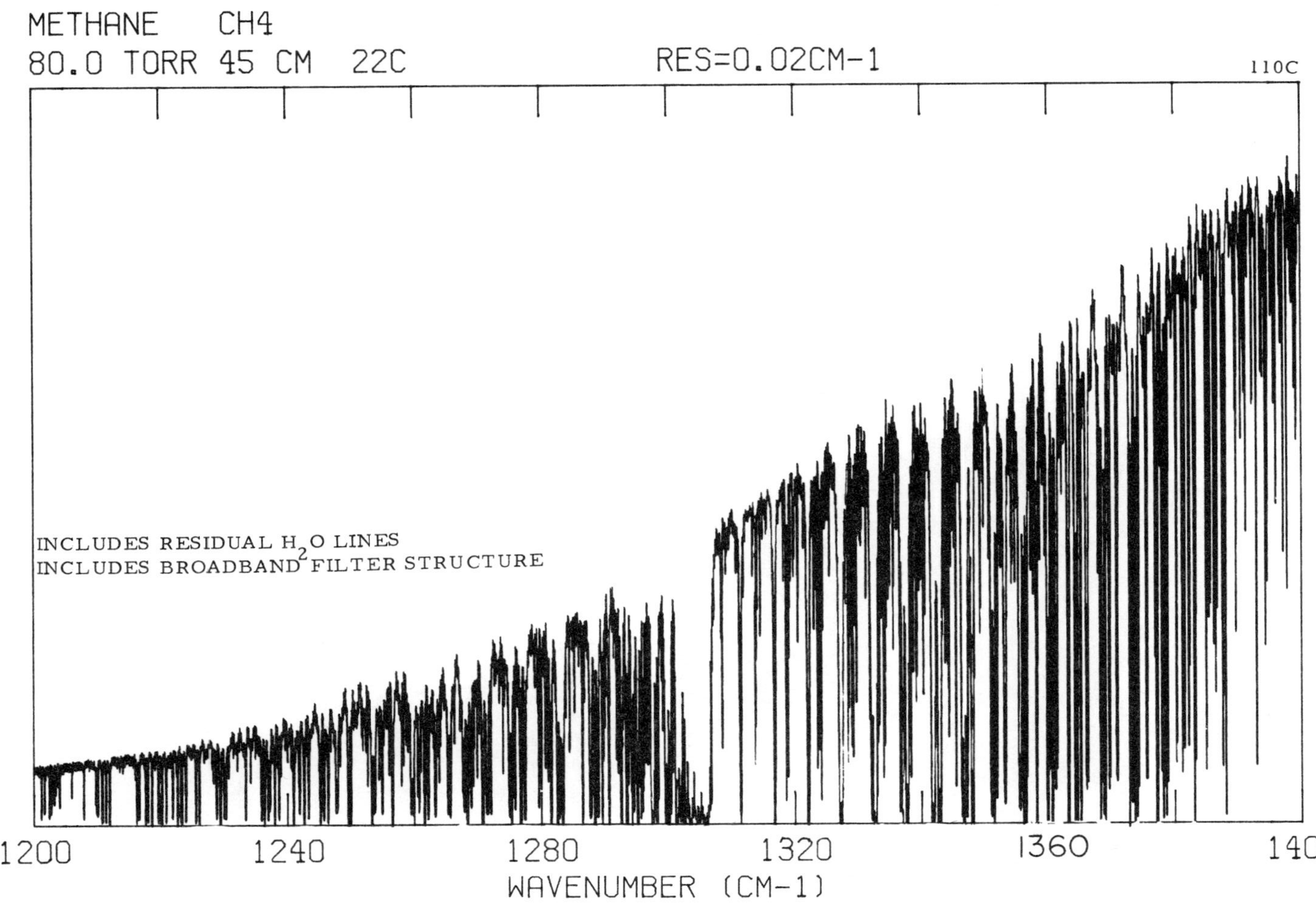

METHANE CH4
80.0 TORR 45 CM 22C
RES=0.02CM-1
110C
INCLUDES RESIDUAL H_2O LINES
INCLUDES BROADBAND FILTER STRUCTURE
1200
1240
1280
1320
1360
1400
WAVENUMBER (CM-1)

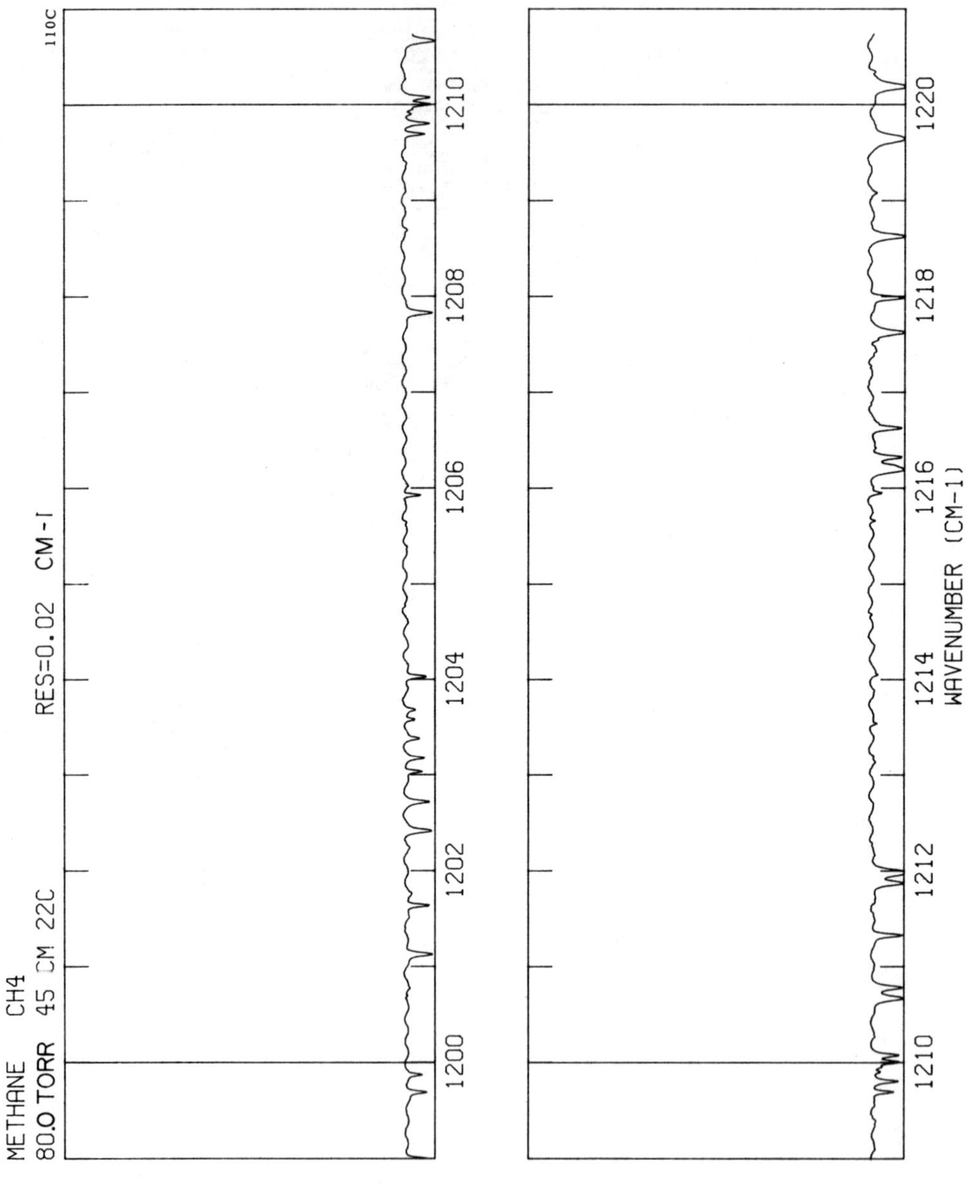
METHANE CH4
80.0 TORR 45 CM 22C
RES=0.02 CM-1
110C
1200
1202
1204
1206
1208
1210
1212
1214
1216
1218
1220
WAVENUMBER (CM-1)

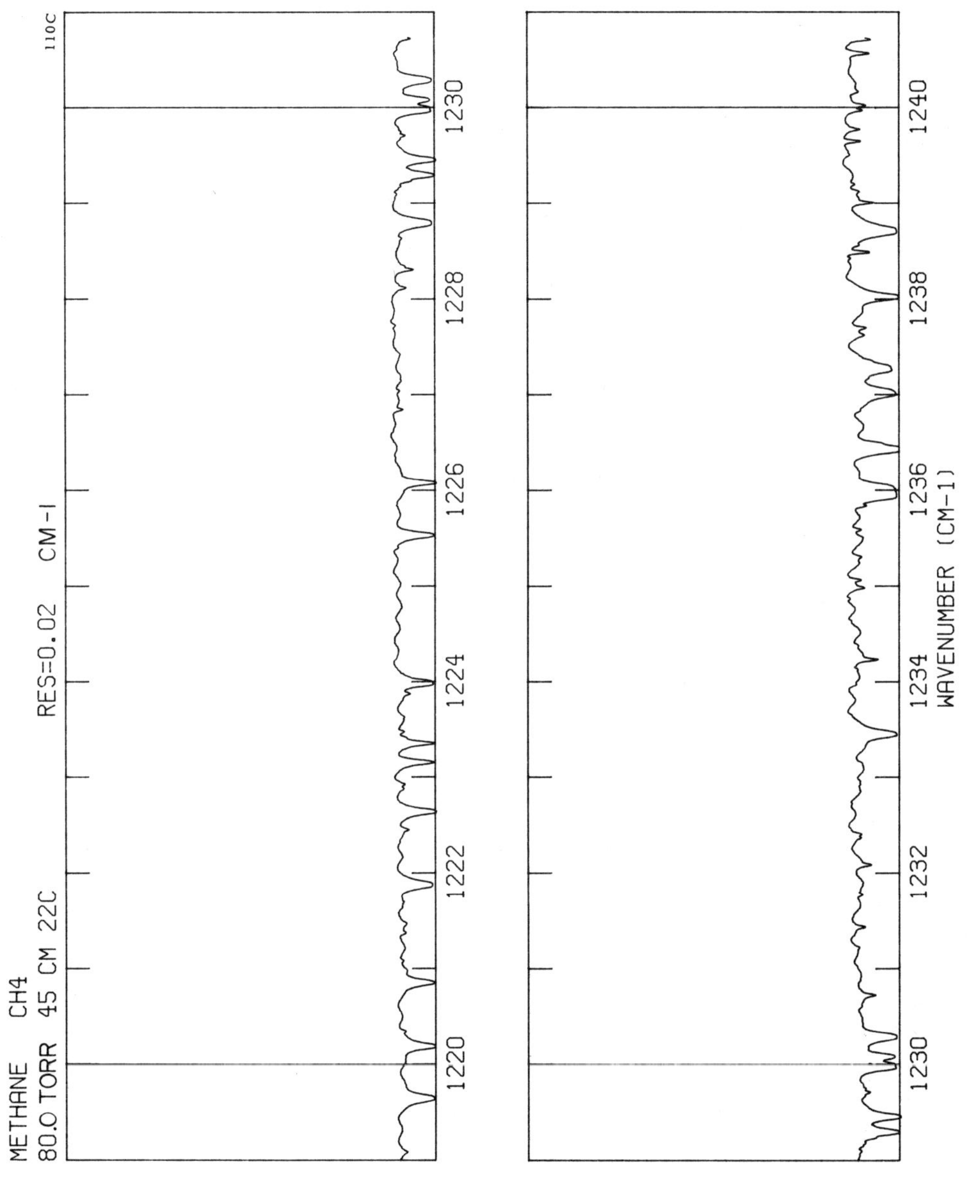
METHANE CH4
80.0 TORR 45 CM 22C
RES=0.02 CM-1
110C
1220
1222
1224
1226
1228
1230
1230
1232
1234
1236
1238
1240
WAVENUMBER (CM-1)

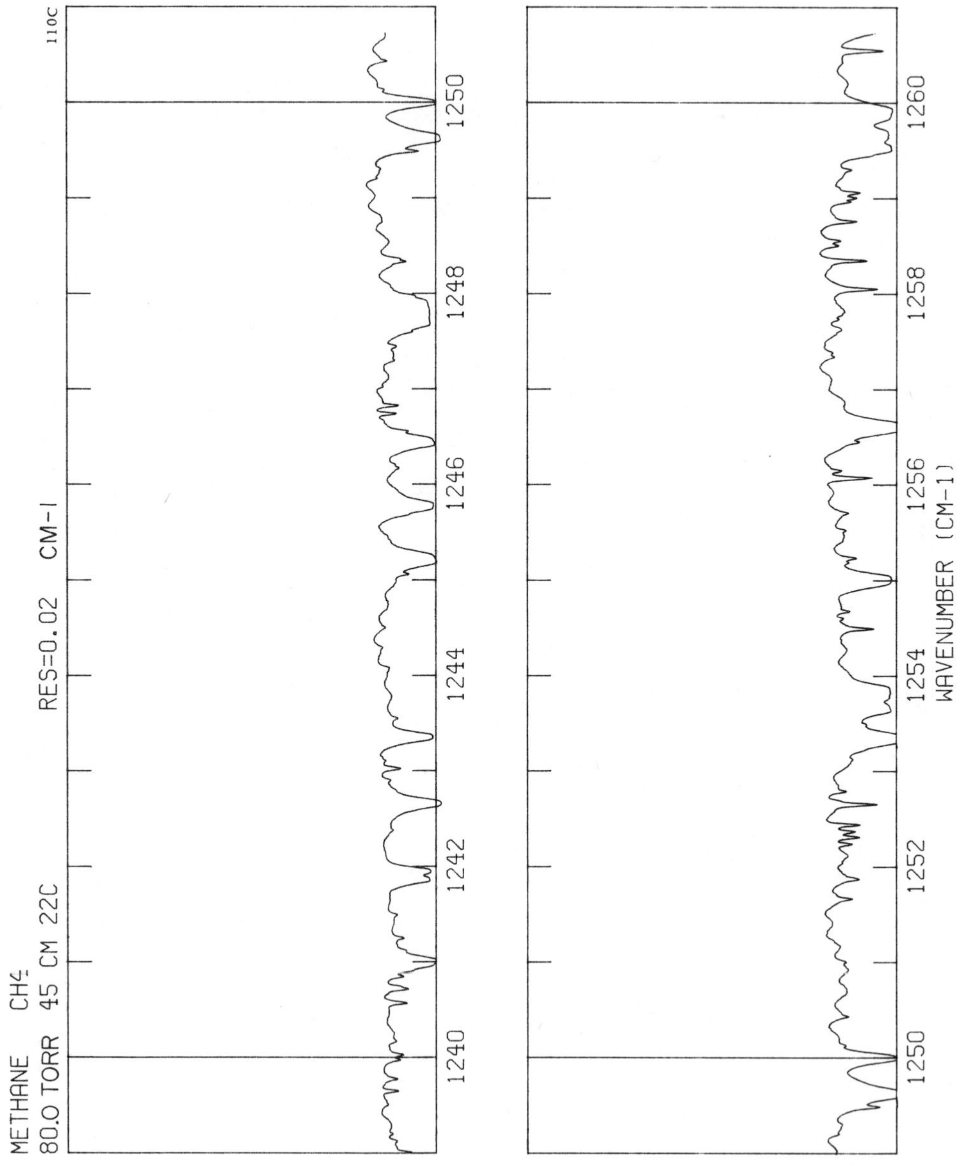

METHANE CH4
80.0 TORR 45 CM 22C
RES=0.02 CM-1
110C
1240
1242
1244
1246
1248
1250
1250
1252
1254
1256
1258
1260
WAVENUMBER (CM-1)

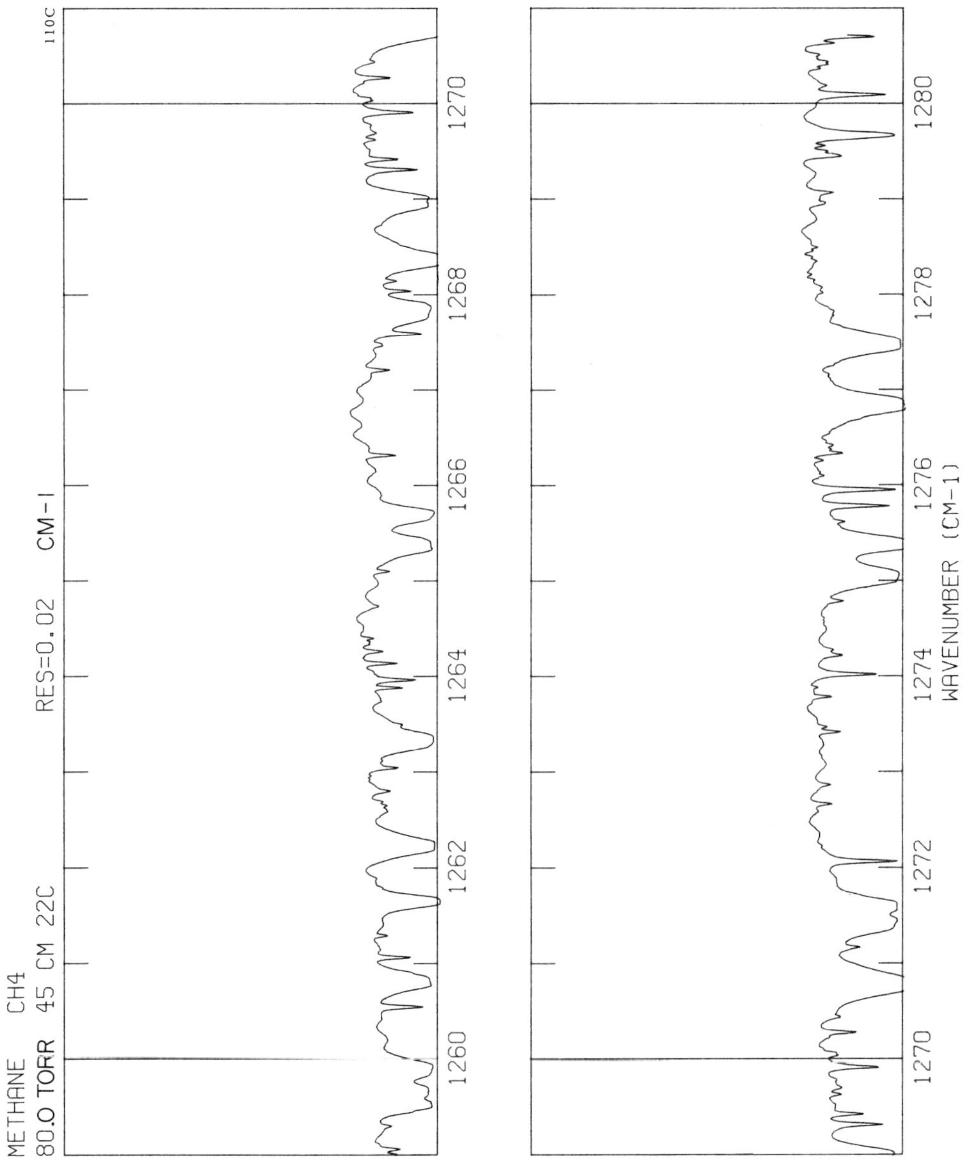
METHANE CH4
80.0 TORR 45 CM 22C
RES=0.02 CM-1
110C
1260
1262
1264
1266
1268
1270
1272
1274
1276
1278
1280
WAVENUMBER (CM-1)

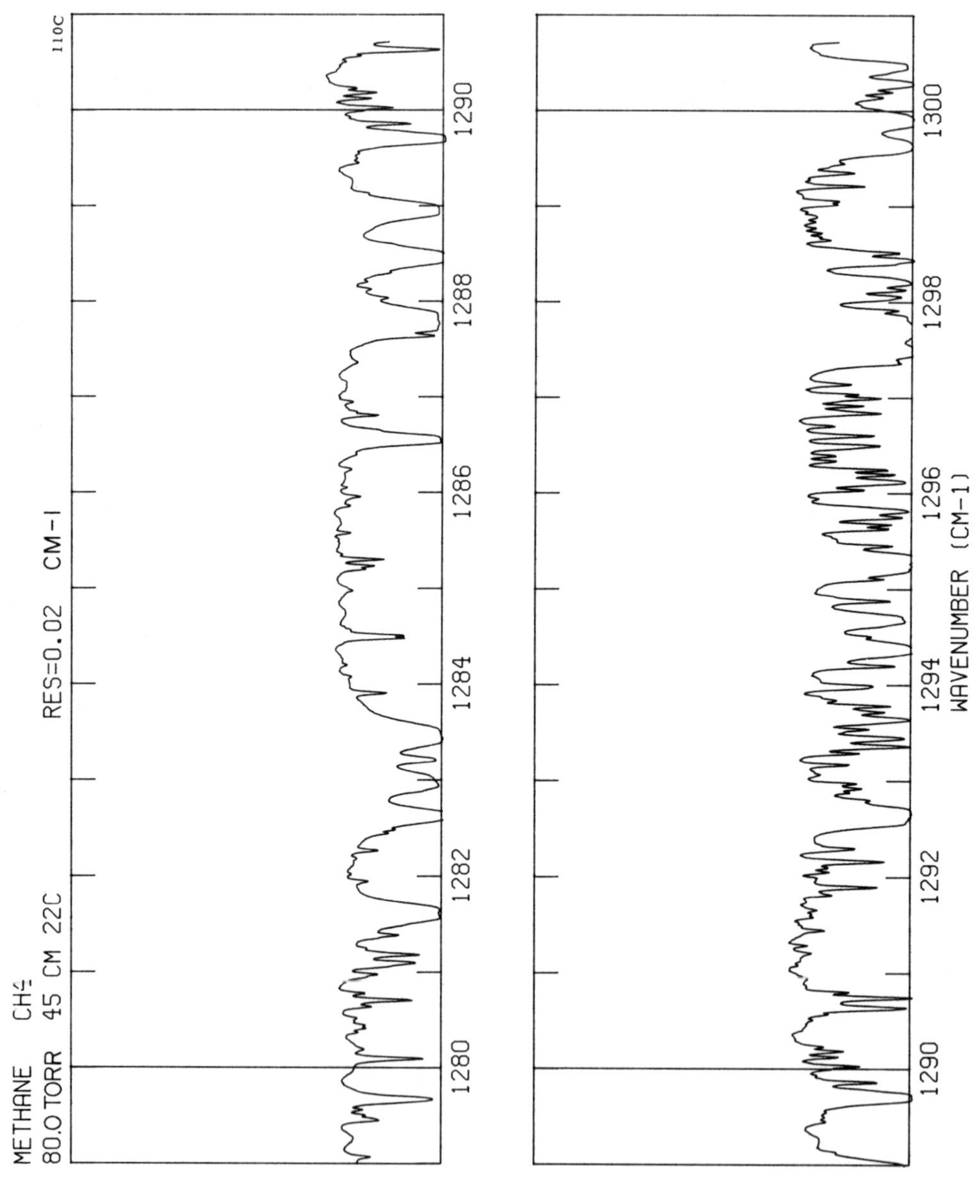
METHANE CH4
80.0 TORR 45 CM 22C
RES=0.02 CM-1
110C
1280
1282
1284
1286
1288
1290
1292
1294
1296
1298
1300
WAVENUMBER (CM-1)

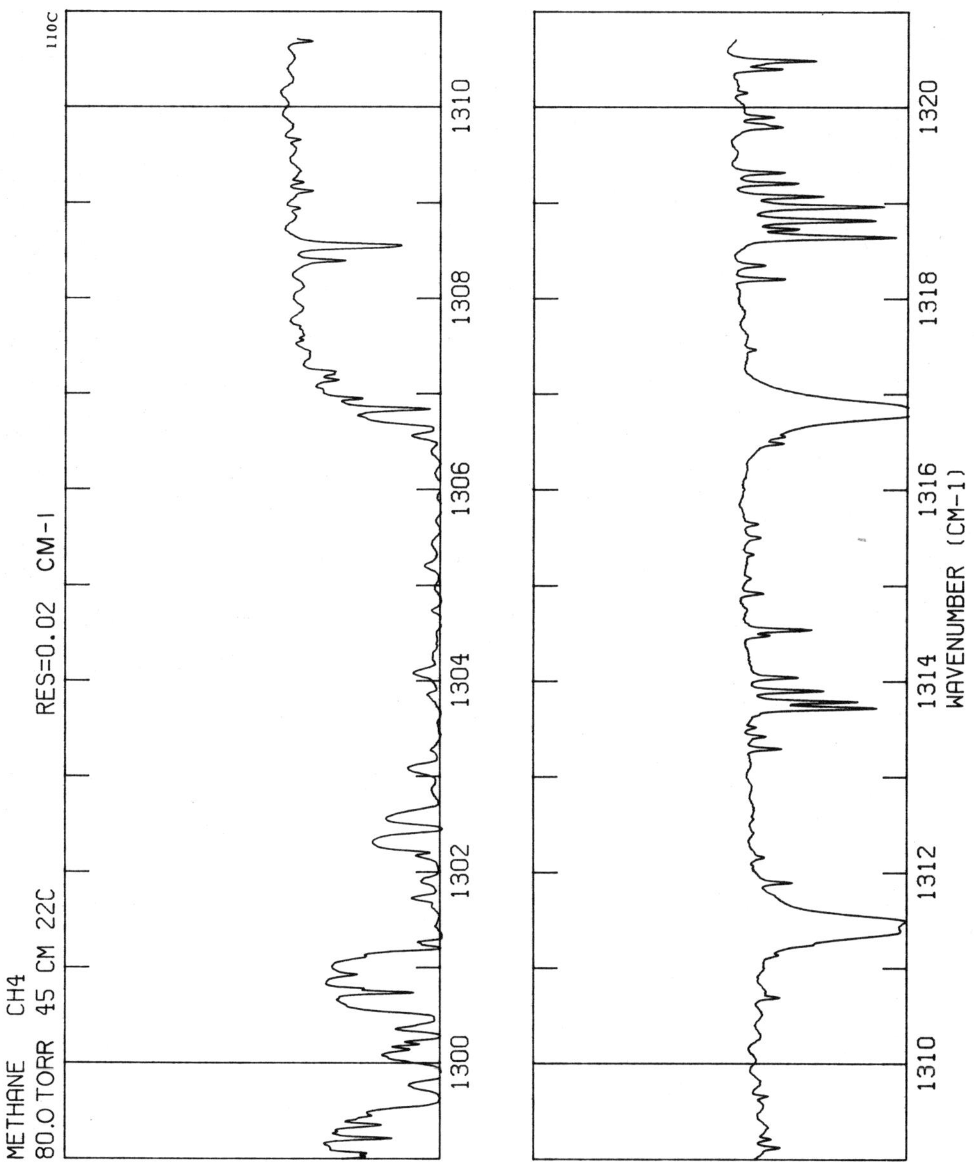
METHANE CH4
80.0 TORR 45 CM 22C
RES=0.02 CM-1
110C
1300
1302
1304
1306
1308
1310
1310
1312
1314
1316
1318
1320
WAVENUMBER (CM-1)

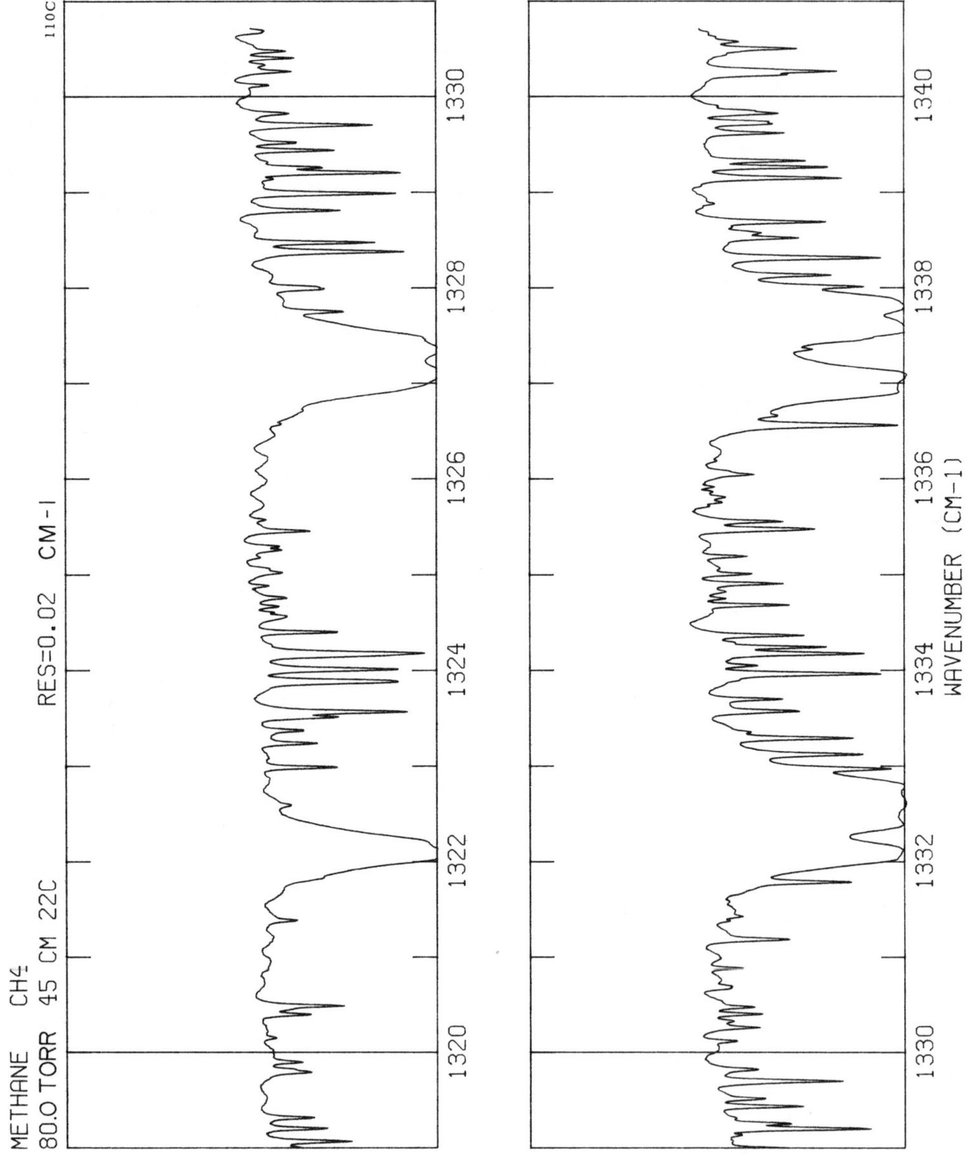
METHANE CH4
80.0 TORR 45 CM 22C
RES=0.02 CM-1
110C
1320
1322
1324
1326
1328
1330
1332
1334
1336
1338
1340
WAVENUMBER (CM-1)

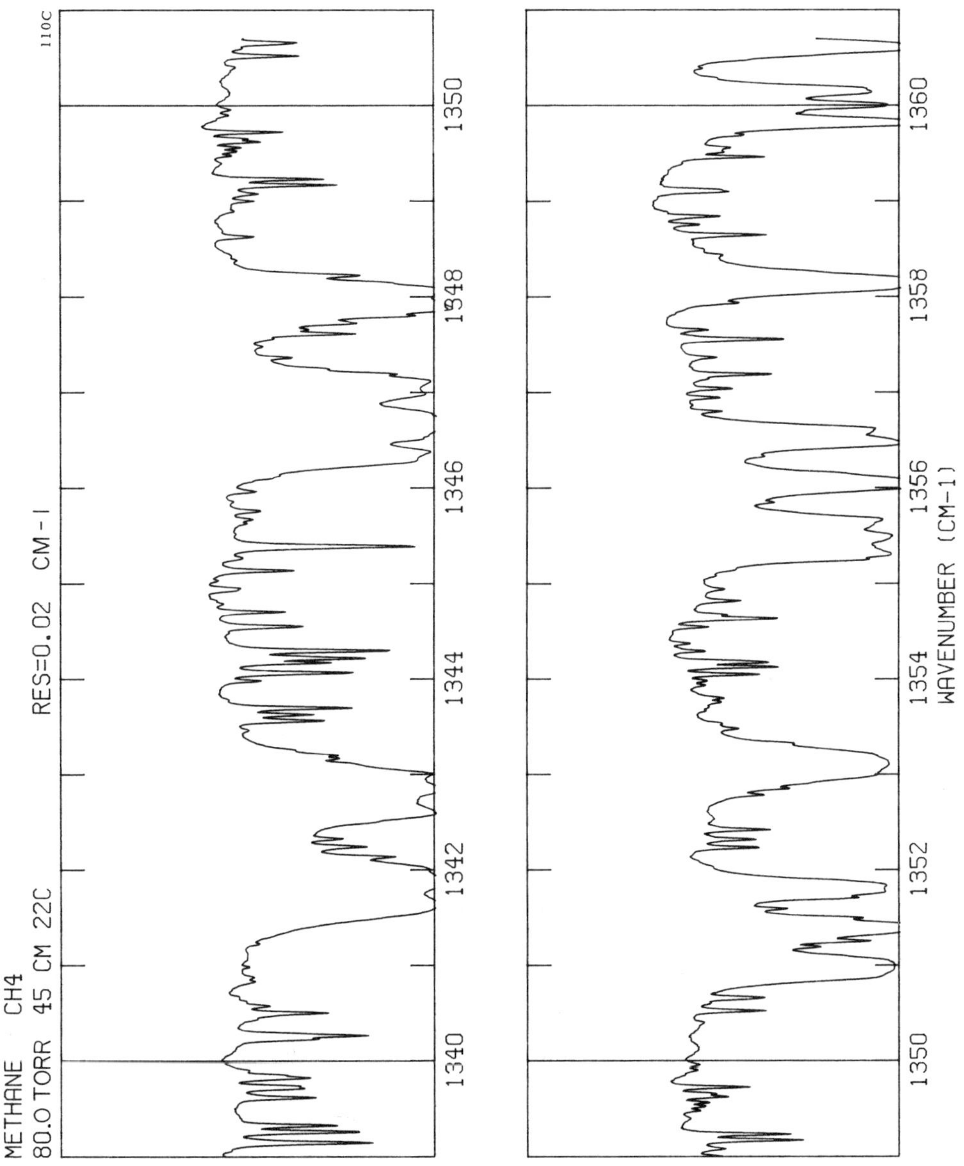
METHANE CH4
80.0 TORR 45 CM 22C
RES=0.02 CM-1
110C
1340
1342
1344
1346
1348
1350
1350
1352
1354
1356
1358
1360
WAVENUMBER (CM-1)

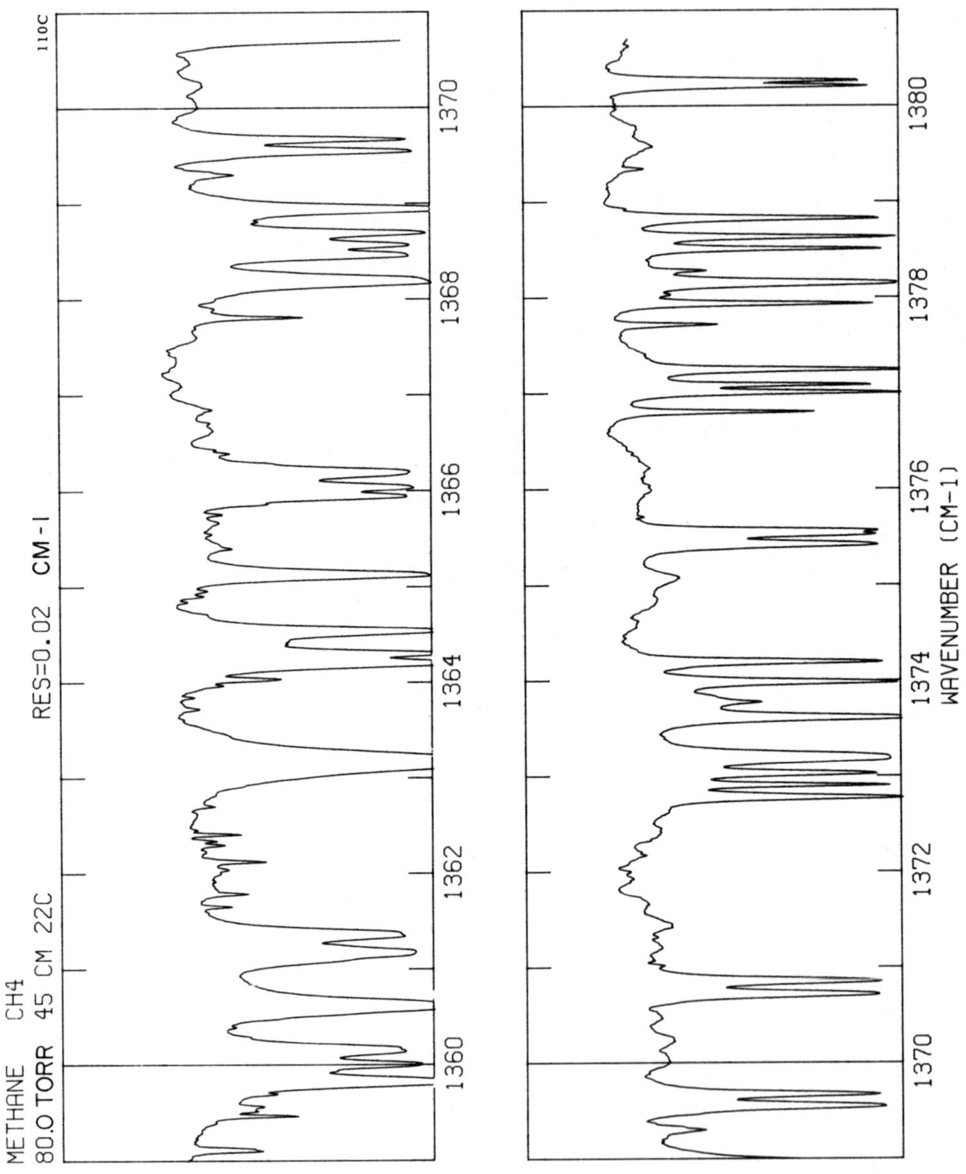
METHANE CH4
80.0 TORR 45 CM 22C
RES=0.02 CM-1
110C
1360
1362
1364
1366
1368
1370
1370
1372
1374
1376
1378
1380
WAVENUMBER (CM-1)

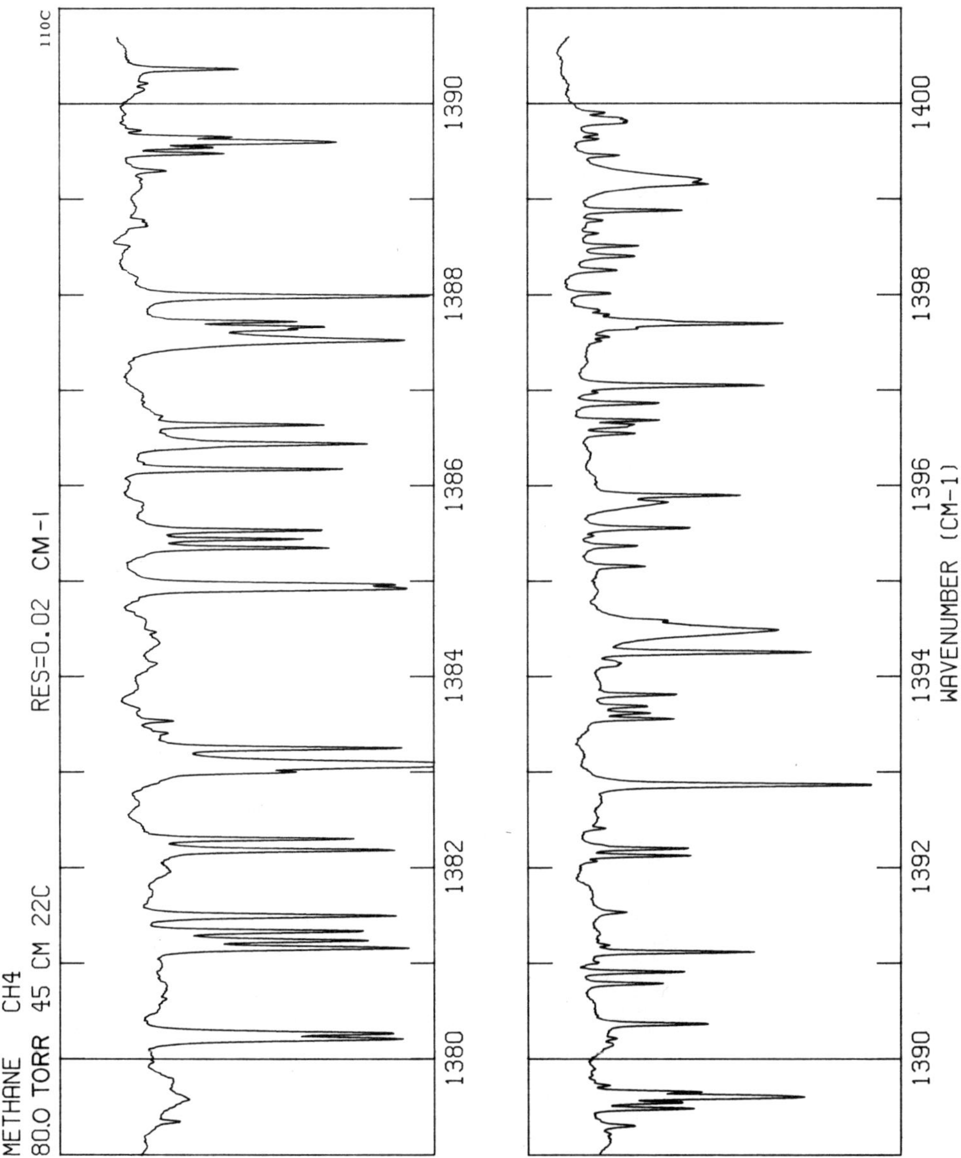

METHANE CH4
80.0 TORR 45 CM 22C
RES=0.02 CM-1
110C
1380
1382
1384
1386
1388
1390
1392
1394
1396
1398
1400
WAVENUMBER (CM-1)

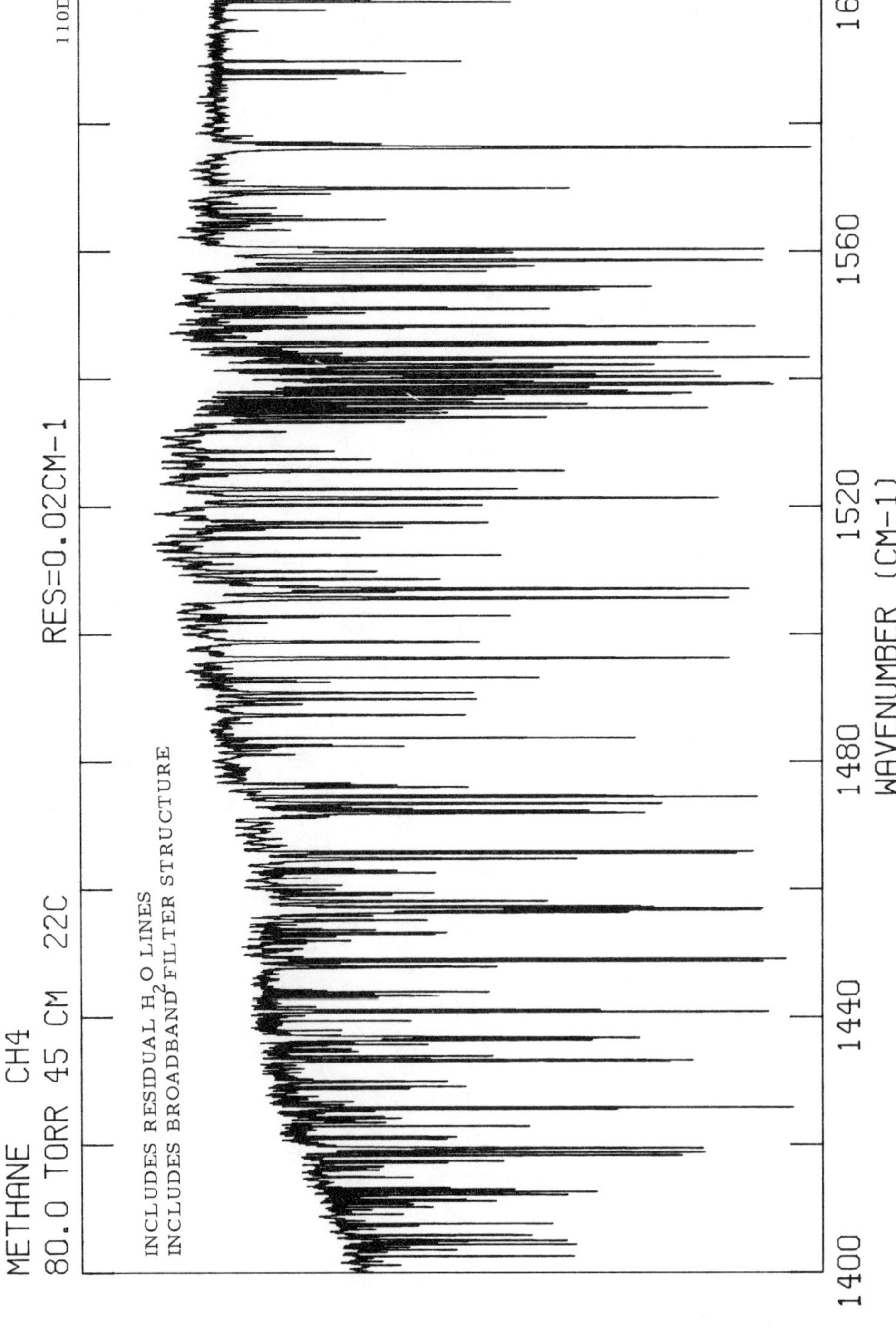
METHANE CH4
80.0 TORR 45 CM 22C
RES=0.02CM-1
110D
INCLUDES RESIDUAL H_2O LINES
INCLUDES BROADBAND FILTER STRUCTURE
1400
1440
1480
1520
1560
1600
WAVENUMBER (CM-1)

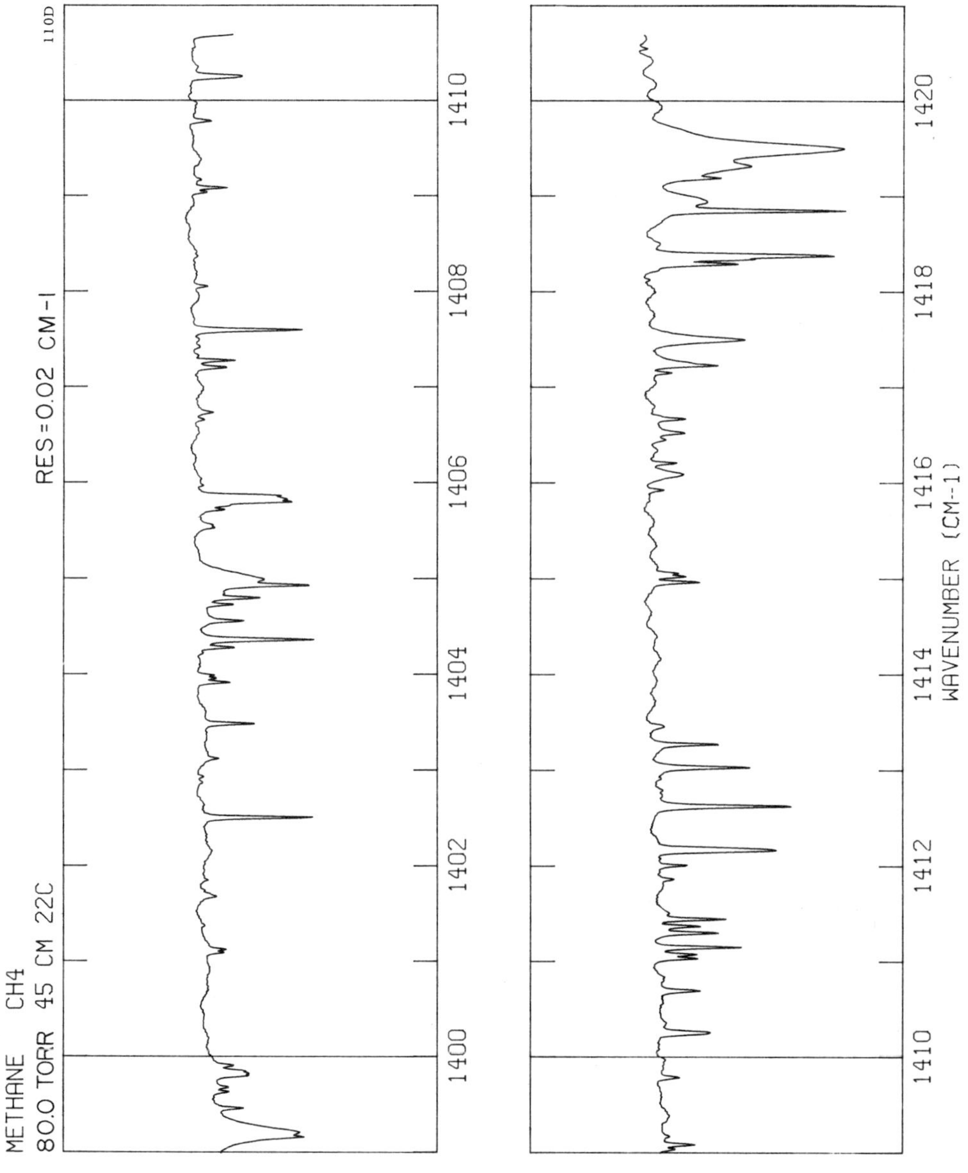
METHANE CH4
80.0 TORR 45 CM 22C
RES=0.02 CM-1
110D
1400
1402
1404
1406
1408
1410
1410
1412
1414
1416
1418
1420
WAVENUMBER (CM-1)

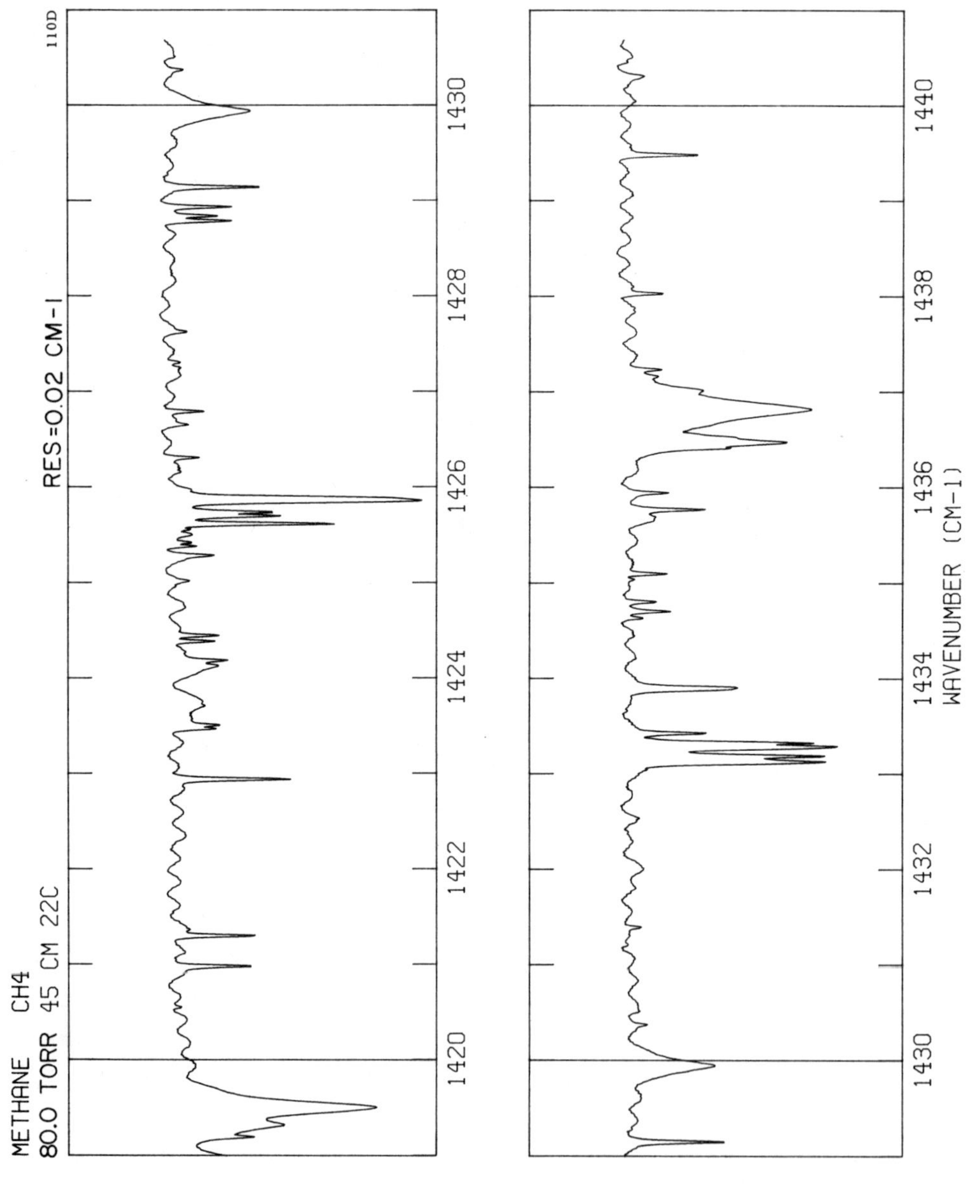
METHANE CH4
80.0 TORR 45 CM 22C
RES=0.02 CM-1
110D
1420
1422
1424
1426
1428
1430
1432
1434
1436
1438
1440
WAVENUMBER (CM-1)

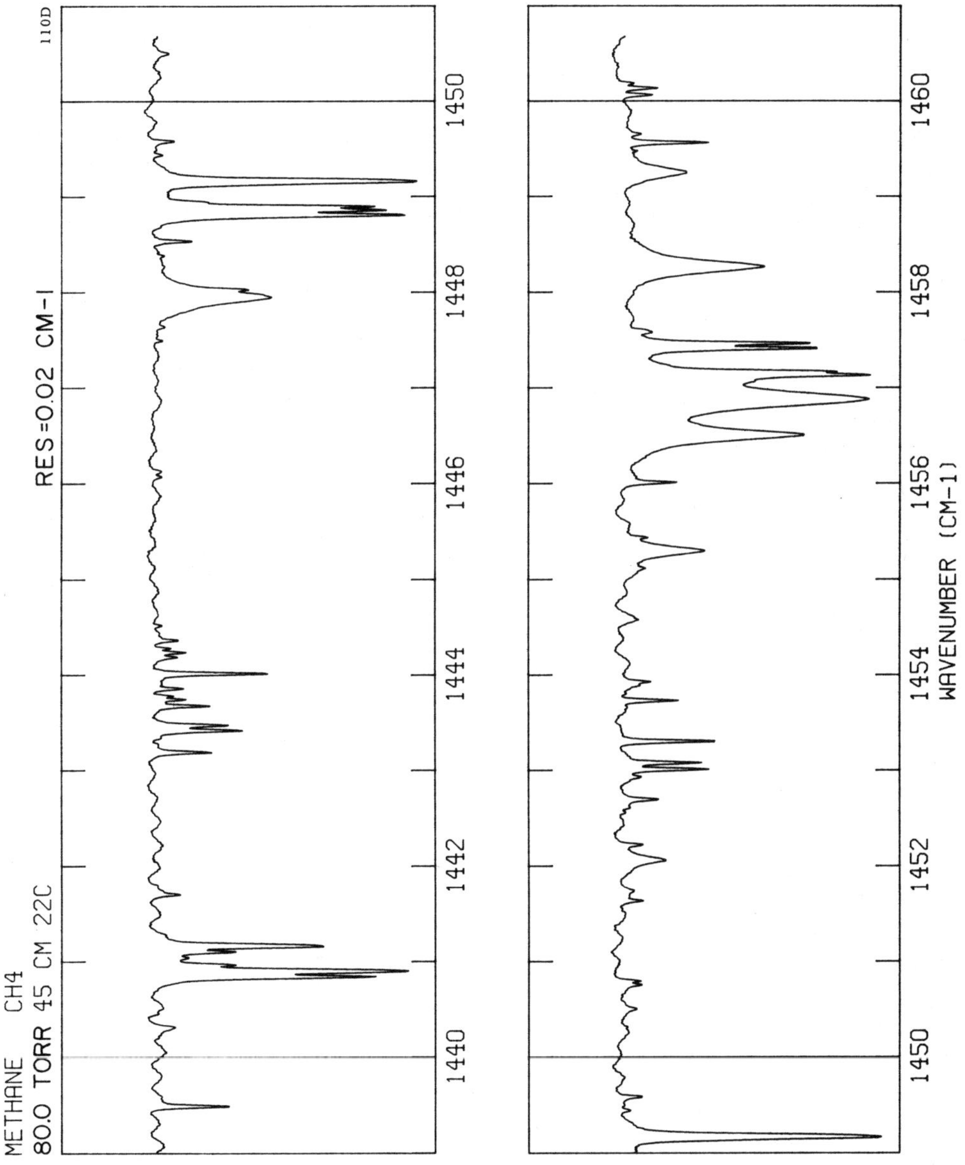

METHANE CH4
80.0 TORR 45 CM 22C
RES=0.02 CM-1
110D
1440
1442
1444
1446
1448
1450
1450
1452
1454
1456
1458
1460
WAVENUMBER (CM-1)

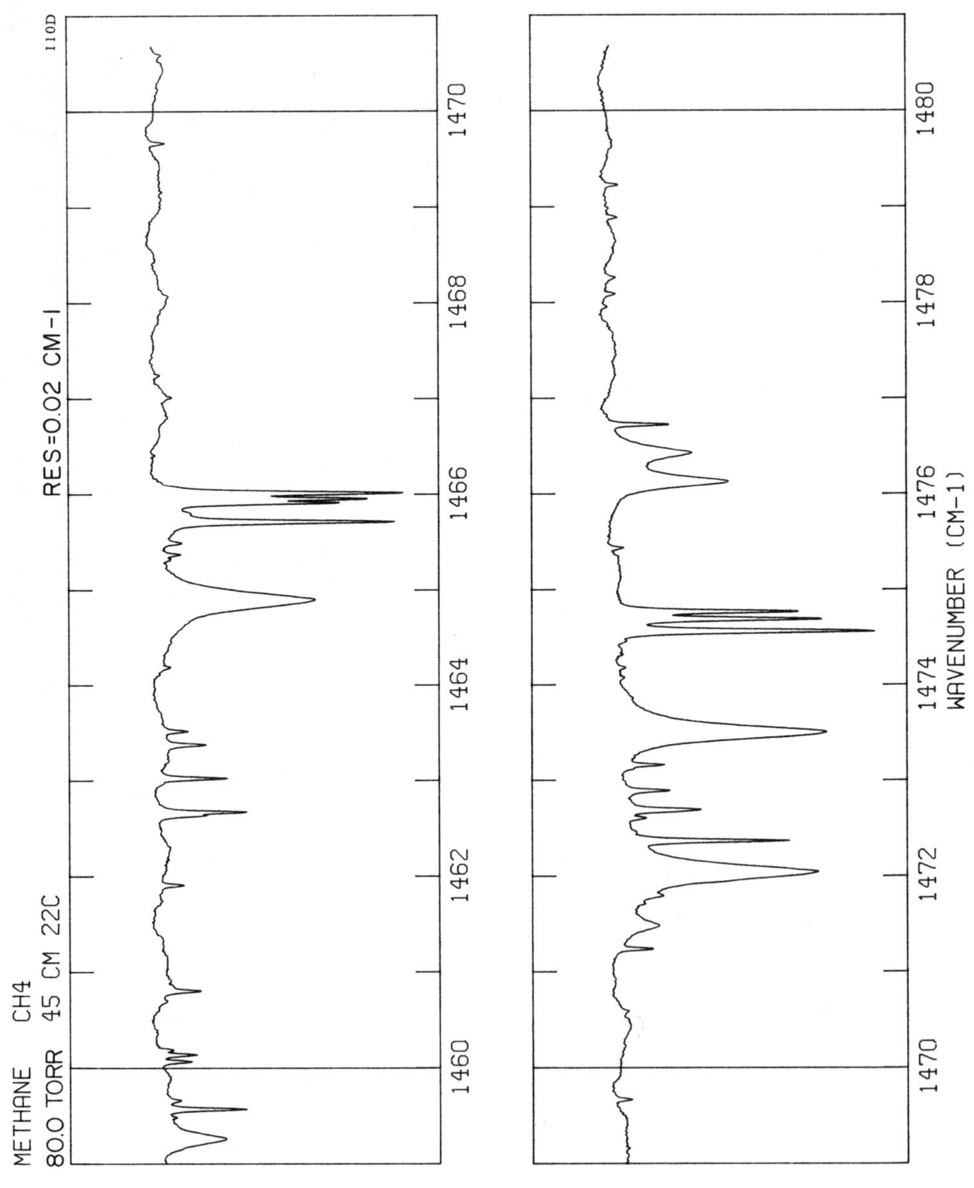
METHANE CH4
80.0 TORR 45 CM 22C
RES=0.02 CM-1
110D
1460
1462
1464
1466
1468
1470
1472
1474
1476
1478
1480
WAVENUMBER (CM-1)

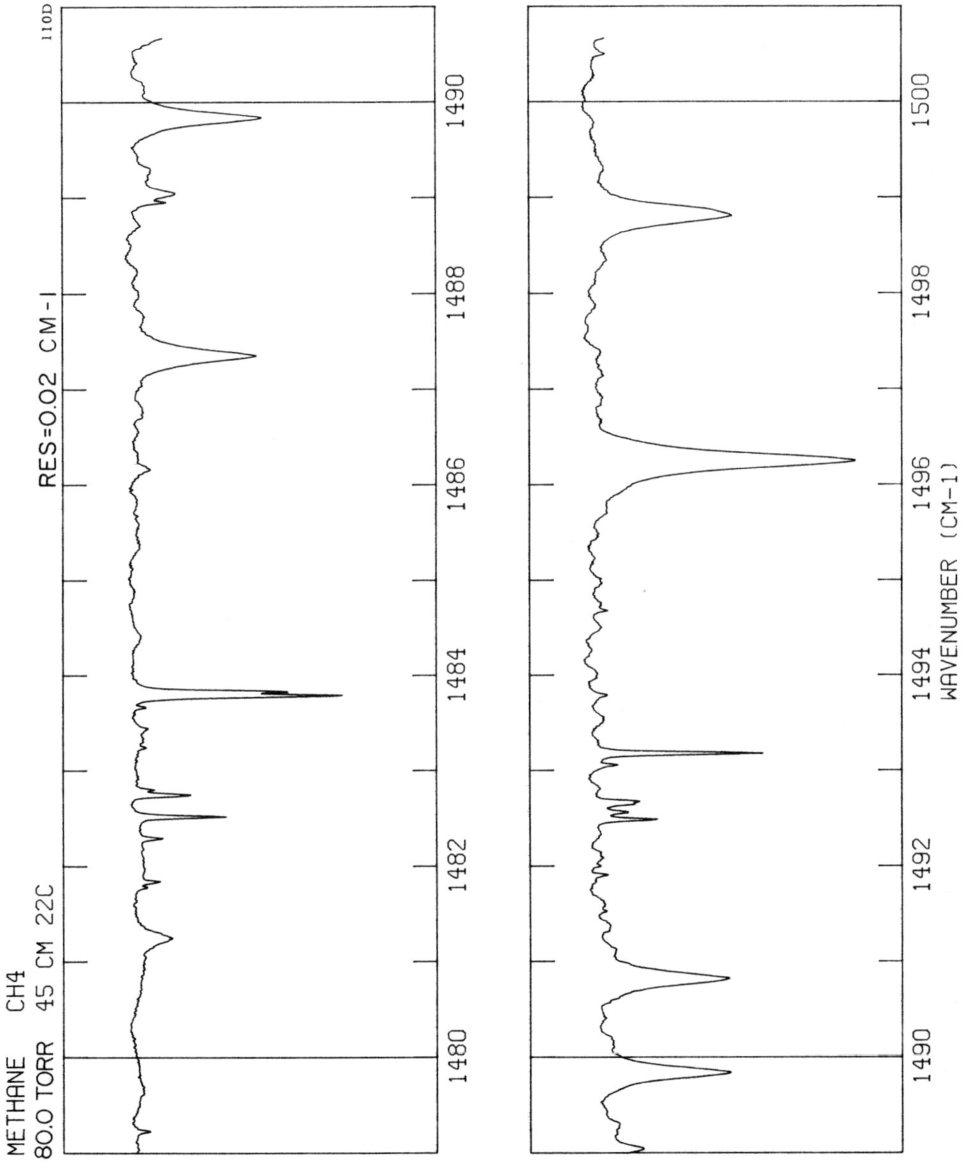

METHANE CH4
80.0 TORR 45 CM 22C
RES=0.02 CM-1
110D
1480
1482
1484
1486
1488
1490
1492
1494
1496
1498
1500
WAVENUMBER (CM-1)

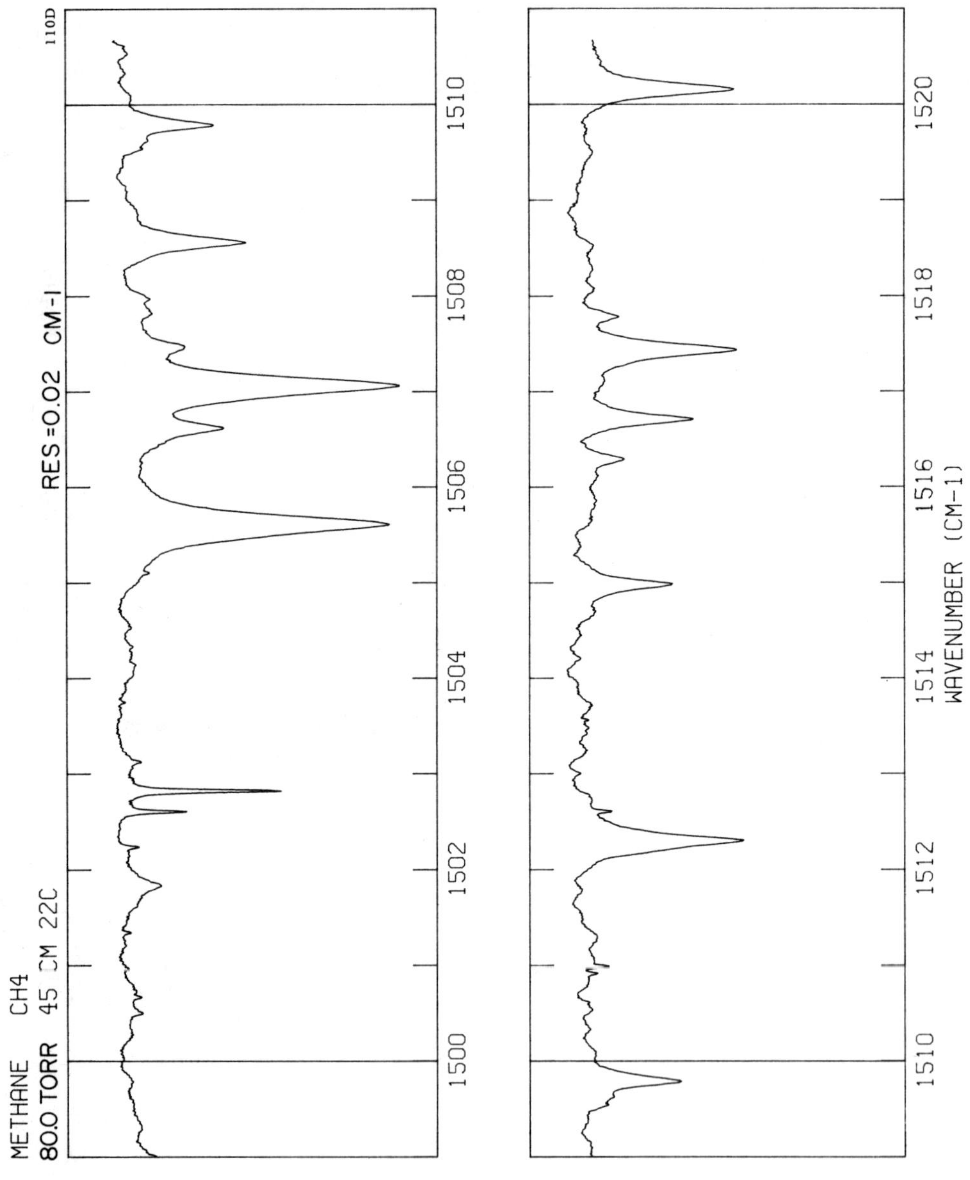
METHANE CH4
80.0 TORR 45 CM 22C
RES =0.02 CM-1
110D
1500
1502
1504
1506
1508
1510
1512
1514
1516
1518
1520
WAVENUMBER (CM-1)

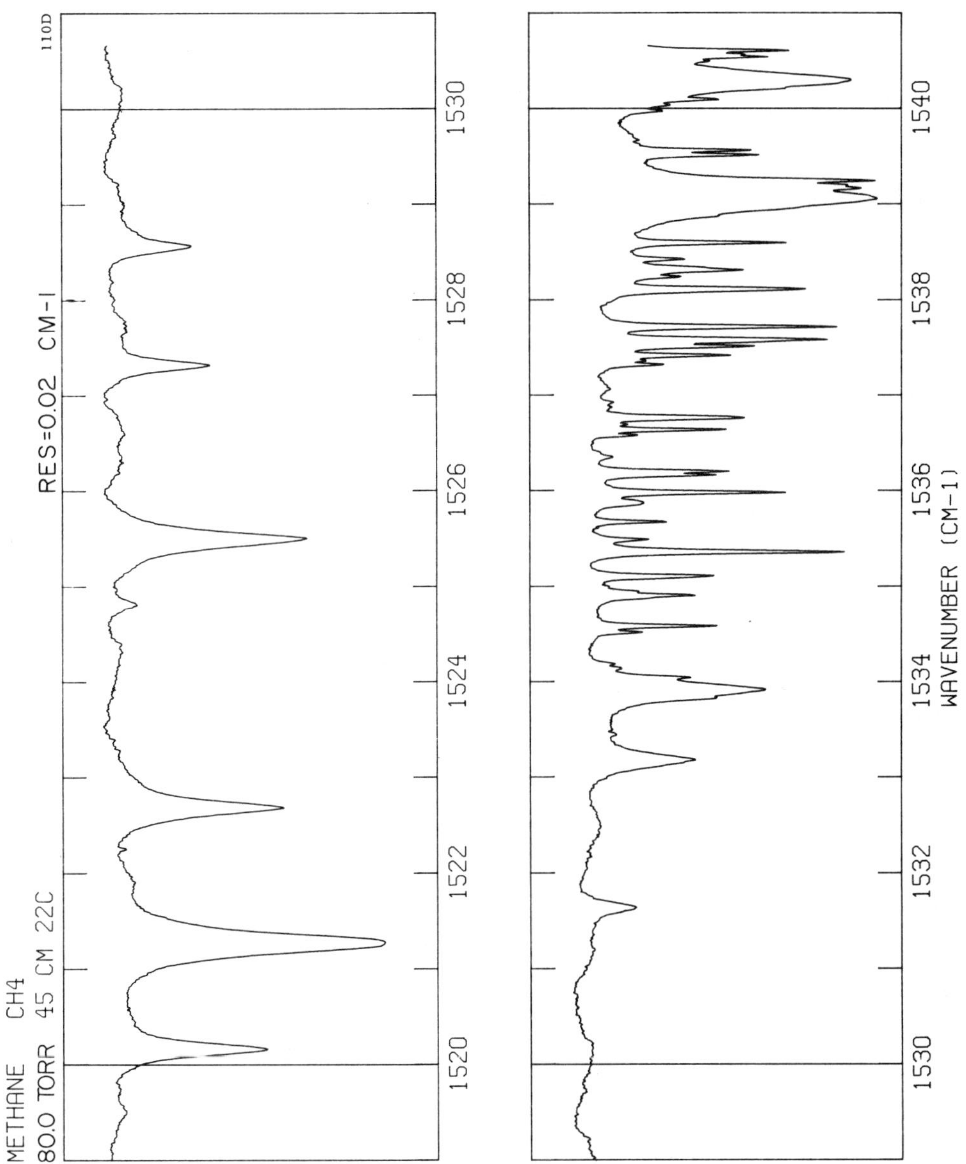
METHANE CH4
80.0 TORR 45 CM 22C
RES=0.02 CM-1
110D
1520
1522
1524
1526
1528
1530
1530
1532
1534
1536
1538
1540
WAVENUMBER (CM-1)

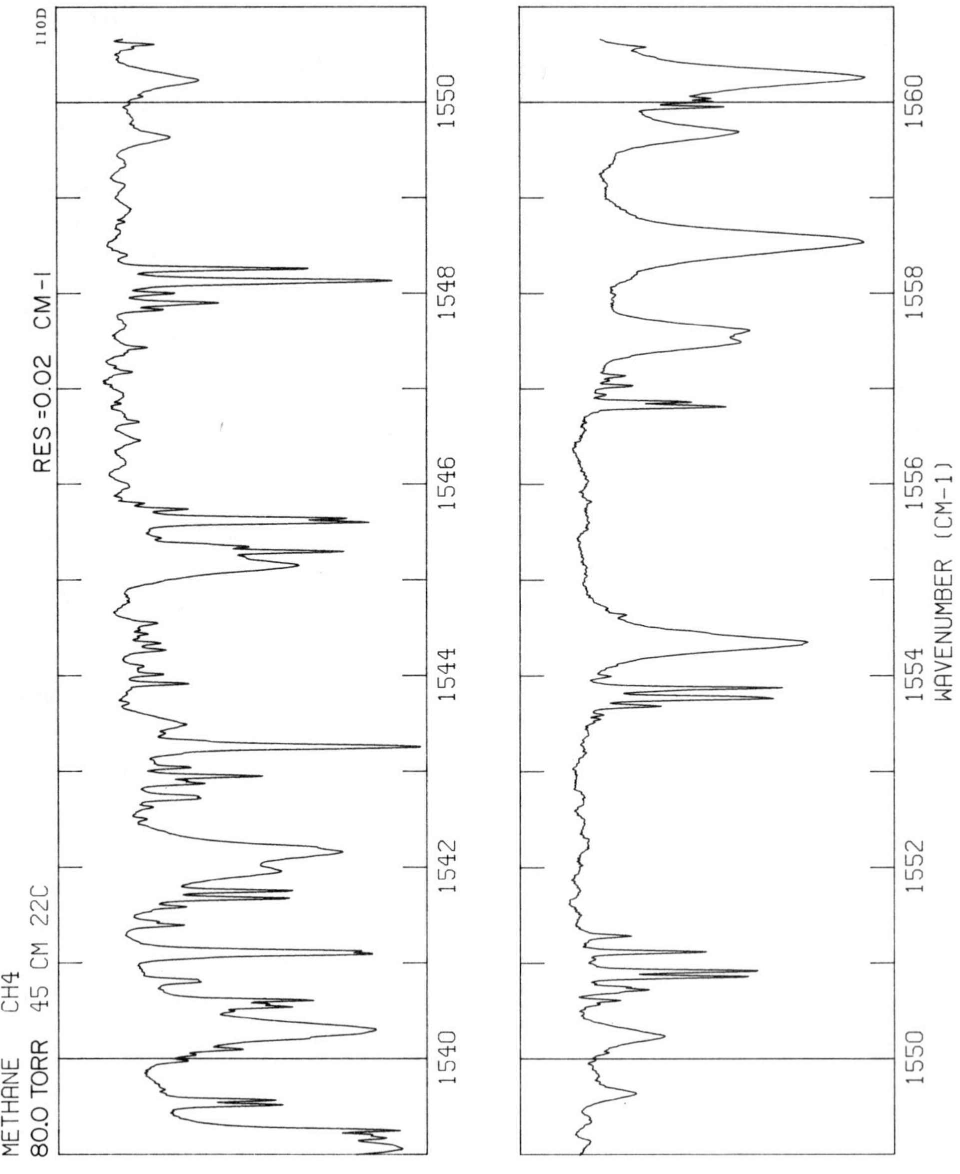

METHANE CH4
80.0 TORR 45 CM 22C
RES =0.02 CM-1
110D
1540
1542
1544
1546
1548
1550
1552
1554
1556
1558
1560
WAVENUMBER (CM-1)

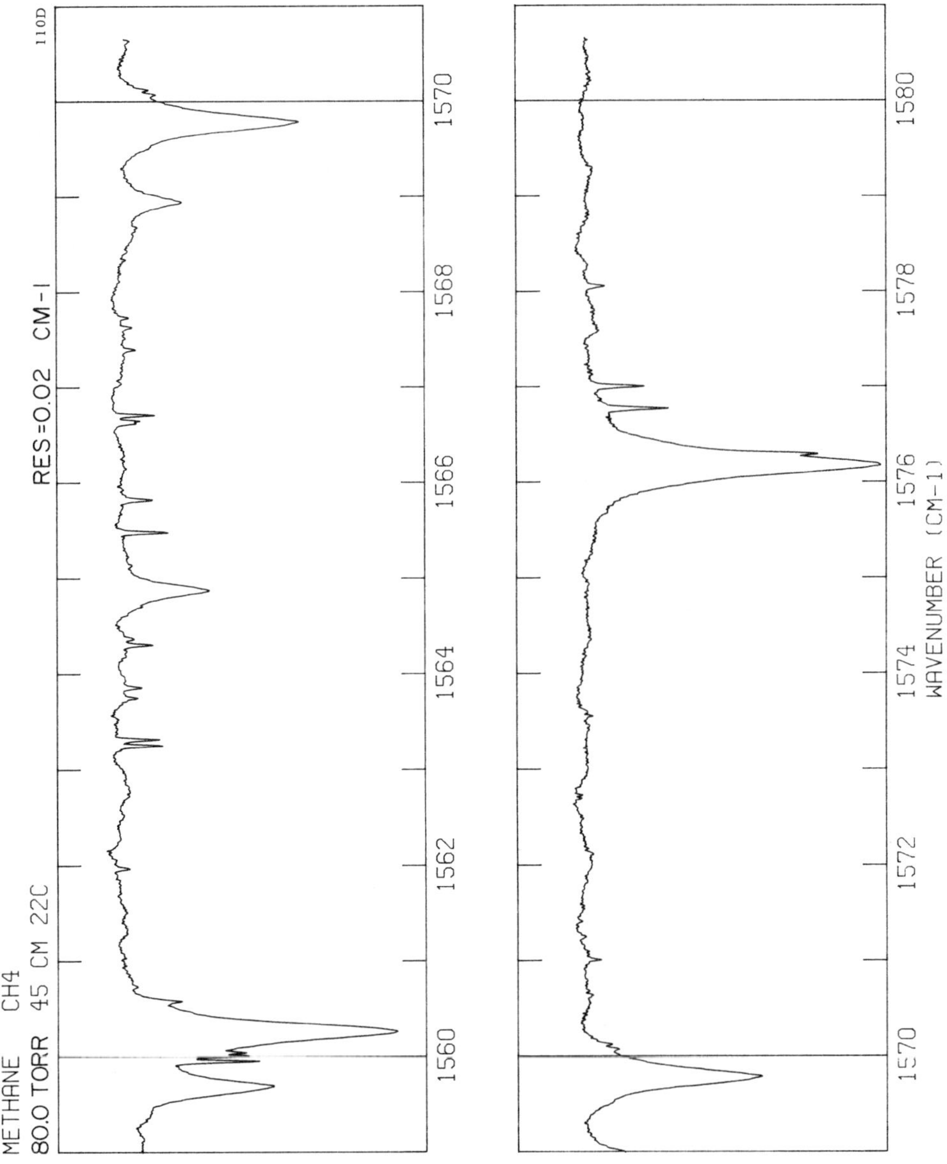

METHANE CH4
80.0 TORR 45 CM 22C
RES=0.02 CM-1
110D
1560
1562
1564
1566
1568
1570
1572
1574
1576
1578
1580
WAVENUMBER (CM-1)

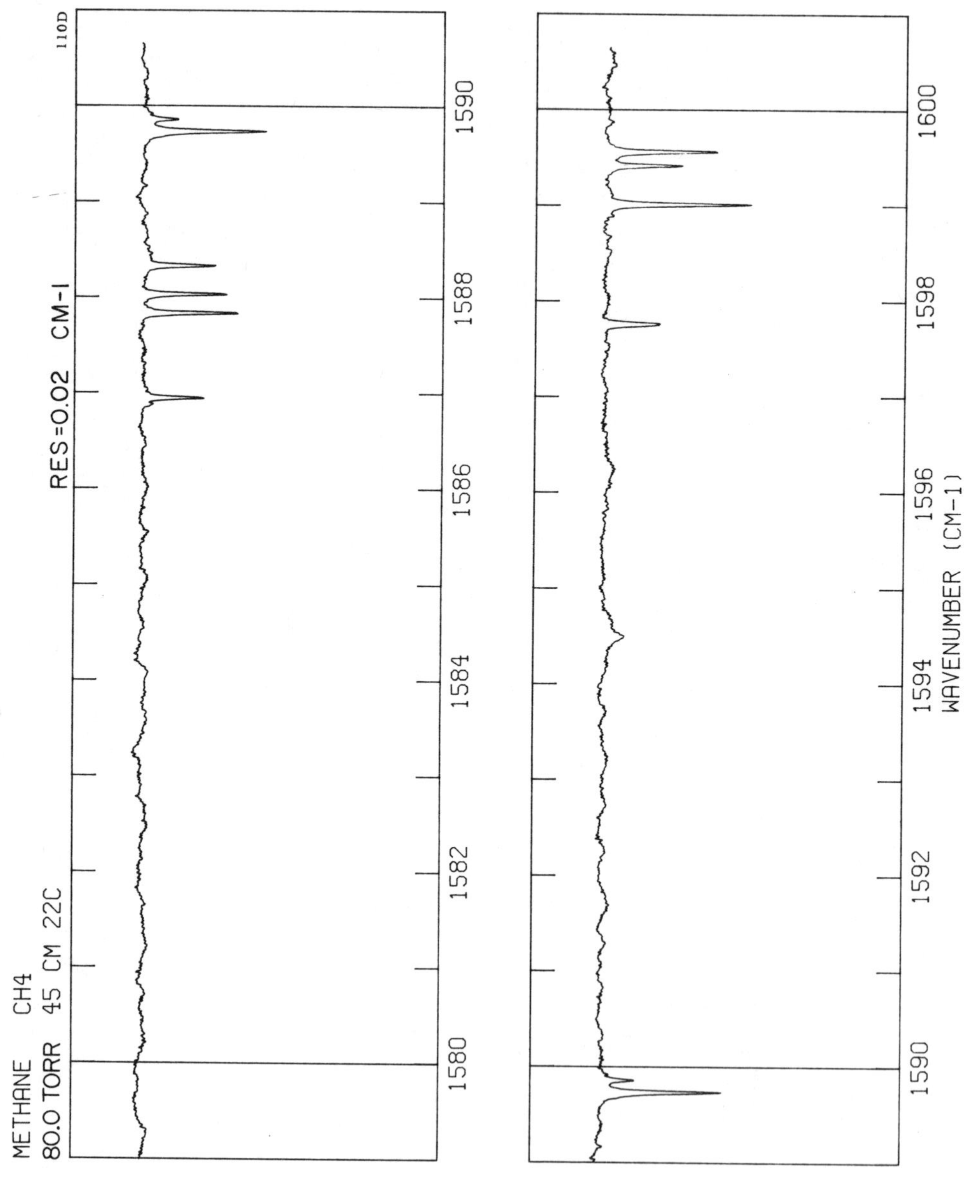
METHANE CH4
80.0 TORR 45 CM 22C
RES=0.02 CM-1
110D
1580
1582
1584
1586
1588
1590
1590
1592
1594
1596
1598
1600
WAVENUMBER (CM-1)

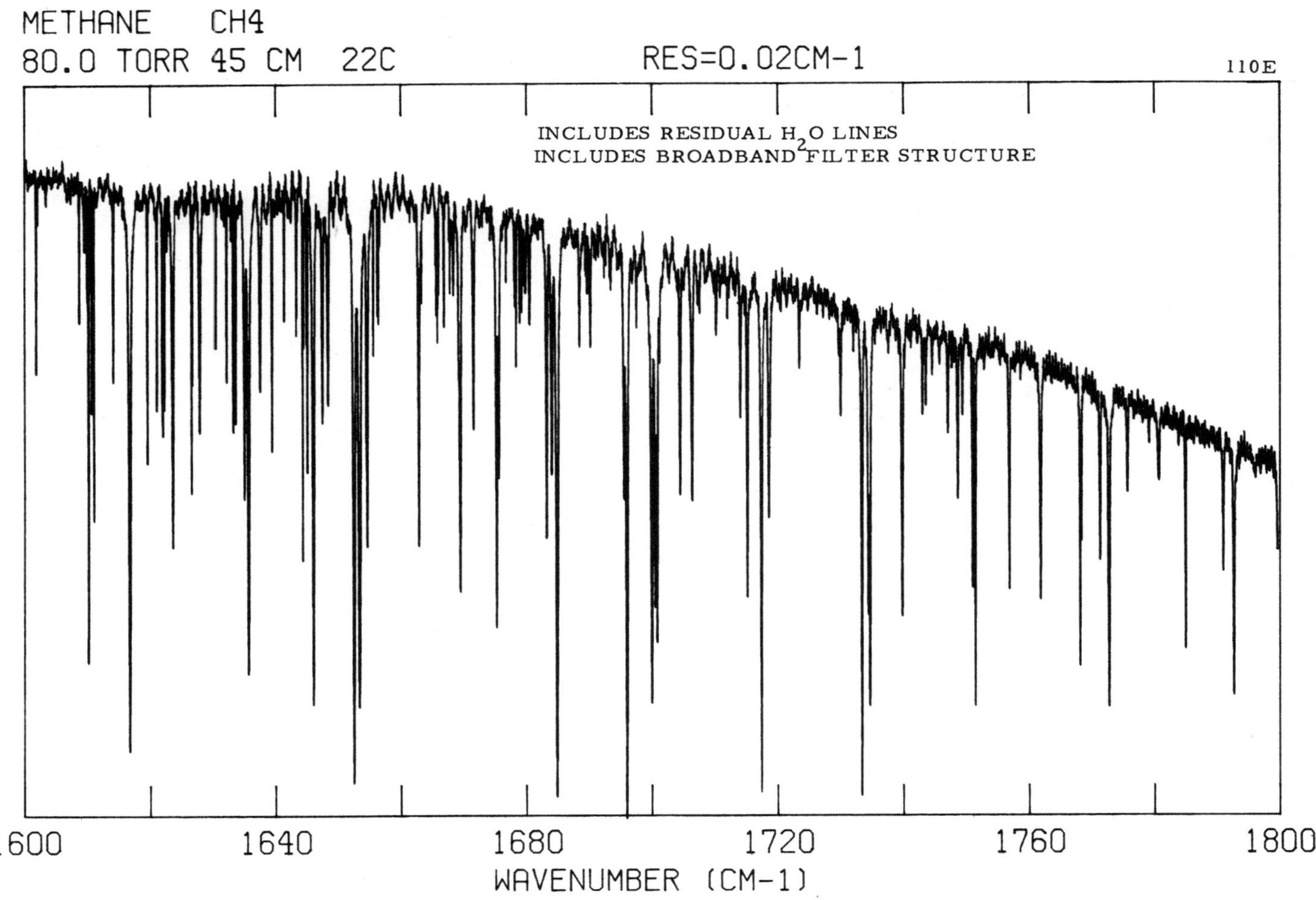
METHANE CH4
80.0 TORR 45 CM 22C
RES=0.02CM-1
110E
INCLUDES RESIDUAL H2O LINES
INCLUDES BROADBAND FILTER STRUCTURE
1600
1640
1680
1720
1760
1800
WAVENUMBER (CM-1)

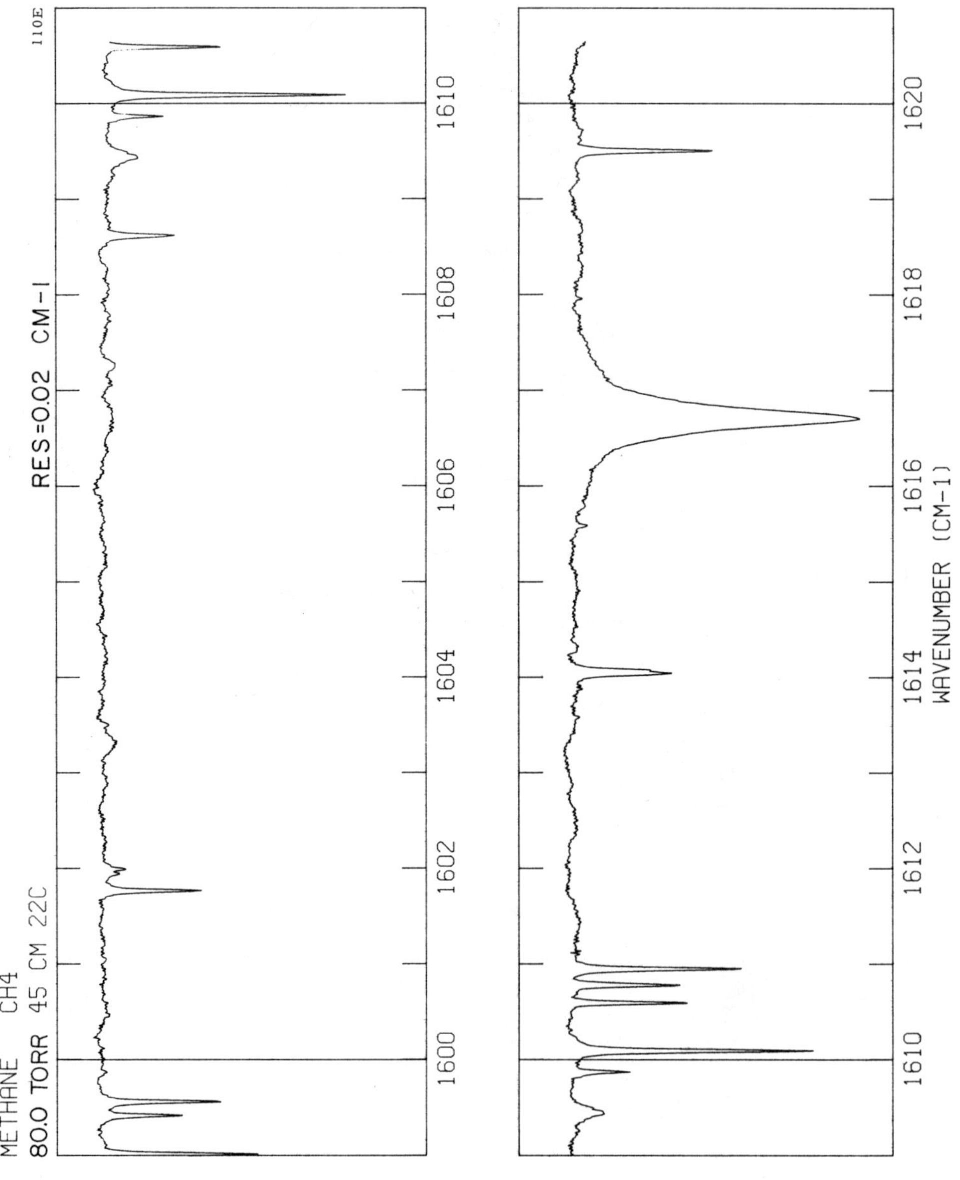
METHANE CH4
80.0 TORR 45 CM 22C
RES=0.02 CM-1
110E
1600
1602
1604
1606
1608
1610
1612
1614
1616
1618
1620
WAVENUMBER (CM-1)

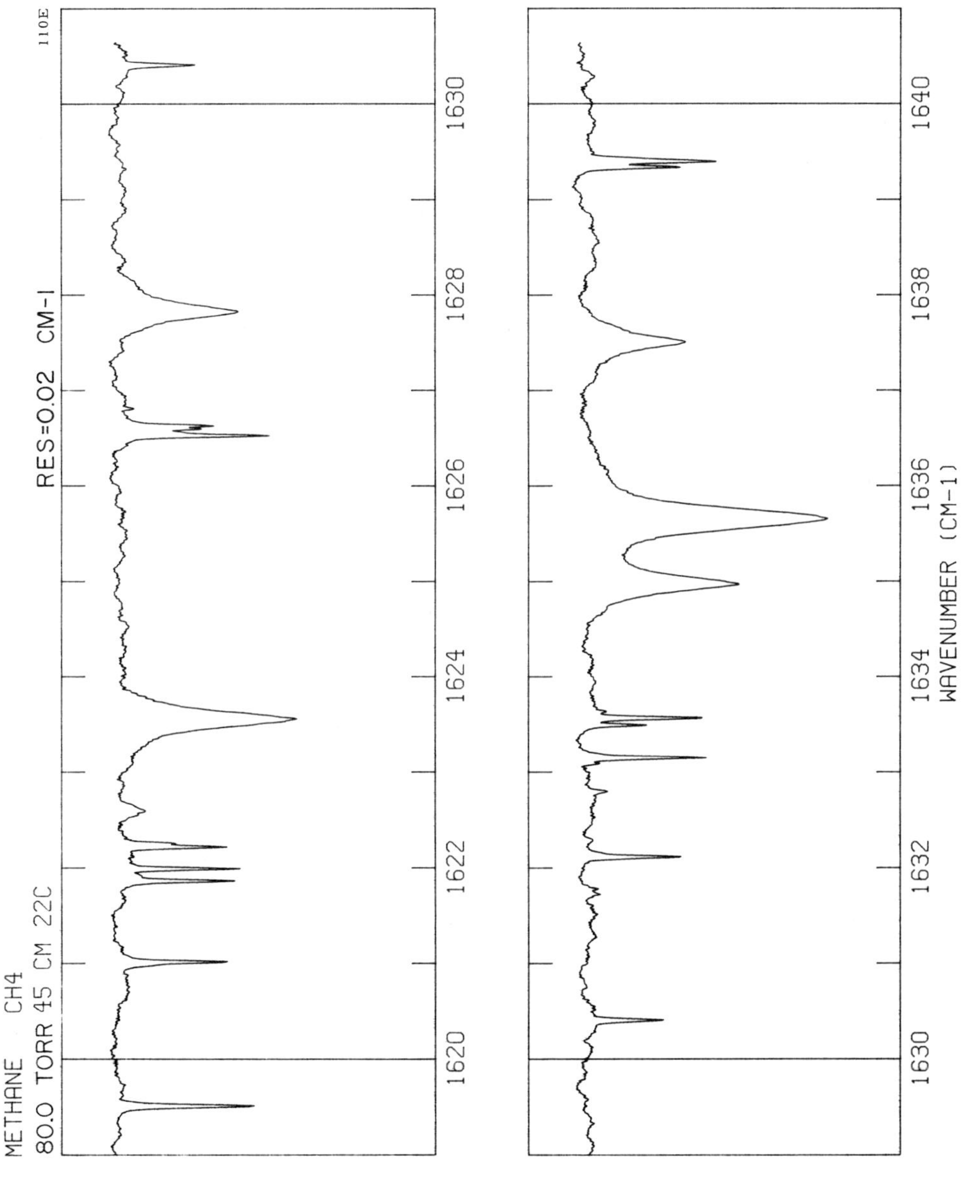
METHANE CH4
80.0 TORR 45 CM 22C
RES=0.02 CM-1
110E
1620
1622
1624
1626
1628
1630
1632
1634
1636
1638
1640
WAVENUMBER (CM-1)

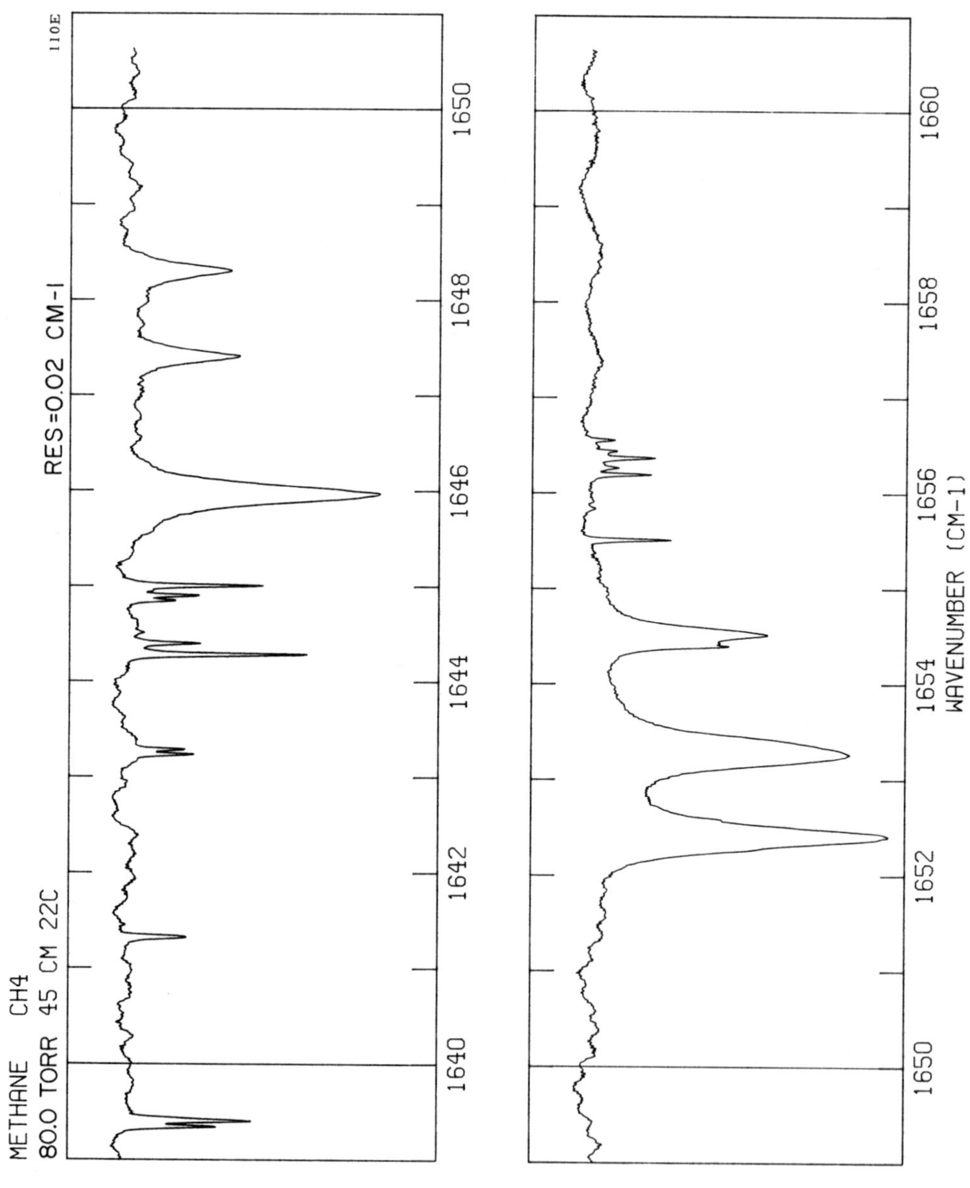
METHANE CH4
80.0 TORR 45 CM 22C
RES=0.02 CM-1
110E
1640
1642
1644
1646
1648
1650
1652
1654
1656
1658
1660
WAVENUMBER (CM-1)

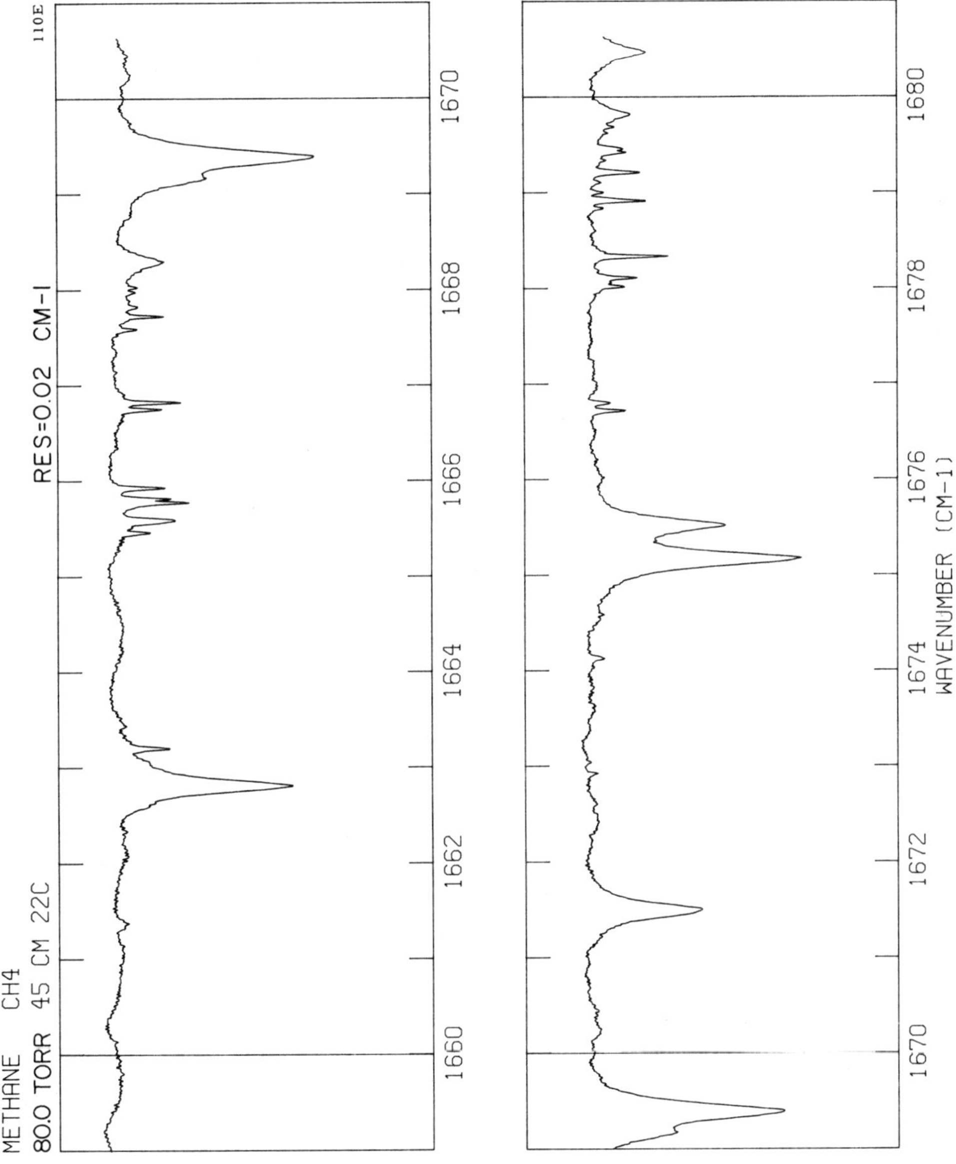

METHANE CH4
80.0 TORR 45 CM 22C
RES=0.02 CM-1
110E
1660
1662
1664
1666
1668
1670
1670
1672
1674
1676
1678
1680
WAVENUMBER (CM-1)

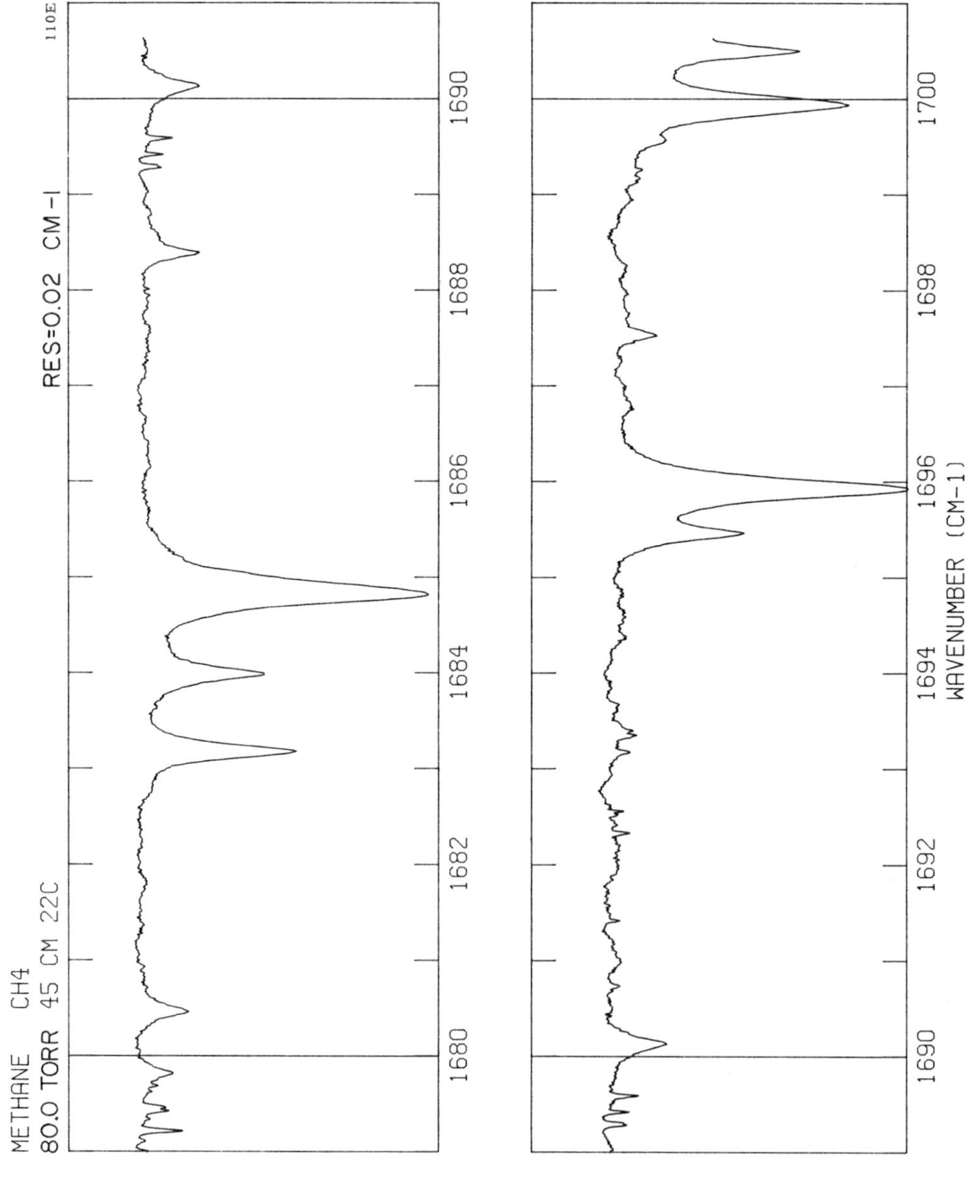
METHANE CH4
80.0 TORR 45 CM 22C
RES=0.02 CM-1
110E
1680
1682
1684
1686
1688
1690
1692
1694
1696
1698
1700
WAVENUMBER (CM-1)

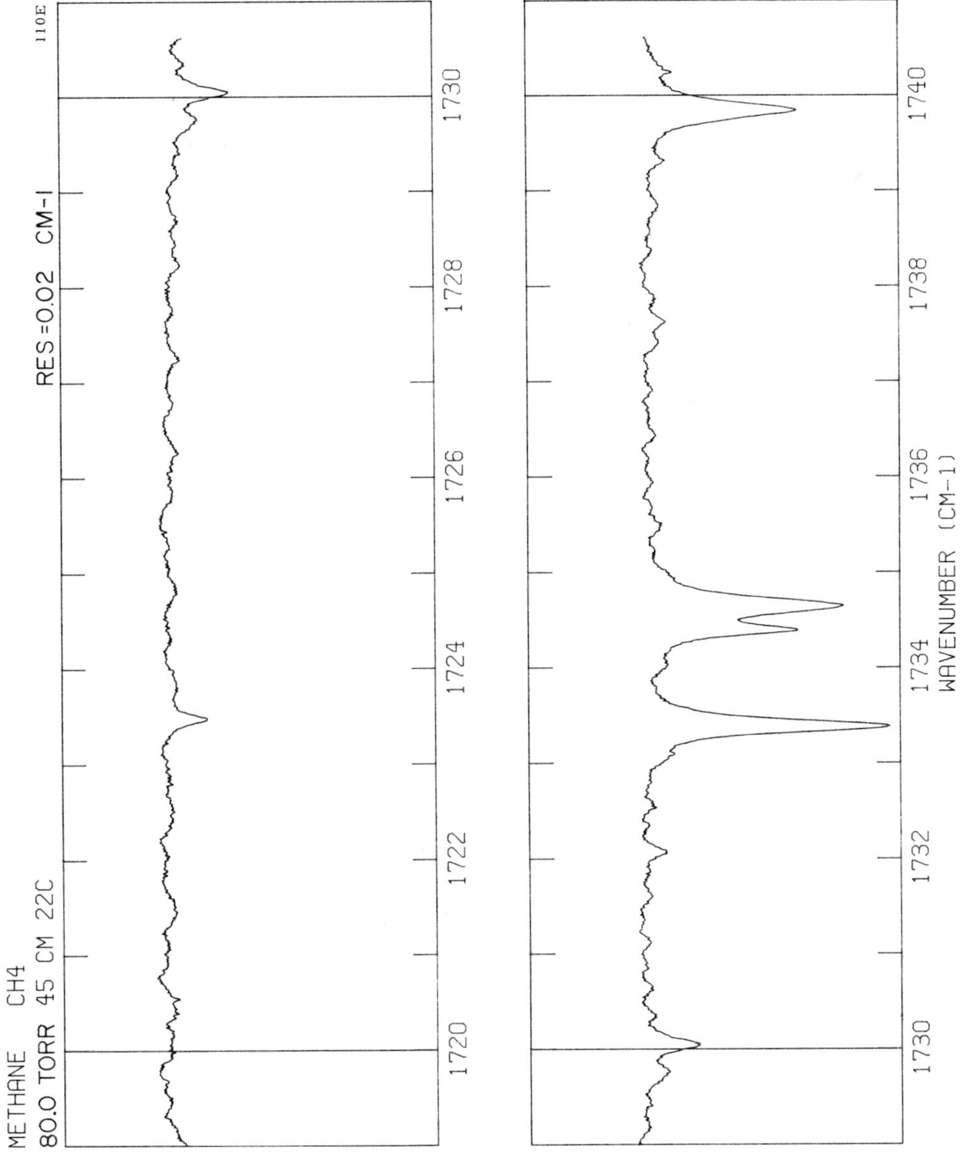
METHANE CH4
80.0 TORR 45 CM 22C
RES=0.02 CM-1
110E
1720
1722
1724
1726
1728
1730
1730
1732
1734
1736
1738
1740
WAVENUMBER (CM-1)

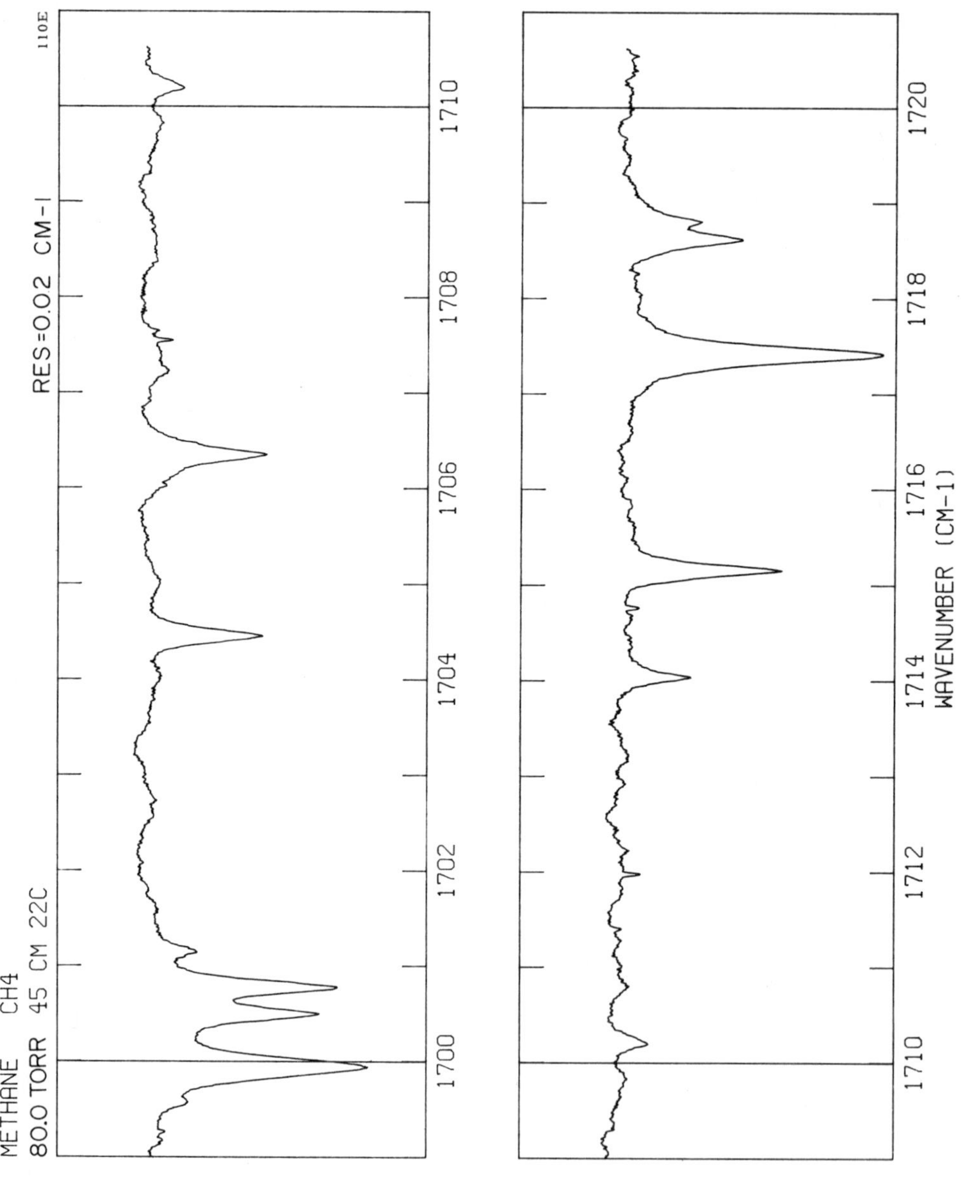
110E
METHANE CH4
80.0 TORR 45 CM 22C
RES=0.02 CM-1
1700
1702
1704
1706
1708
1710
1712
1714
1716
1718
1720
WAVENUMBER (CM-1)

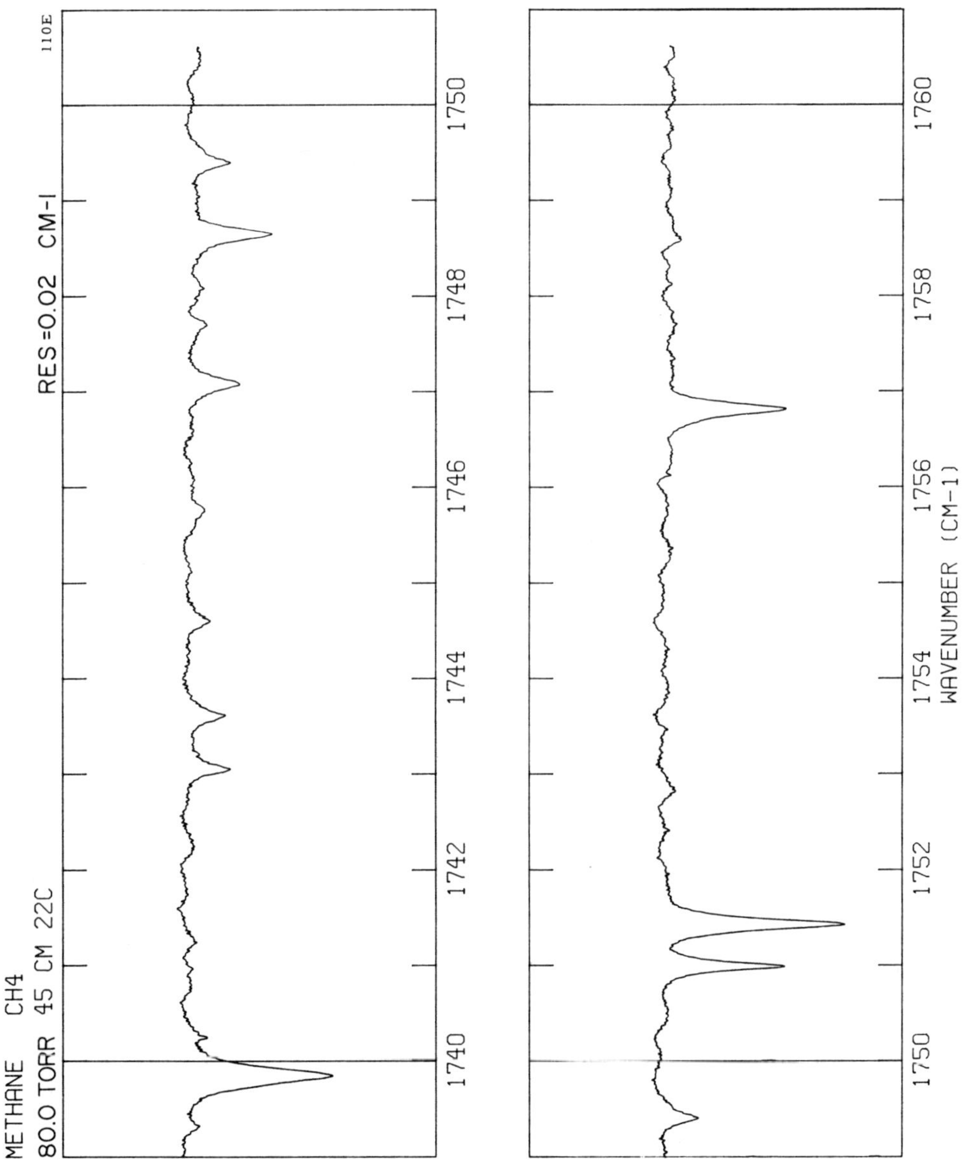
METHANE CH4
80.0 TORR 45 CM 22C
RES=0.02 CM-1
110E
1740
1742
1744
1746
1748
1750
1752
1754
1756
1758
1760
WAVENUMBER (CM-1)

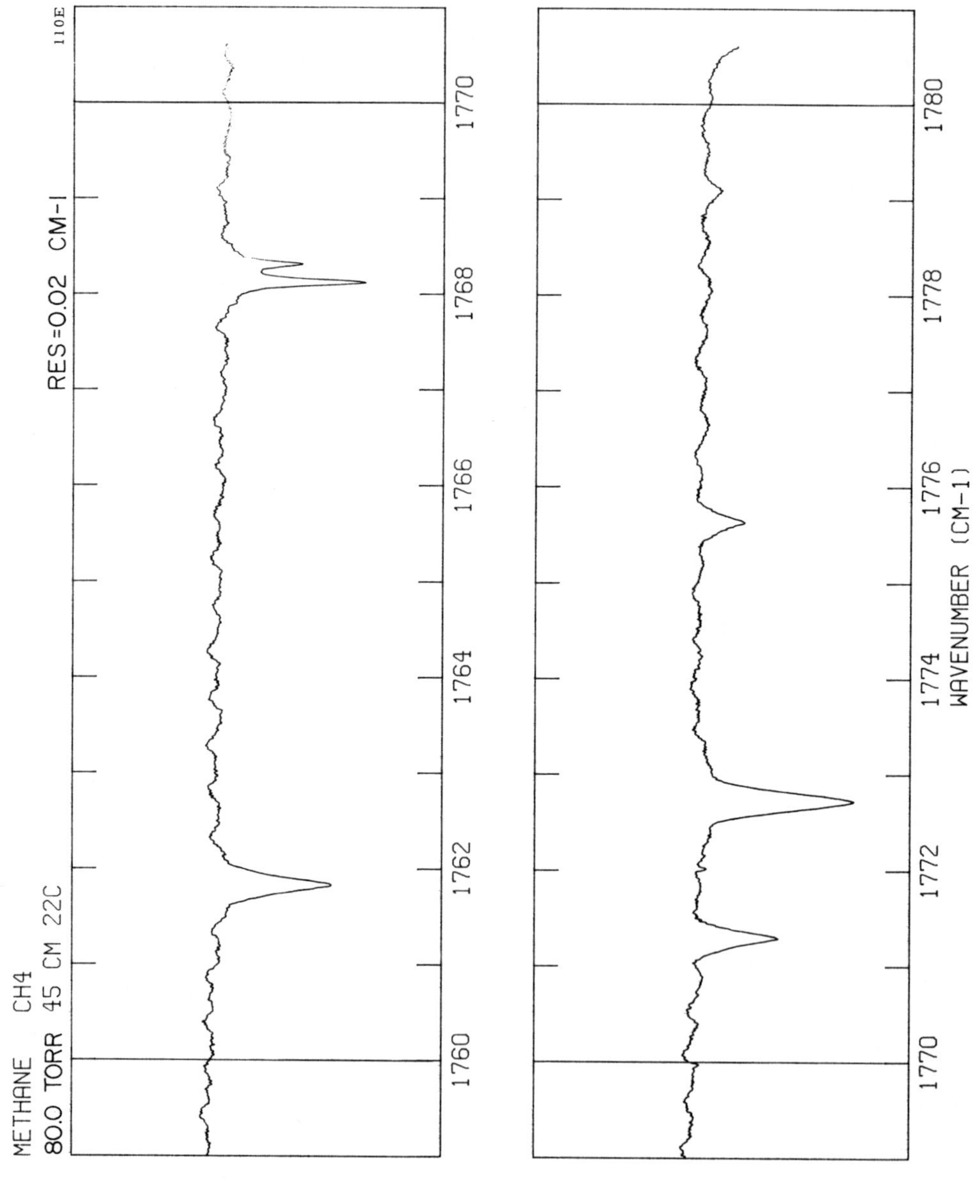
METHANE CH4
80.0 TORR 45 CM 22C
RES=0.02 CM-1
110E
1760
1762
1764
1766
1768
1770
1770
1772
1774
1776
1778
1780
WAVENUMBER (CM-1)

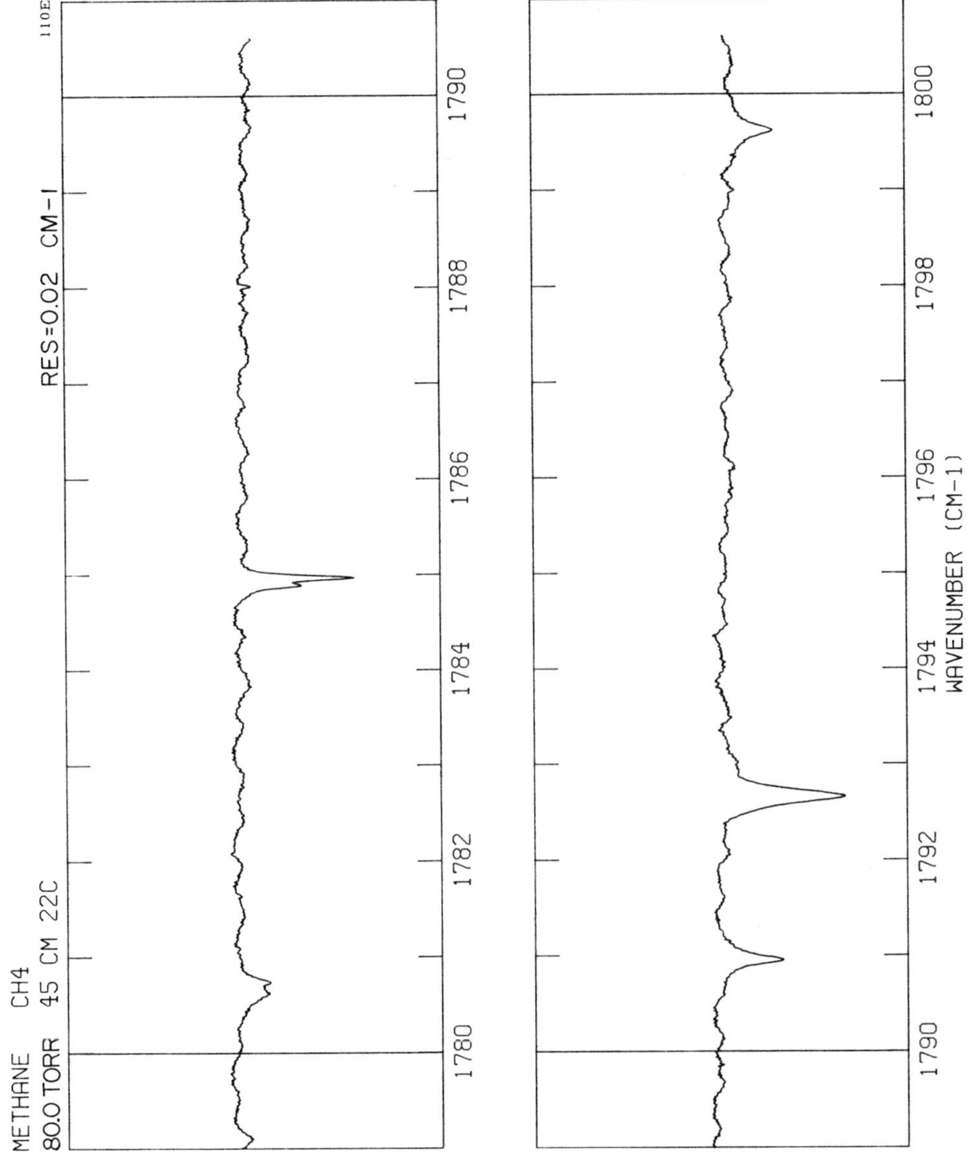
METHANE CH4
80.0 TORR 45 CM 22C
RES=0.02 CM-1
110E
1780
1782
1784
1786
1788
1790
1790
1792
1794
1796
1798
1800
WAVENUMBER (CM-1)

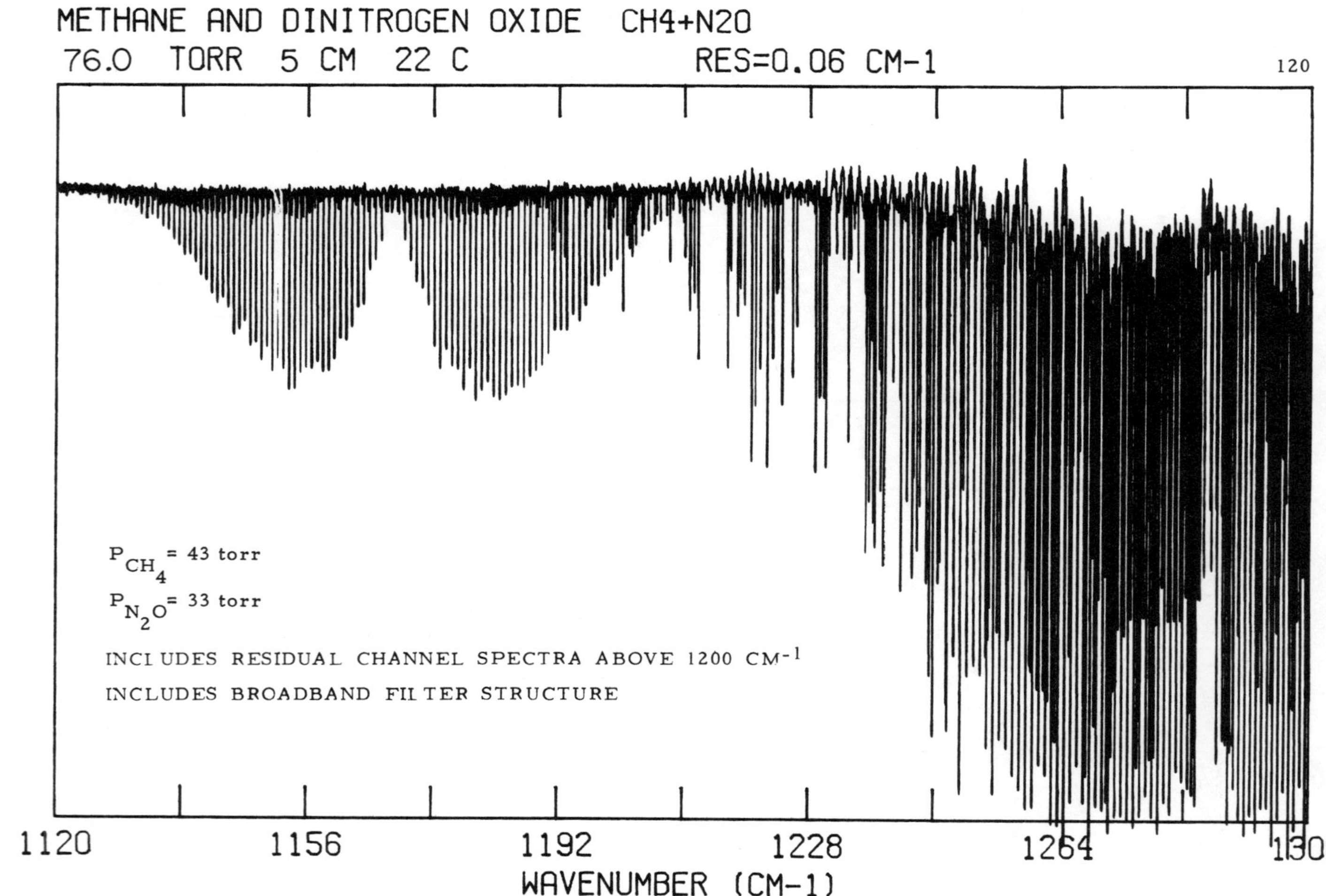
METHANE AND DINITROGEN OXIDE CH4+N2O
76.0 TORR 5 CM 22 C RES=0.06 CM-1
120
P_{CH_4} = 43 torr
P_{N_2O} = 33 torr
INCLUDES RESIDUAL CHANNEL SPECTRA ABOVE 1200 CM^{-1}
INCLUDES BROADBAND FILTER STRUCTURE
1120
1156
1192
1228
1264
1300
WAVENUMBER (CM-1)

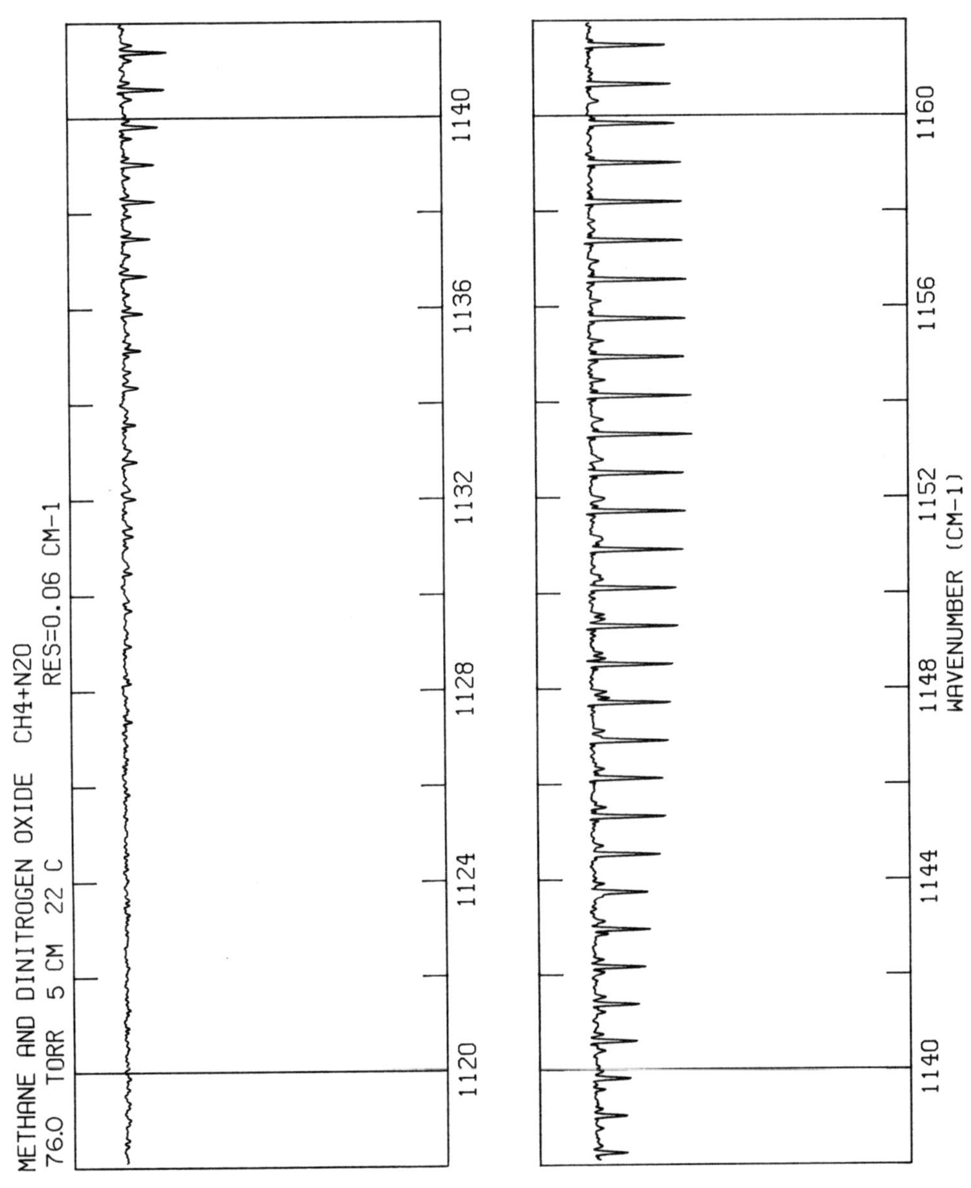

METHANE AND DINITROGEN OXIDE CH4+N2O
76.0 TORR 5 CM 22 C
RES=0.06 CM-1
1120
1124
1128
1132
1136
1140
1140
1144
1148
1152
1156
1160
WAVENUMBER (CM-1)

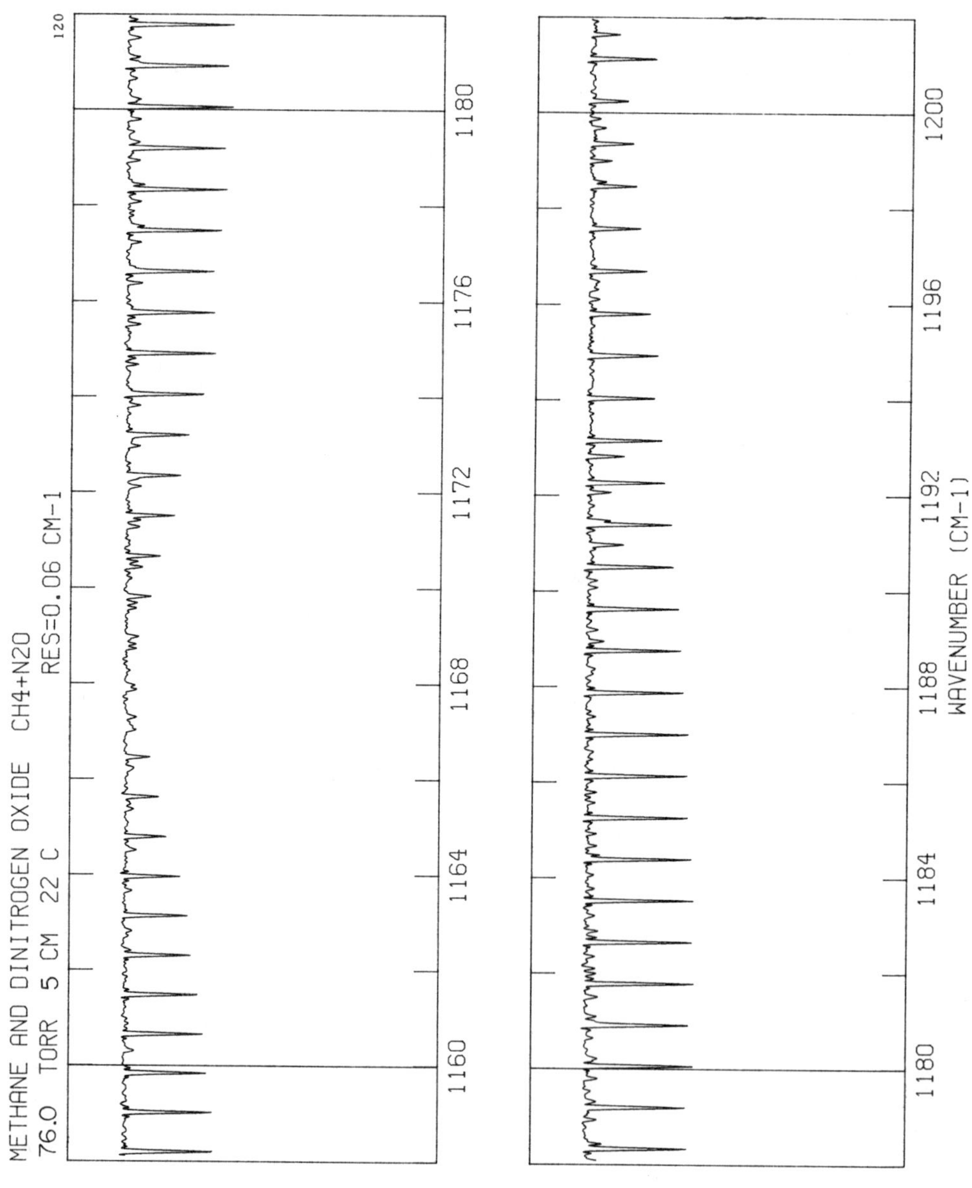
METHANE AND DINITROGEN OXIDE CH4+N2O
76.0 TORR 5 CM 22 C RES=0.06 CM-1
120
1160
1164
1168
1172
1176
1180
1180
1184
1188
1192
1196
1200
WAVENUMBER (CM-1)

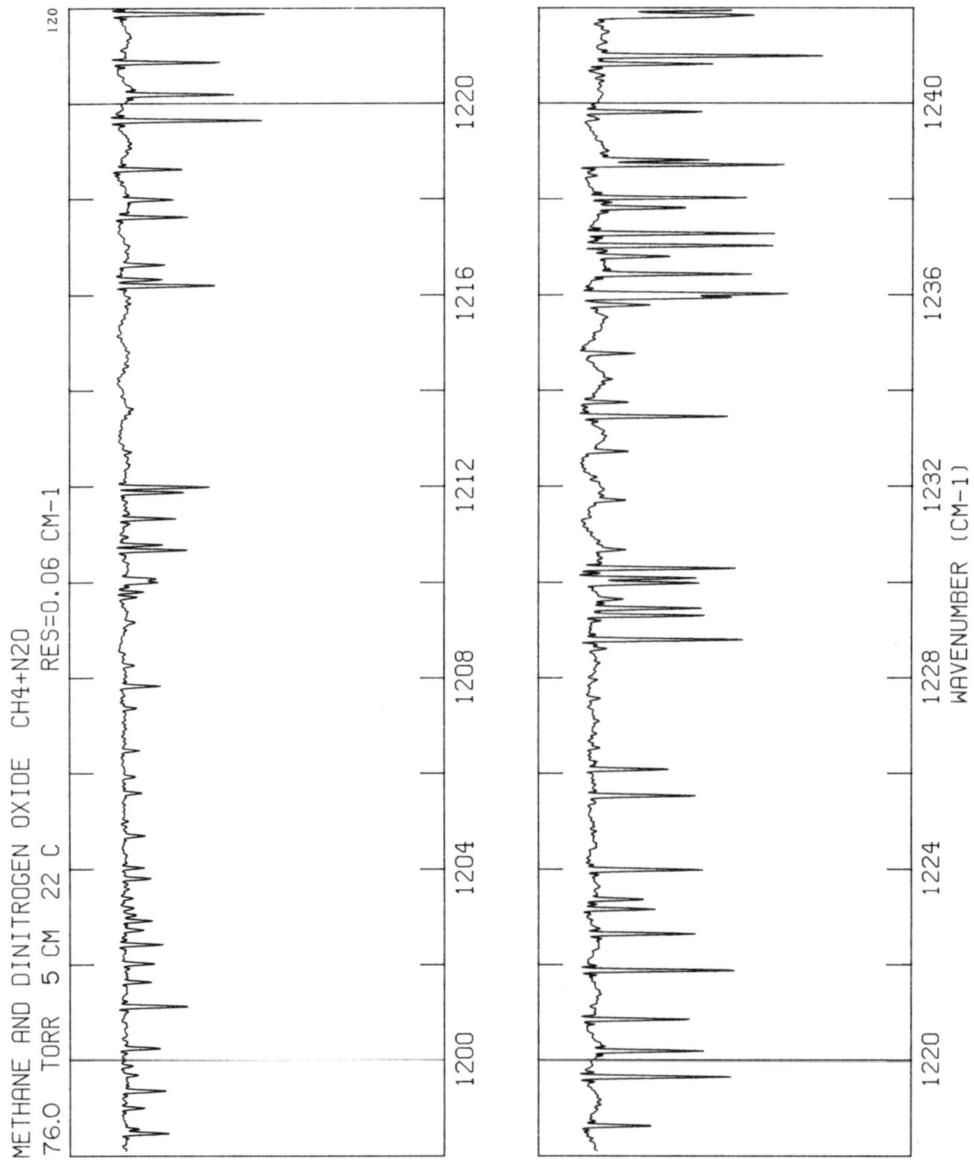
METHANE AND DINITROGEN OXIDE CH4+N2O
76.0 TORR 5 CM 22 C
RES=0.06 CM-1
120
1200
1204
1208
1212
1216
1220
1220
1224
1228
1232
1236
1240
WAVENUMBER (CM-1)

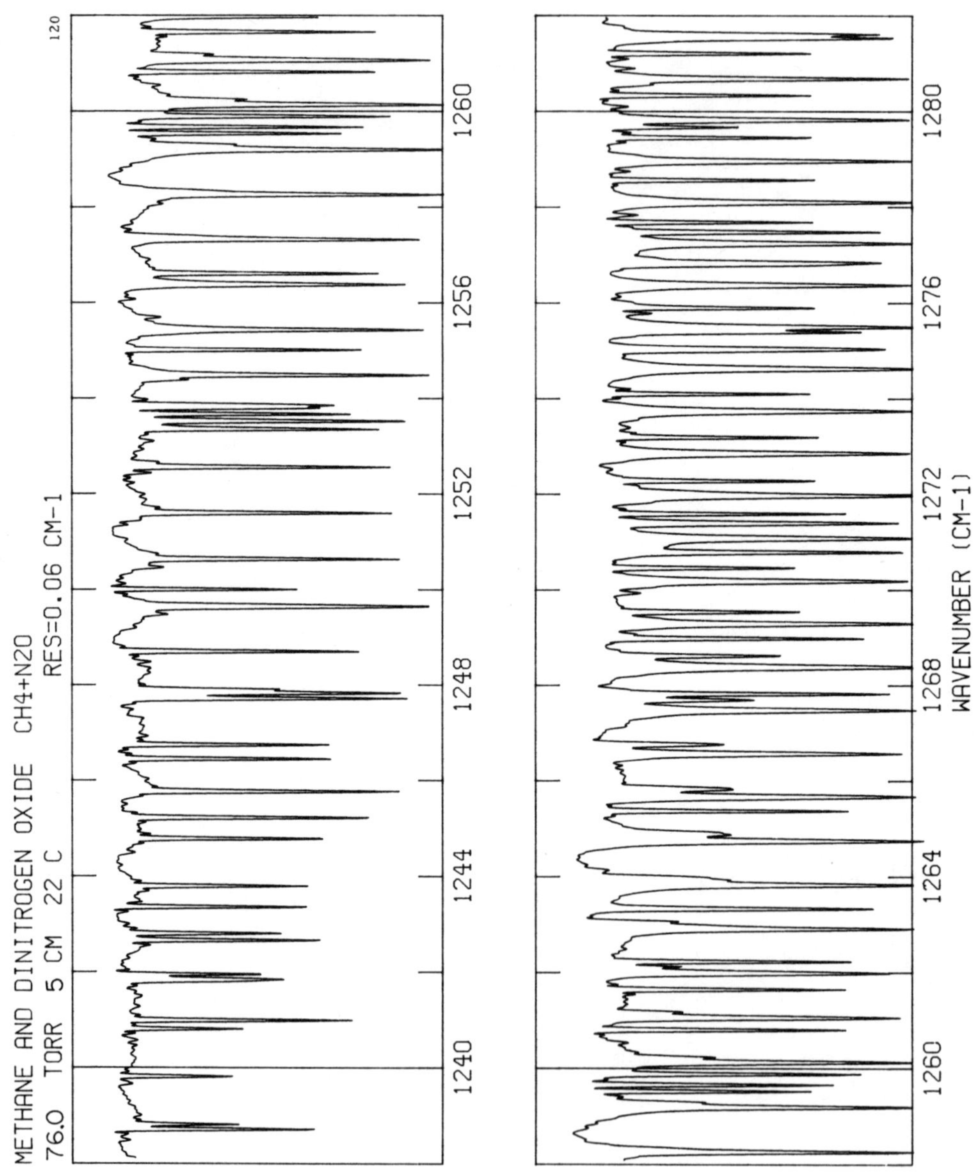
METHANE AND DINITROGEN OXIDE CH4+N2O
76.0 TORR 5 CM 22 C RES=0.06 CM-1
120
1240
1244
1248
1252
1256
1260
1260
1264
1268
1272
1276
1280
WAVENUMBER (CM-1)

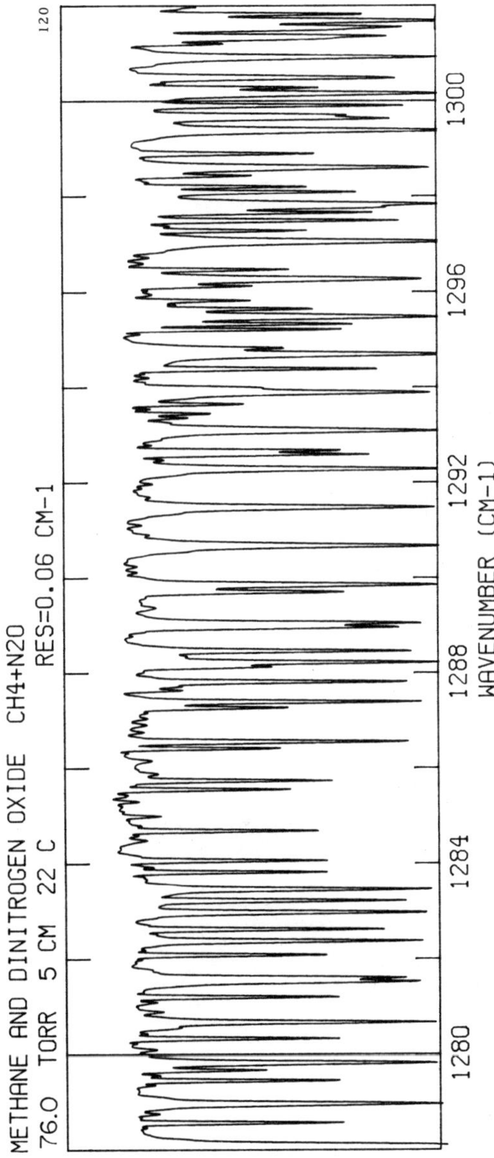
METHANE AND DINITROGEN OXIDE CH4+N2O
76.0 TORR 5 CM 22 C RES=0.06 CM-1
1280
1284
1288
1292
1296
1300
WAVENUMBER (CM-1)
120

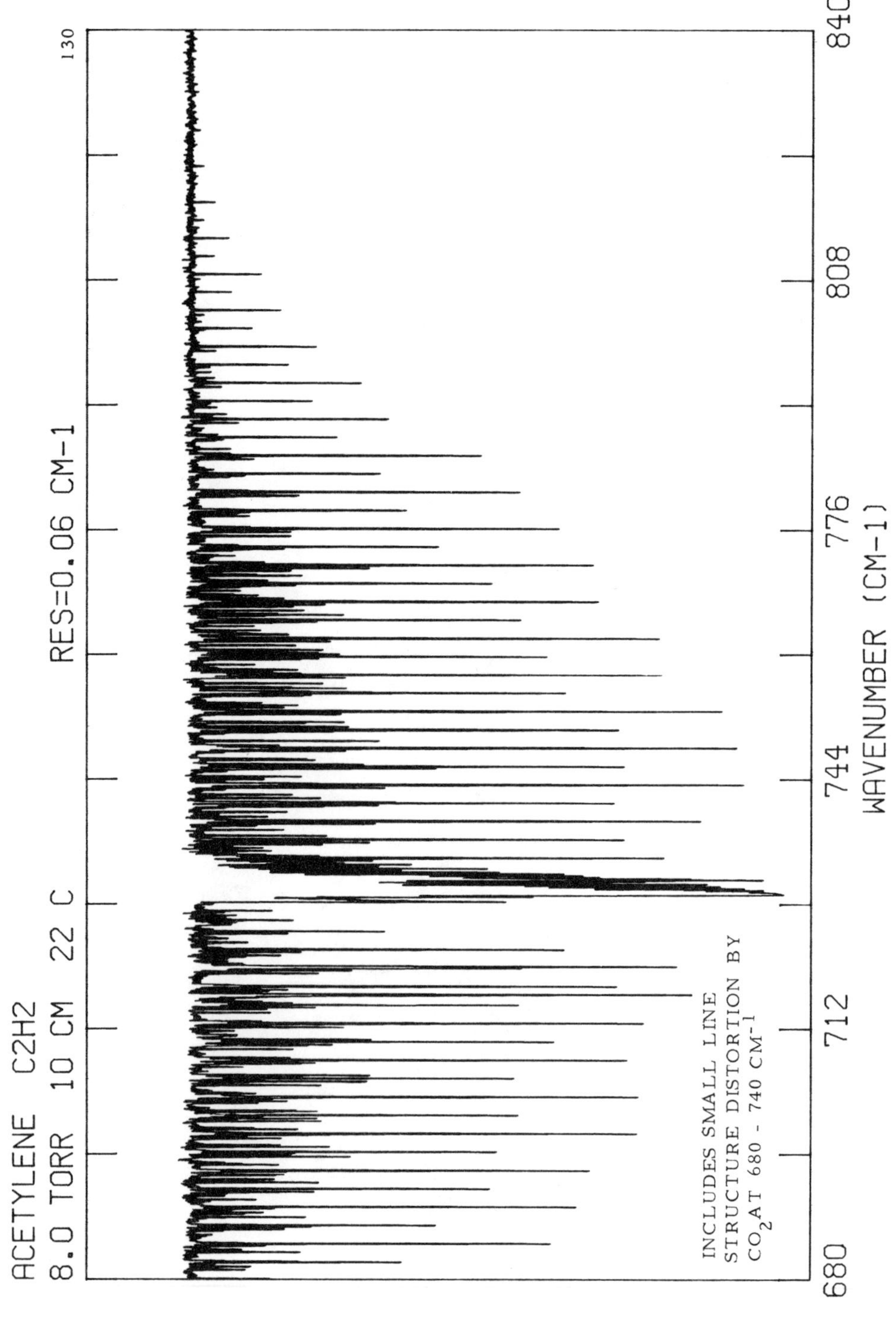
ACETYLENE C2H2
8.0 TORR 10 CM 22 C
RES=0.06 CM-1
130
INCLUDES SMALL LINE
STRUCTURE DISTORTION BY
CO_2 AT 680 - 740 CM^{-1}
680
712
744
776
808
840
WAVENUMBER (CM-1)

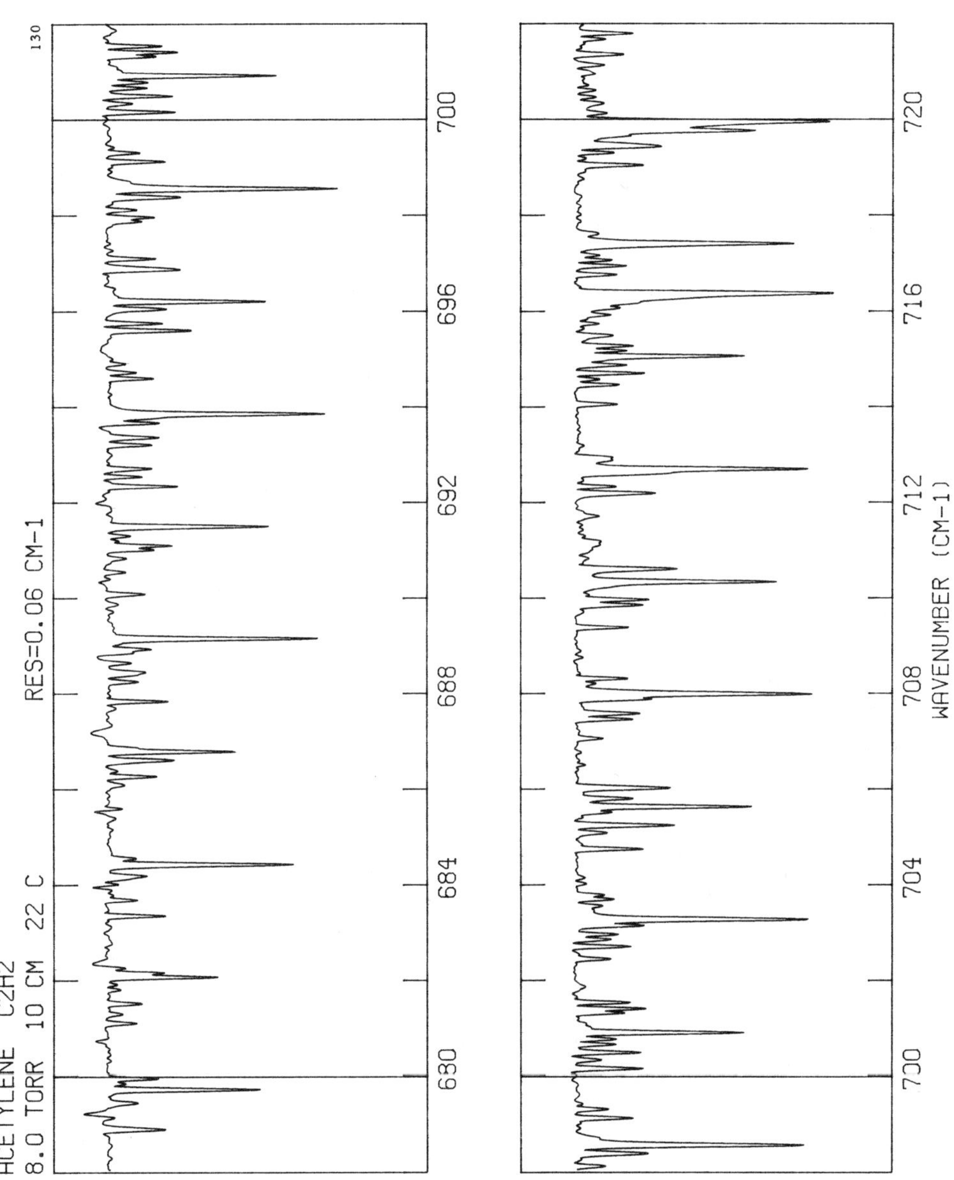
ACETYLENE C2H2
8.0 TORR 10 CM 22 C
RES=0.06 CM-1
130
680
684
688
692
696
700
700
704
708
712
716
720
WAVENUMBER (CM-1)

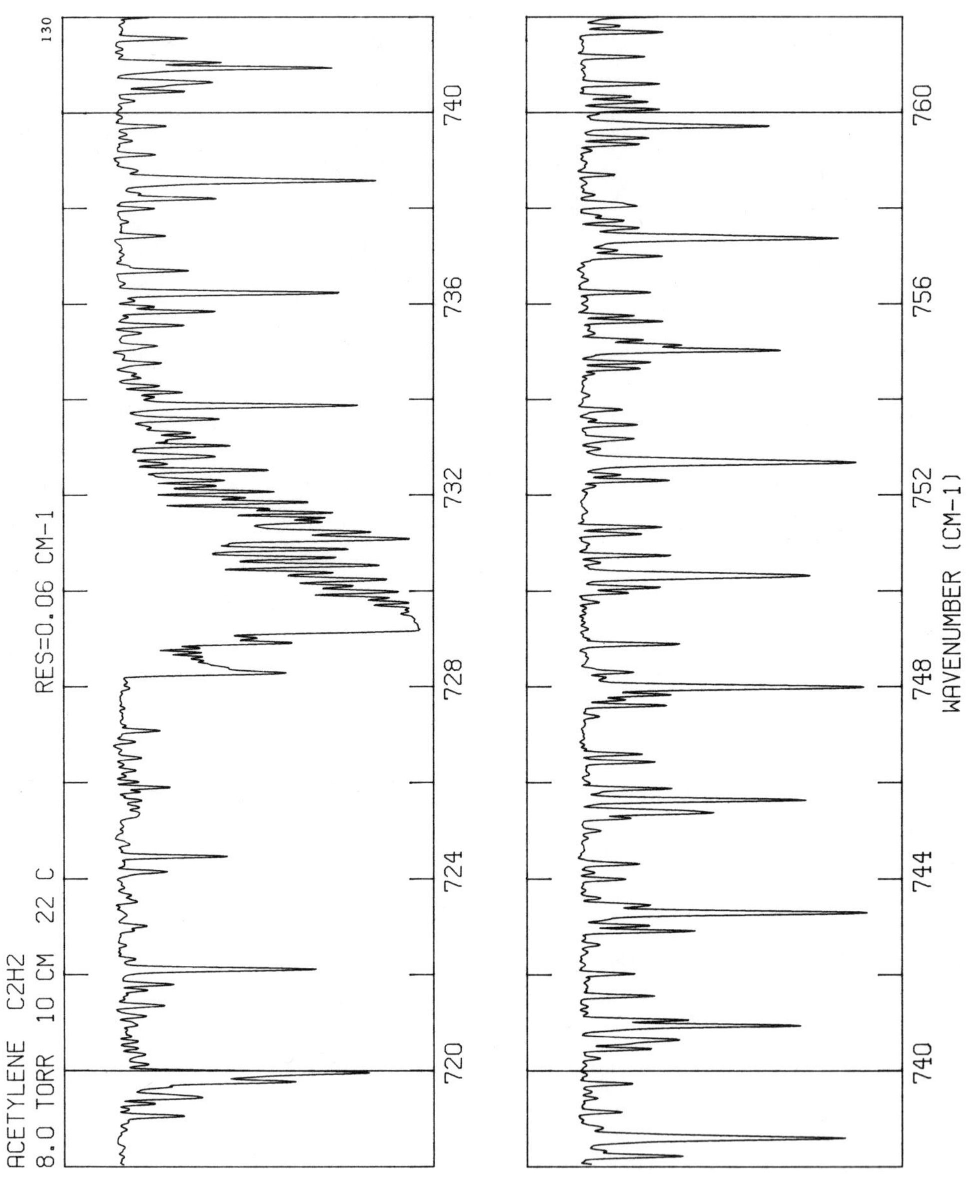

ACETYLENE C2H2
8.0 TORR 10 CM 22 C
RES=0.06 CM-1
130
720
724
728
732
736
740
740
744
748
752
756
760
WAVENUMBER (CM-1)

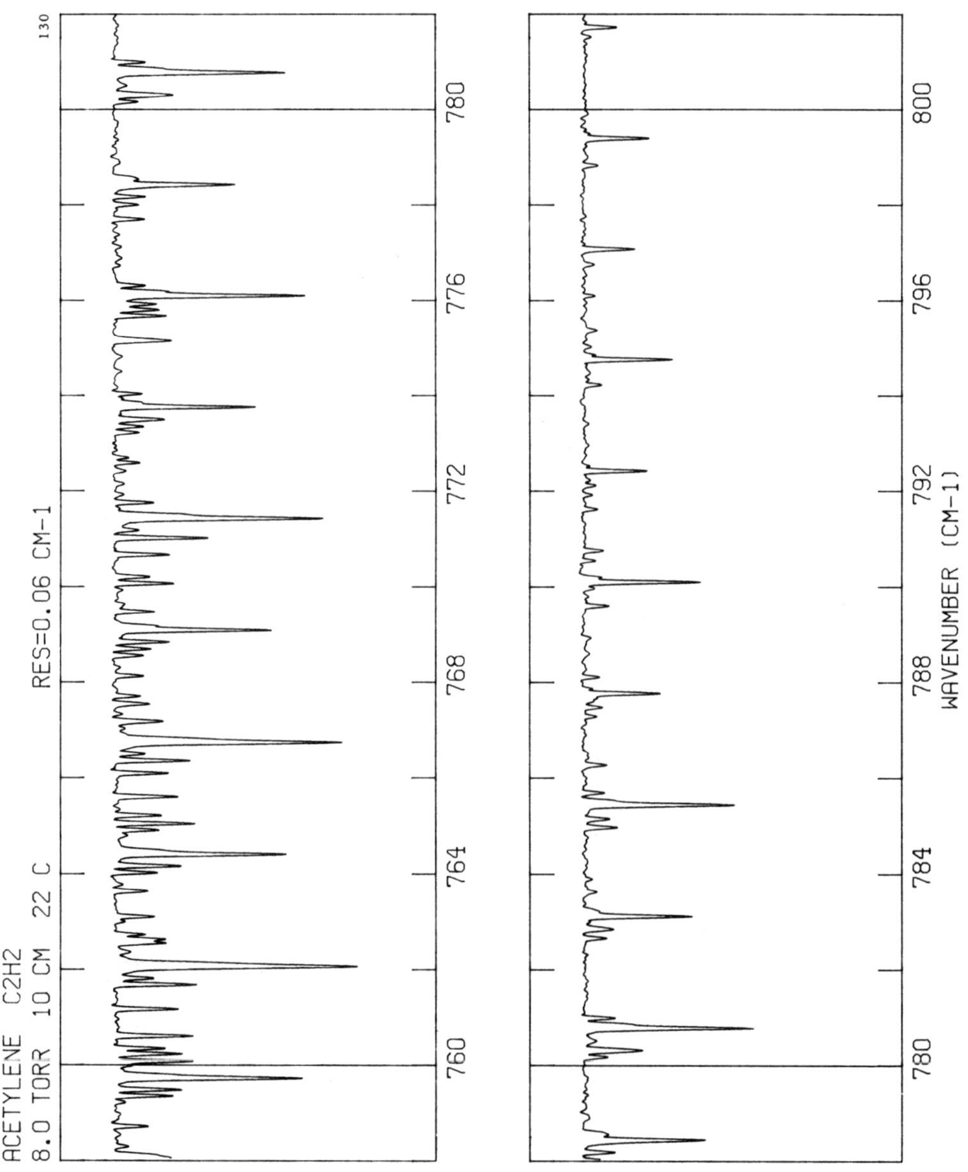
ACETYLENE C2H2
8.0 TORR 10 CM 22 C
RES=0.06 CM-1
130
760
764
768
772
776
780
784
788
792
796
800
WAVENUMBER (CM-1)

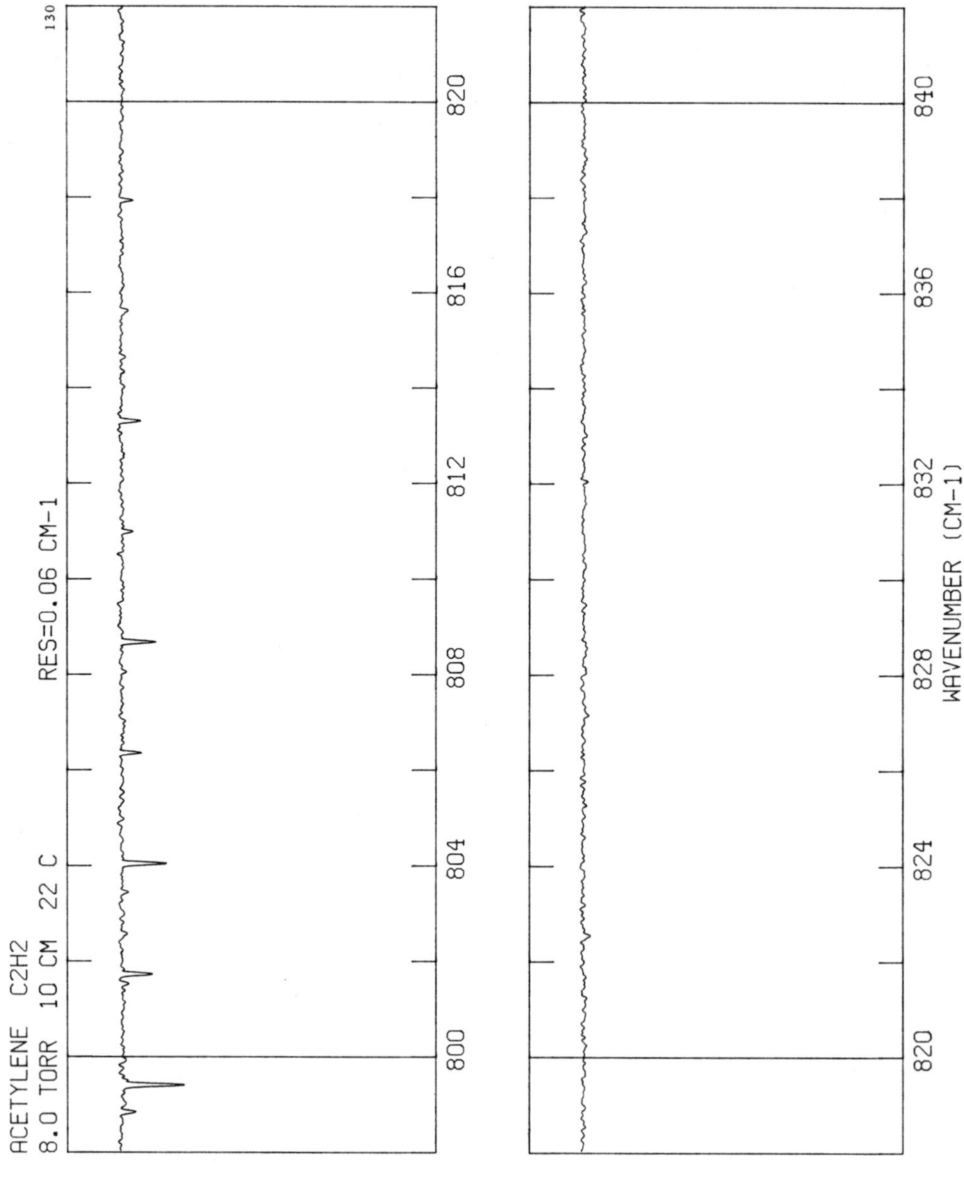
ACETYLENE C2H2
8.0 TORR 10 CM 22 C
RES=0.06 CM-1
130
800
804
808
812
816
820
824
828
832
836
840
WAVENUMBER (CM-1)

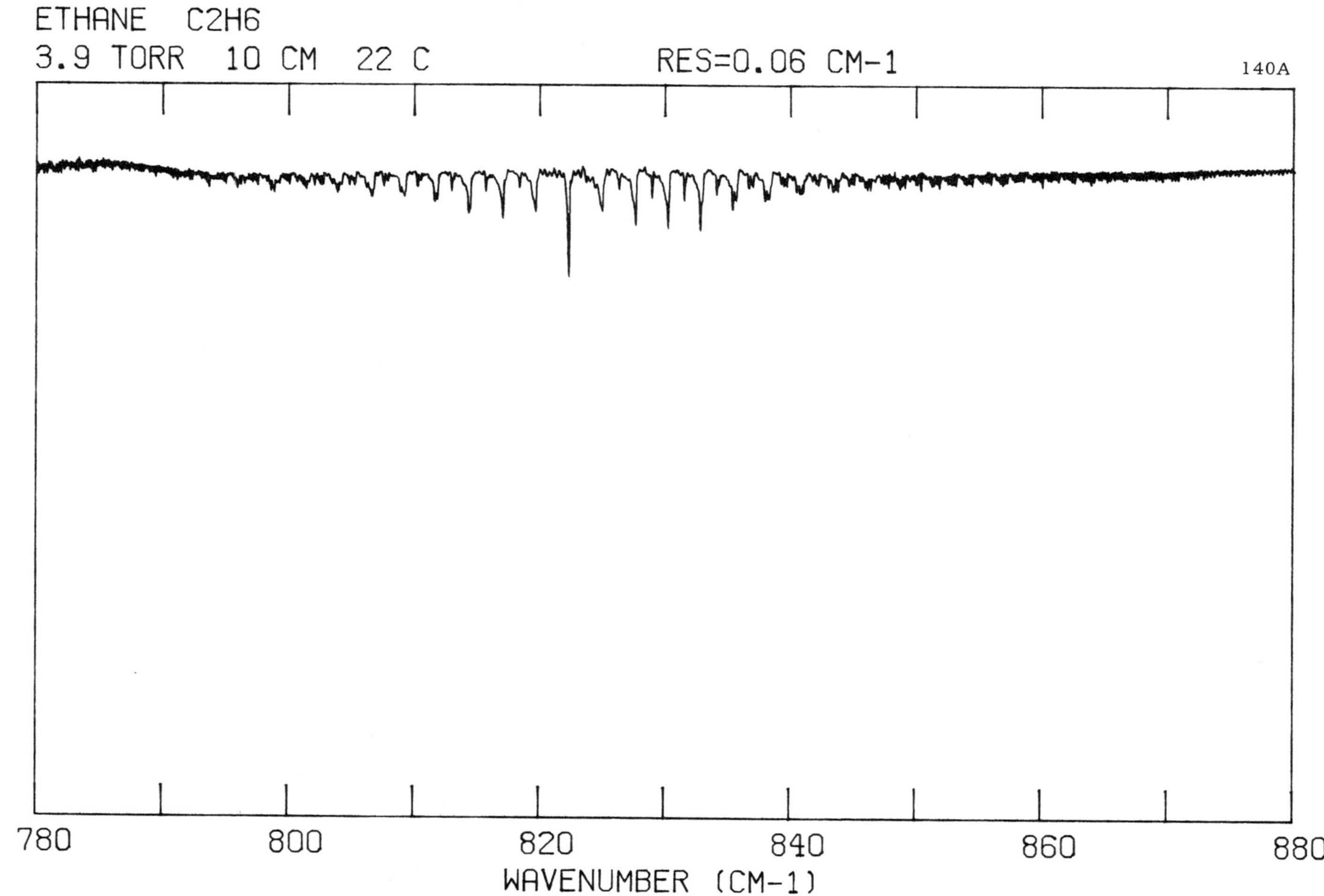
ETHANE C2H6
3.9 TORR 10 CM 22 C
RES=0.06 CM-1
140A
780
800
820
840
860
880
WAVENUMBER (CM-1)

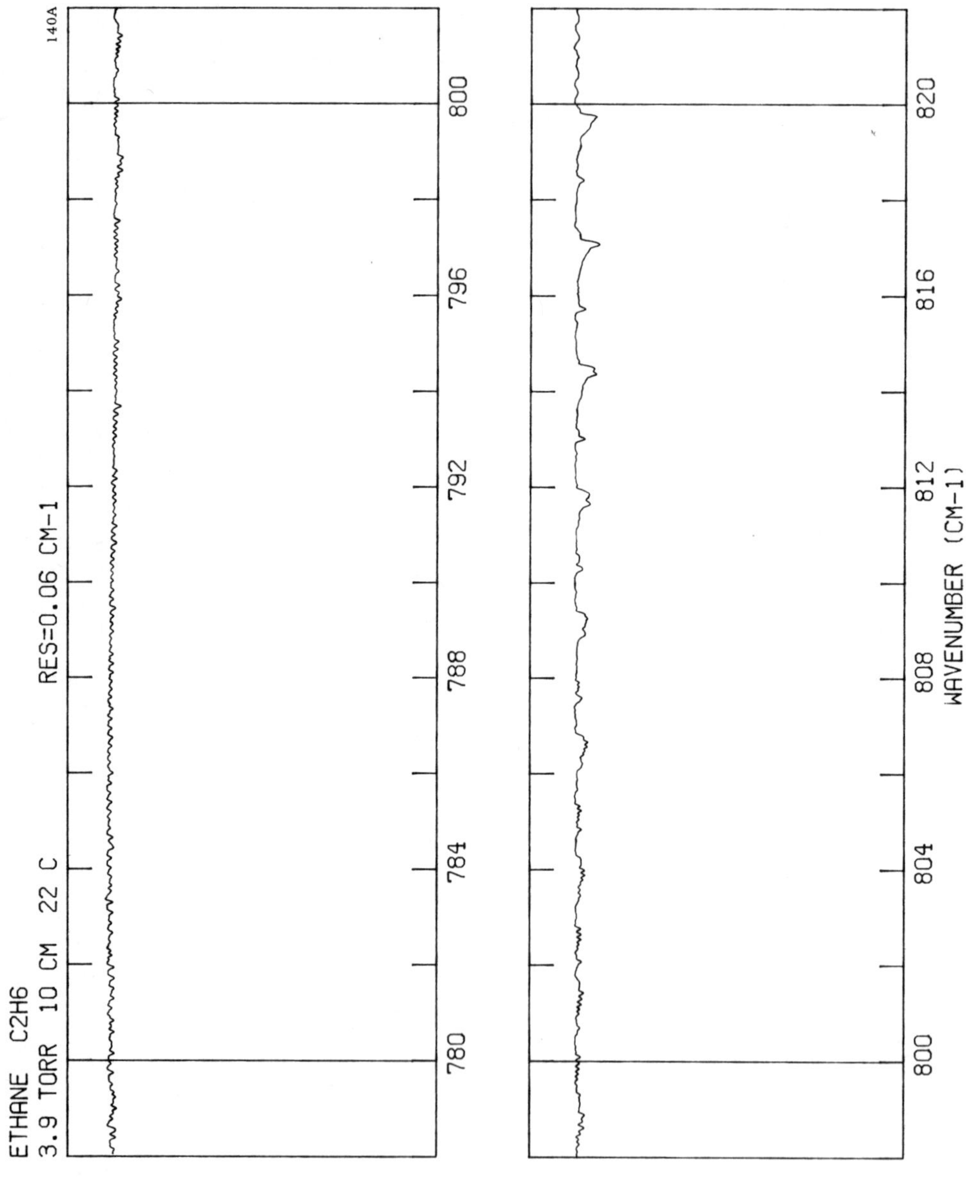
140A
ETHANE C2H6
3.9 TORR 10 CM 22 C
RES=0.06 CM-1
780
784
788
792
796
800
800
804
808
812
816
820
WAVENUMBER (CM-1)

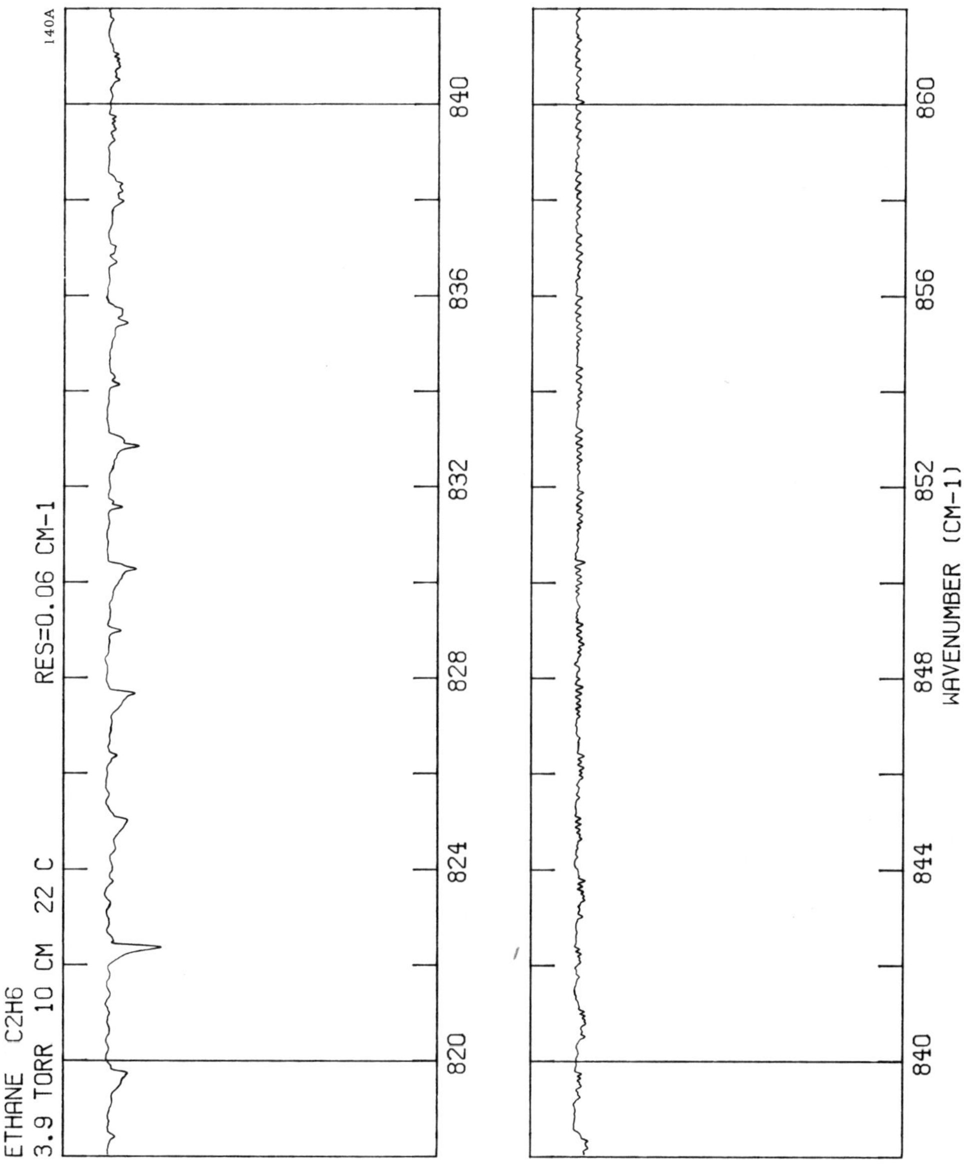
ETHANE C2H6
3.9 TORR 10 CM 22 C
RES=0.06 CM-1
140A
820
824
828
832
836
840
840
844
848
852
856
860
WAVENUMBER (CM-1)

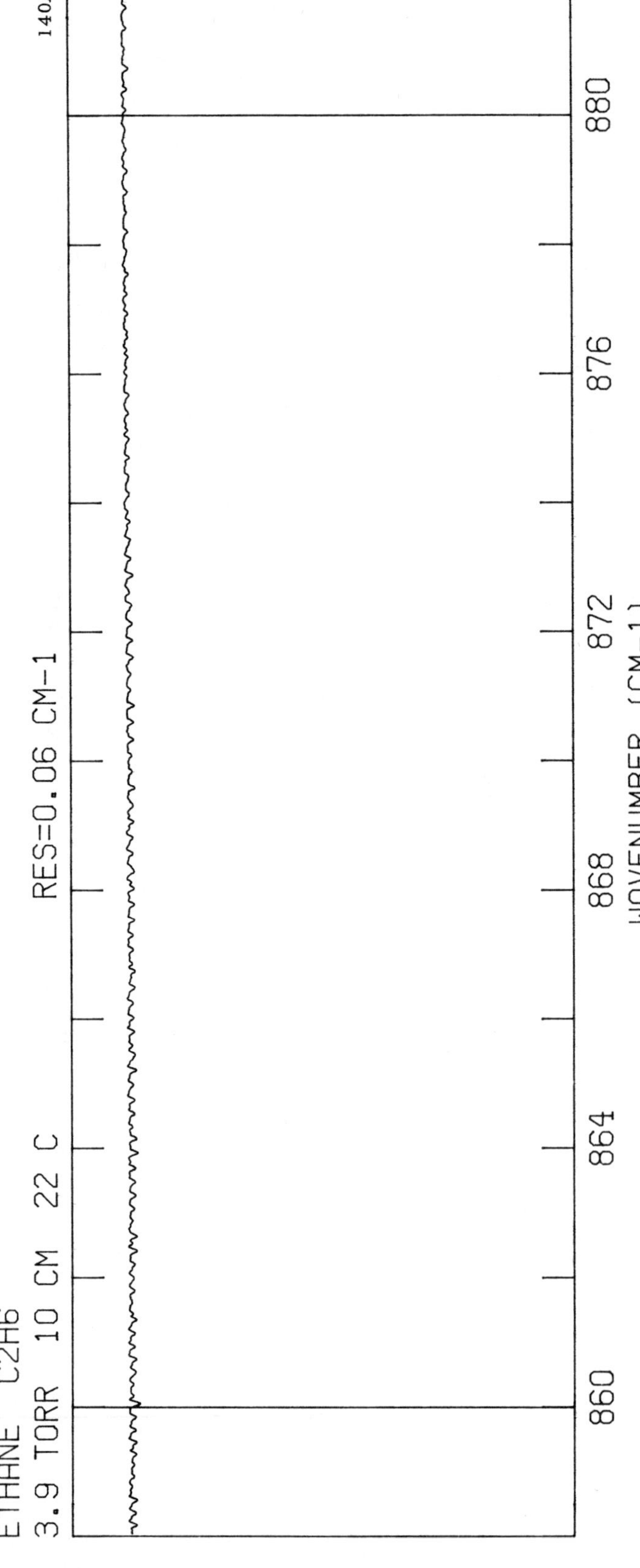
ETHANE C2H6
3.9 TORR 10 CM 22 C
RES=0.06 CM-1
140A
860
864
868
872
876
880
WAVENUMBER (CM-1)

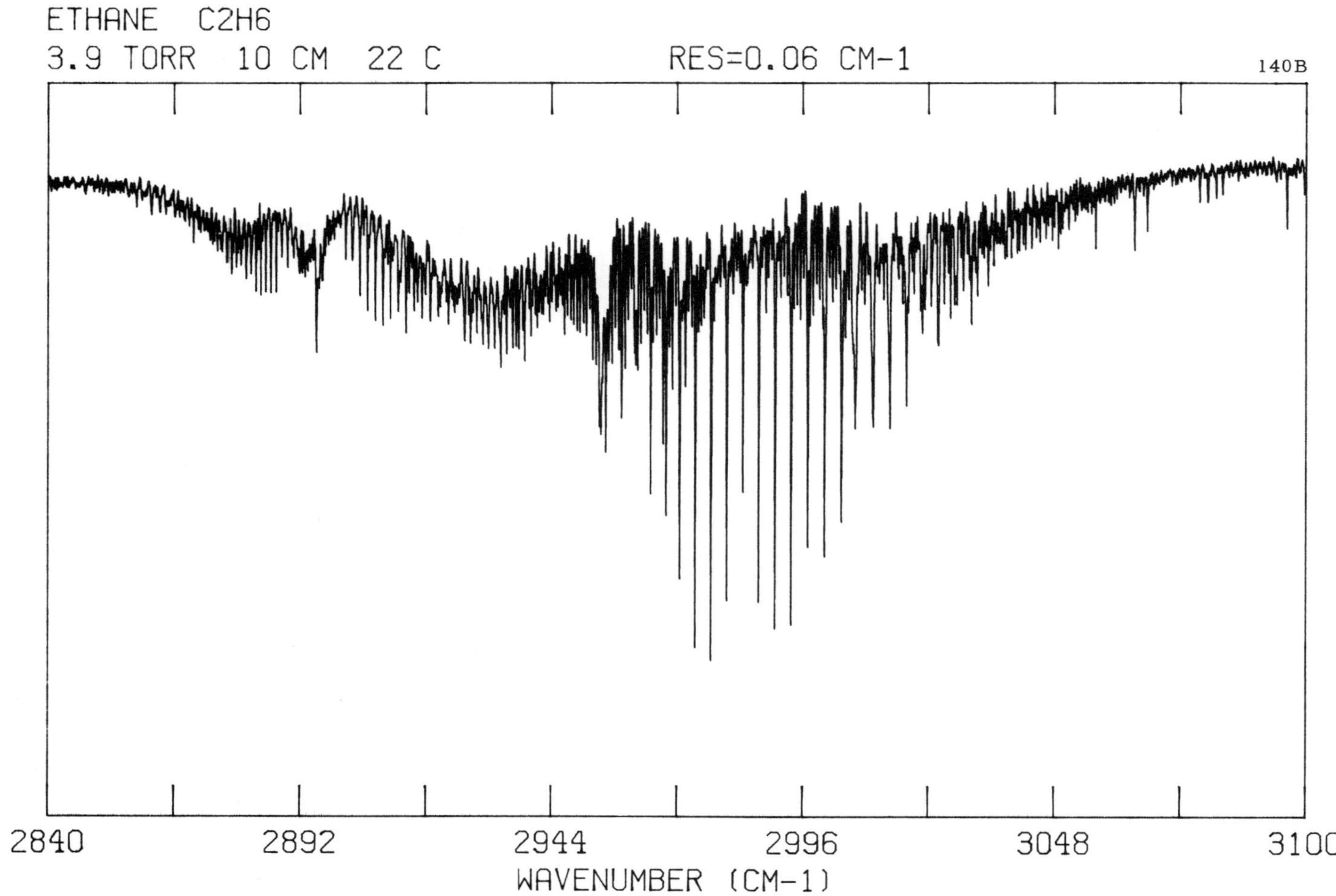
ETHANE C2H6
3.9 TORR 10 CM 22 C
RES=0.06 CM-1
140B
2840
2892
2944
2996
3048
3100
WAVENUMBER (CM-1)

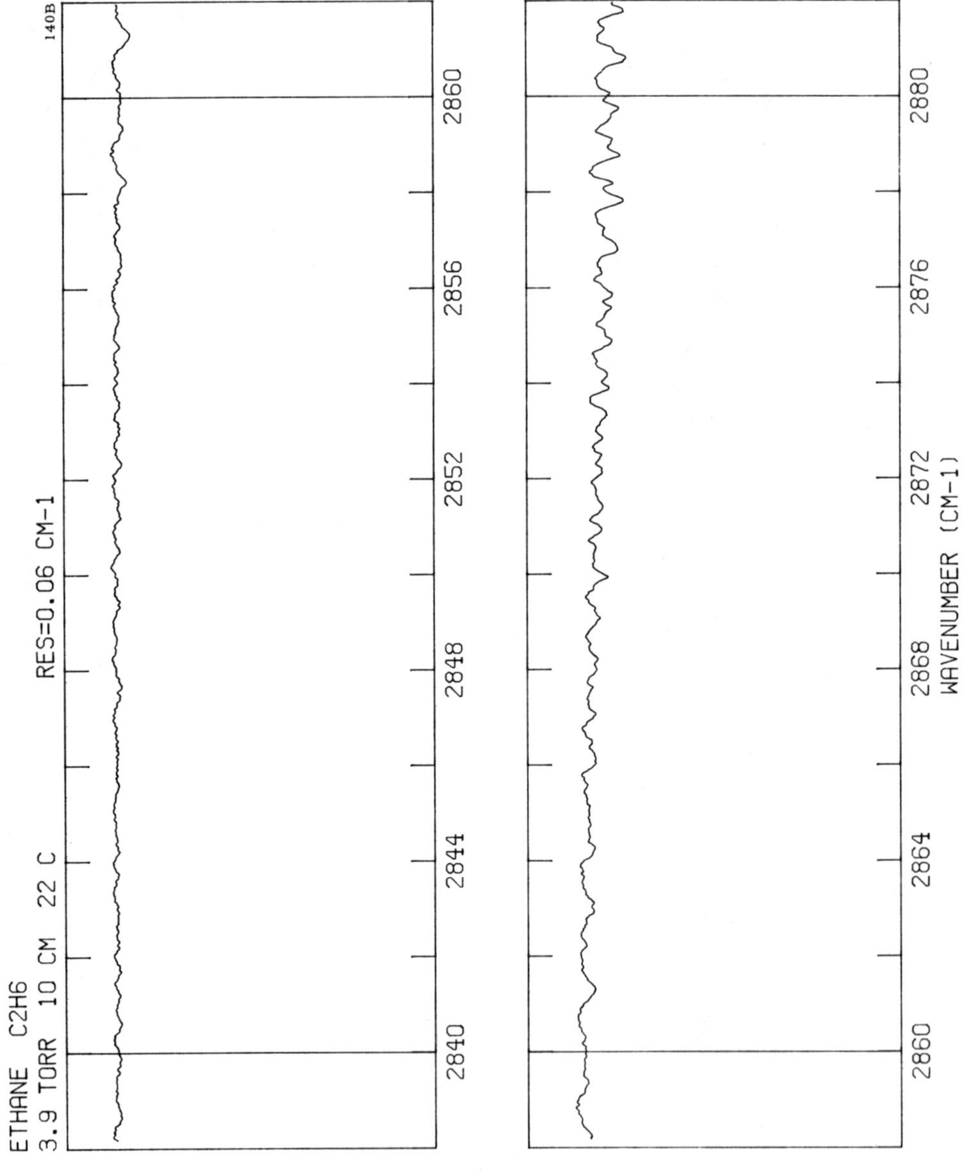
140B
ETHANE C2H6
3.9 TORR 10 CM 22 C
RES=0.06 CM-1
2840
2844
2848
2852
2856
2860
2864
2868
2872
2876
2880
WAVENUMBER (CM-1)

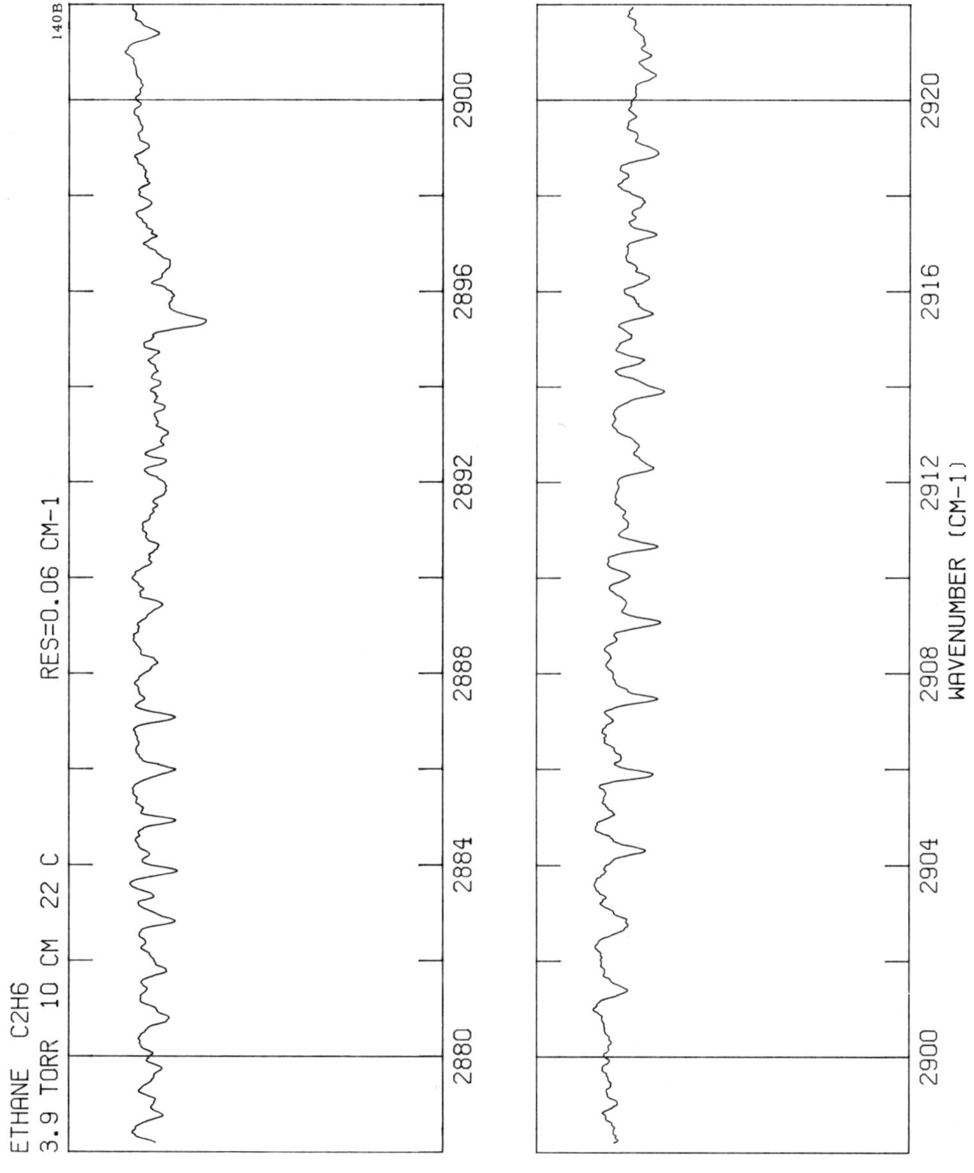
ETHANE C2H6
3.9 TORR 10 CM 22 C
RES=0.06 CM-1
140B
2880
2884
2888
2892
2896
2900
2900
2904
2908
2912
2916
2920
WAVENUMBER (CM-1)

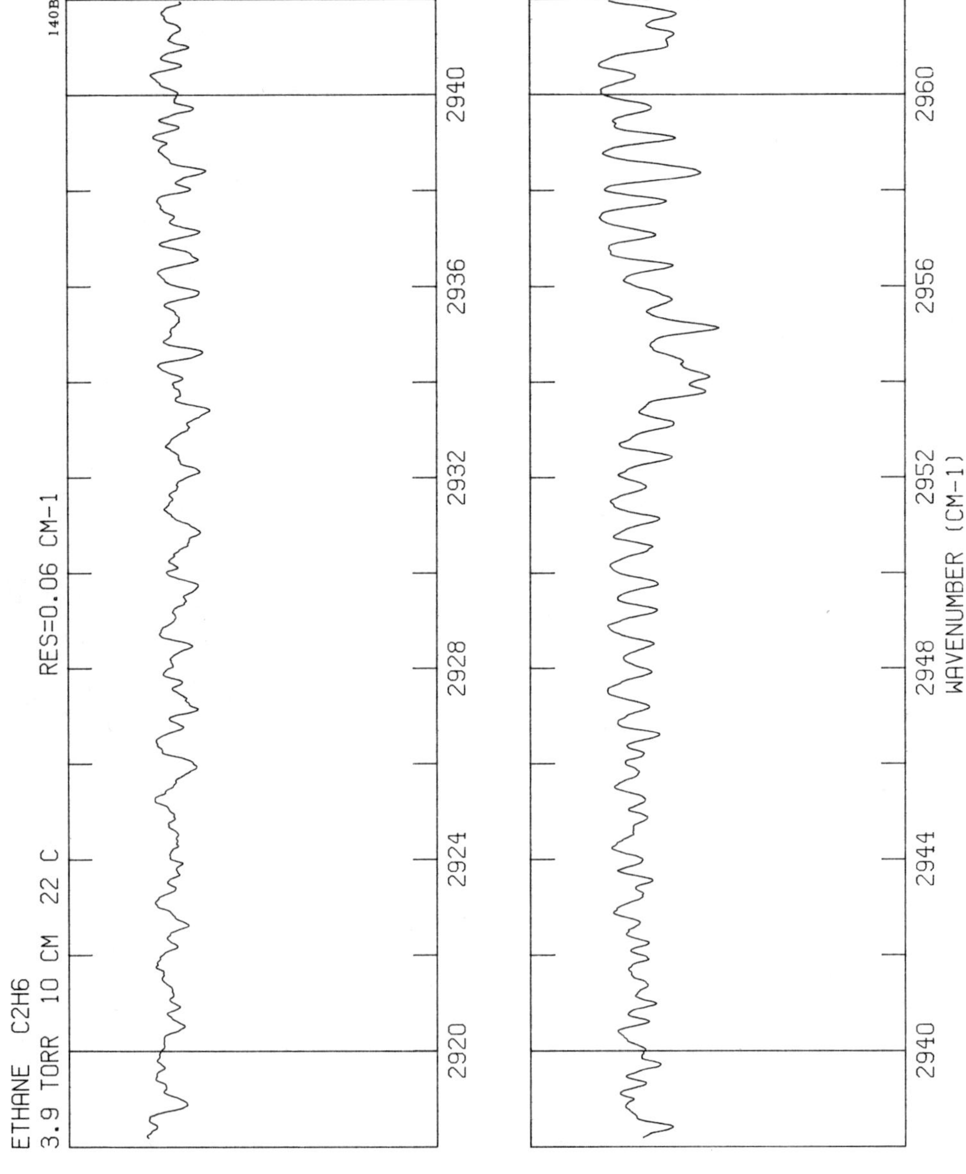
ETHANE C2H6
3.9 TORR 10 CM 22 C
RES=0.06 CM-1
140B
2920
2924
2928
2932
2936
2940
2940
2944
2948
2952
2956
2960
WAVENUMBER (CM-1)

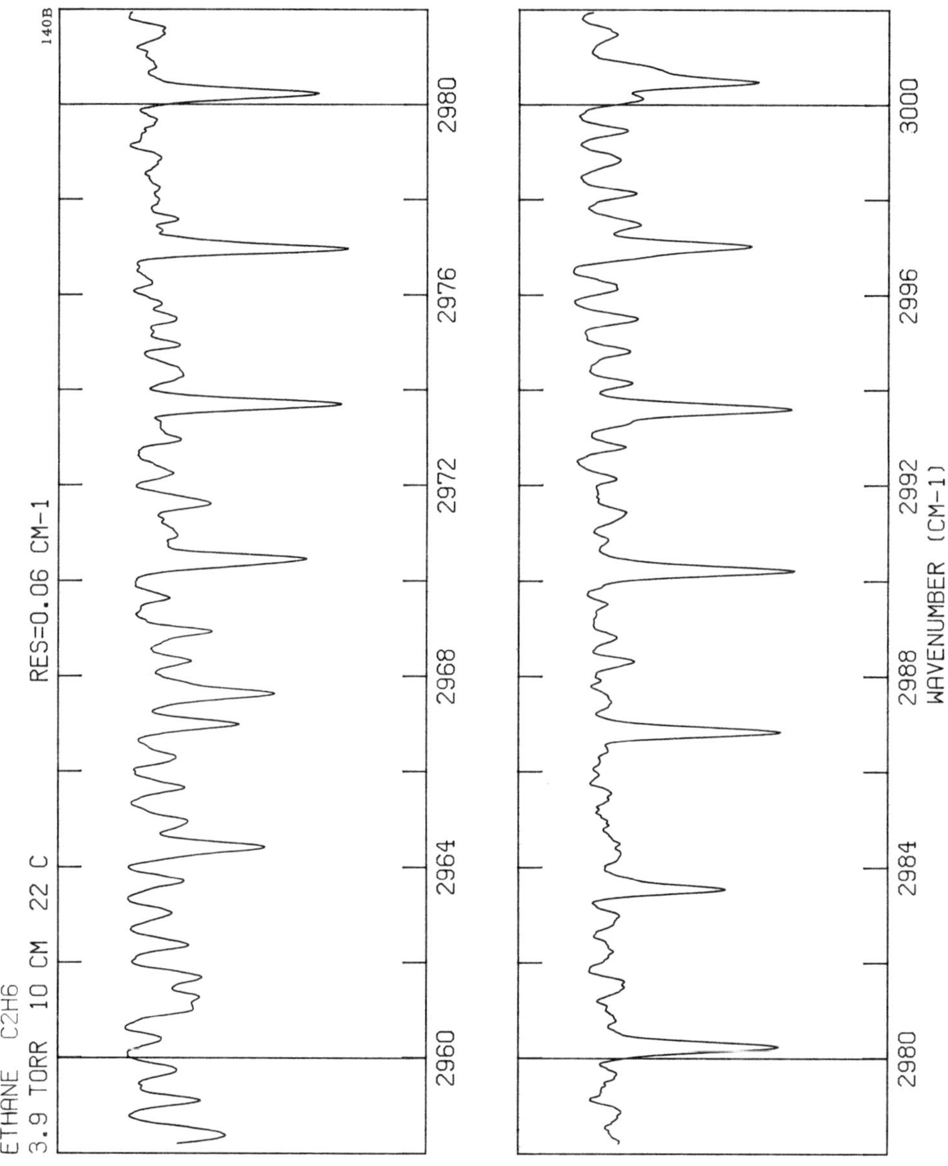
ETHANE C2H6
3.9 TORR 10 CM 22 C
RES=0.06 CM-1
140B
2960
2964
2968
2972
2976
2980
2980
2984
2988
2992
2996
3000
WAVENUMBER (CM-1)

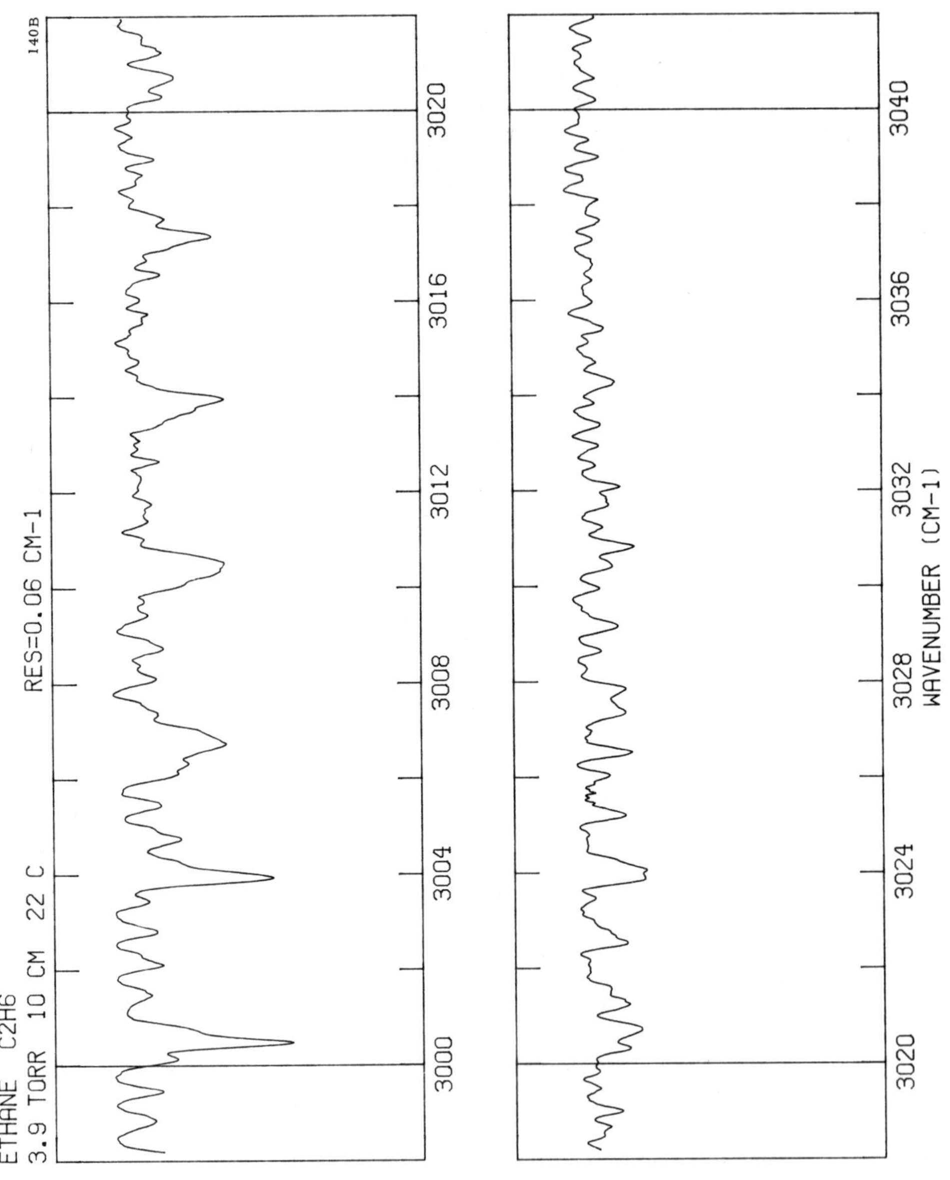
ETHANE C2H6
3.9 TORR 10 CM 22 C
RES=0.06 CM-1
140B
3000
3004
3008
3012
3016
3020
3020
3024
3028
3032
3036
3040
WAVENUMBER (CM-1)

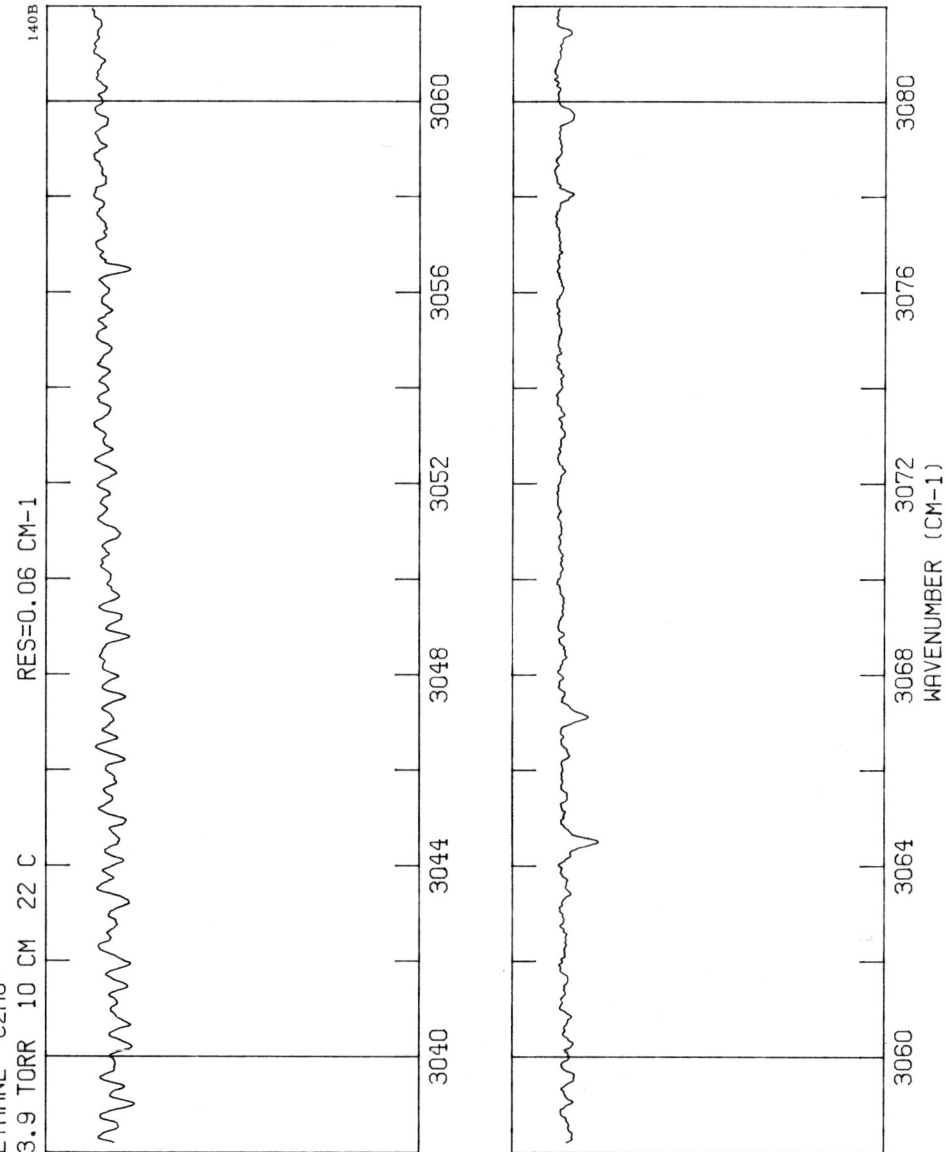
ETHANE C2H6
3.9 TORR 10 CM 22 C
RES=0.06 CM-1
140B
3040
3044
3048
3052
3056
3060
3060
3064
3068
3072
3076
3080
WAVENUMBER (CM-1)

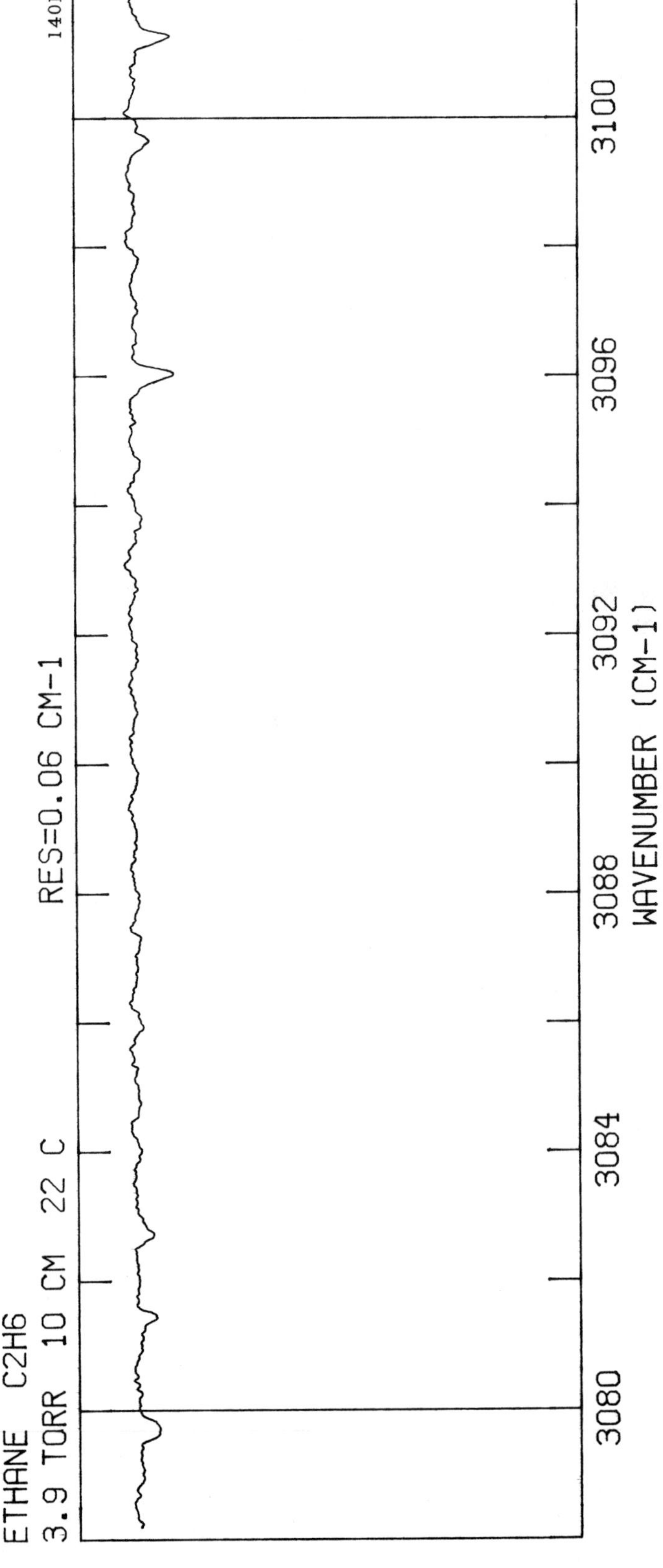
ETHANE C2H6
3.9 TORR 10 CM 22 C
RES=0.06 CM-1
140B
3080
3084
3088
3092
3096
3100
WAVENUMBER (CM-1)

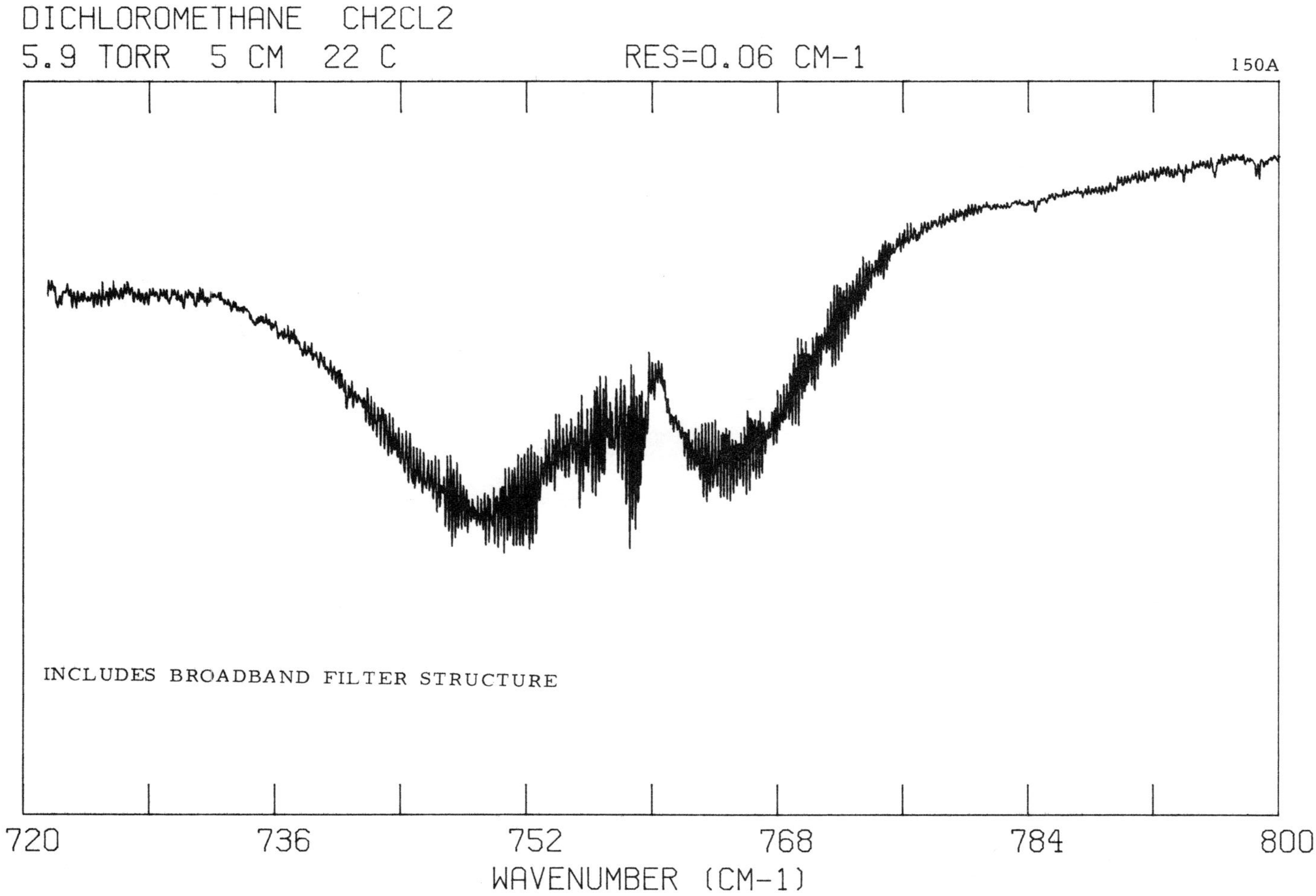

DICHLOROMETHANE CH2CL2
5.9 TORR 5 CM 22 C
RES=0.06 CM-1
150A
INCLUDES BROADBAND FILTER STRUCTURE
720
736
752
768
784
800
WAVENUMBER (CM-1)

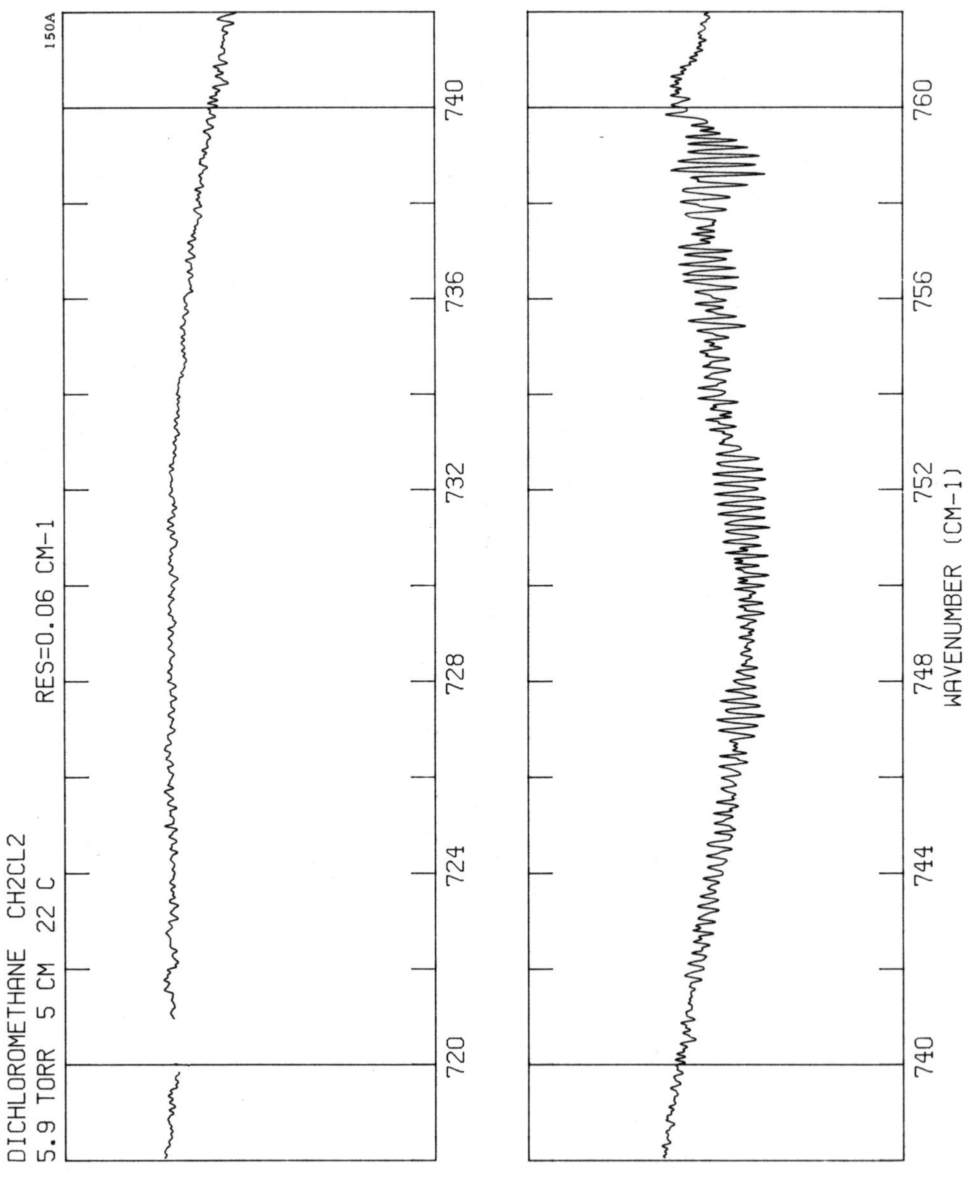
DICHLOROMETHANE CH2CL2
5.9 TORR 5 CM 22 C
RES=0.06 CM-1
150A
720
724
728
732
736
740
740
744
748
752
756
760
WAVENUMBER (CM-1)

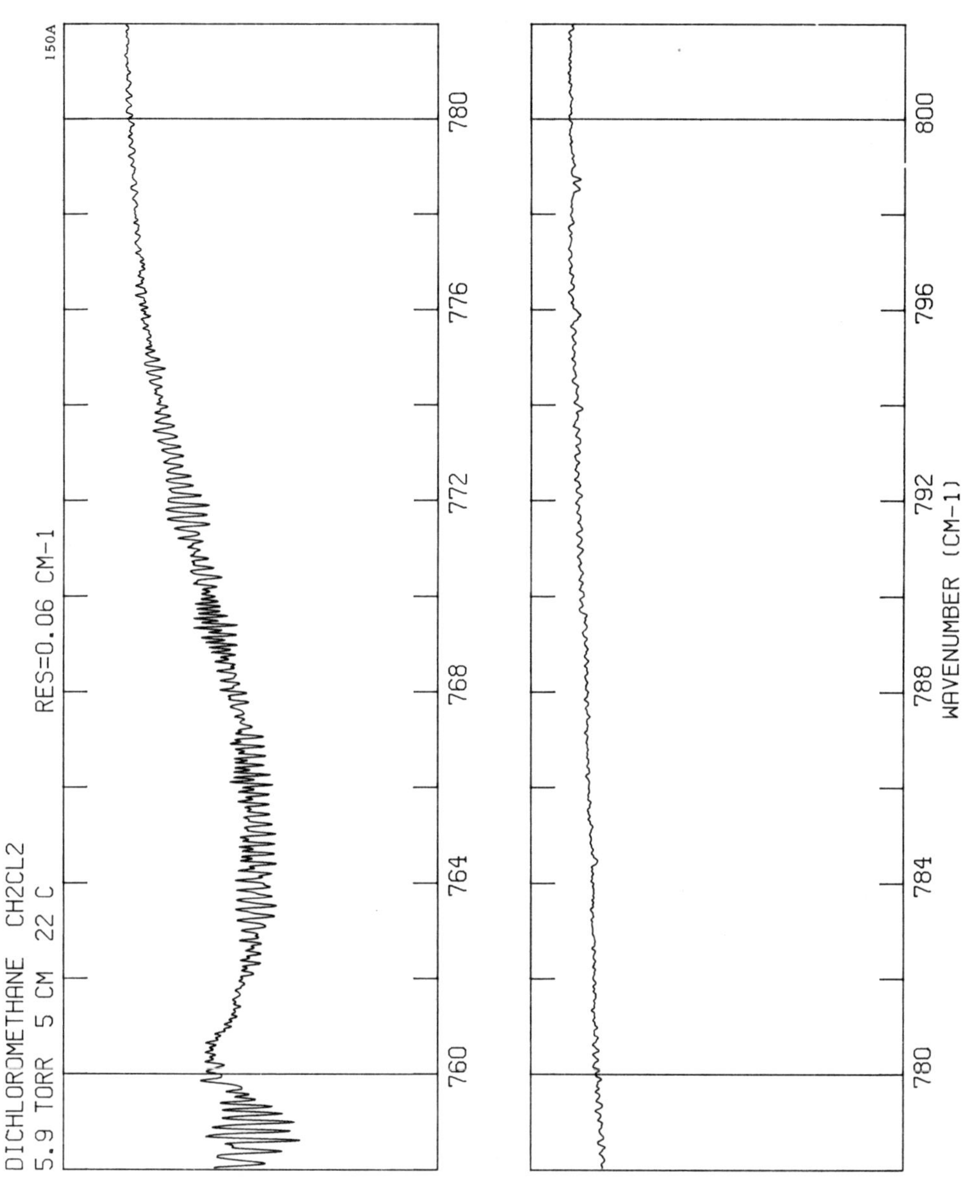
DICHLOROMETHANE CH2CL2
5.9 TORR 5 CM 22 C
RES=0.06 CM-1
150A
760
764
768
772
776
780
780
784
788
792
796
800
WAVENUMBER (CM-1)

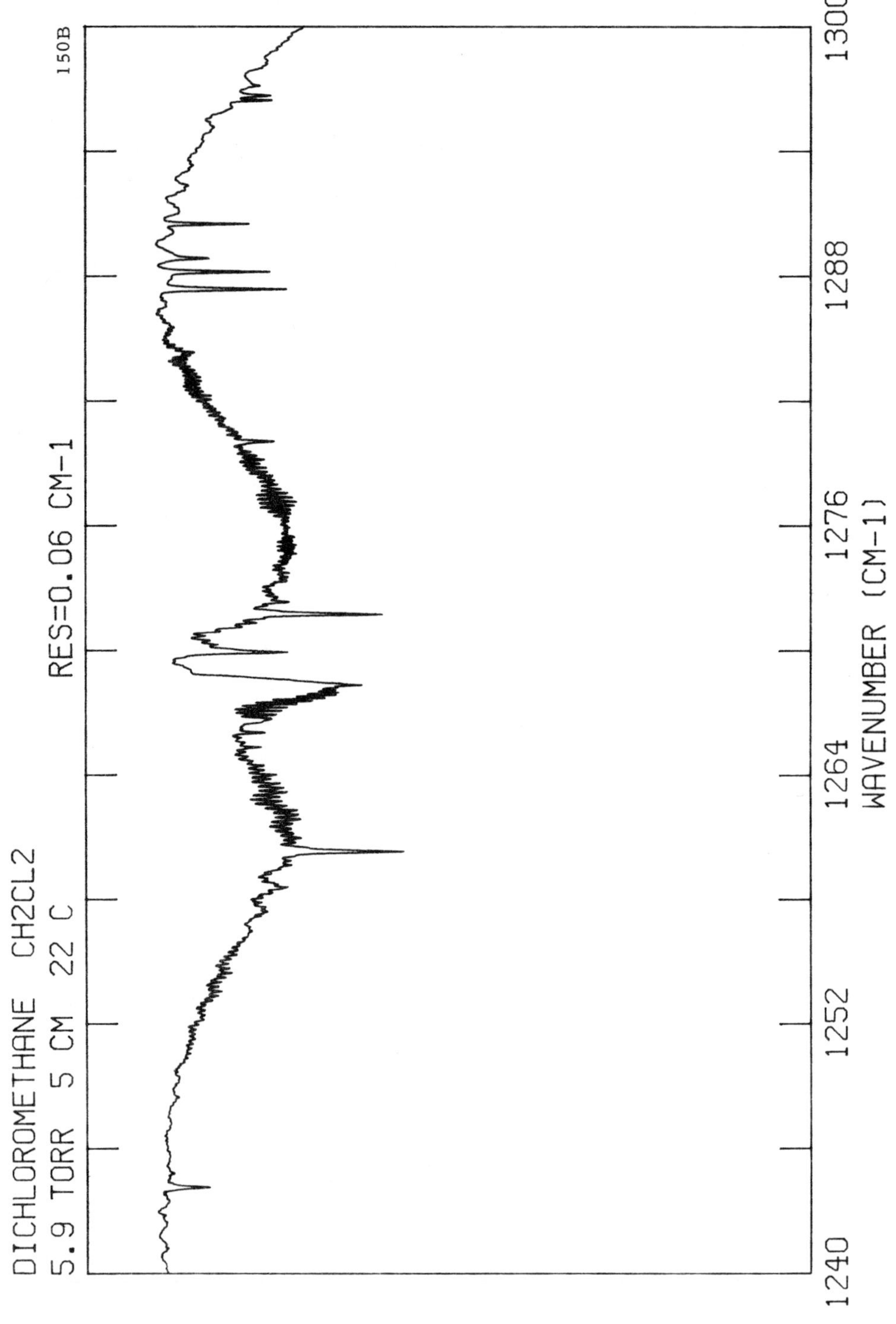

DICHLOROMETHANE CH2CL2
5.9 TORR 5 CM 22 C
RES=0.06 CM-1
150B
1240
1252
1264
1276
1288
1300
WAVENUMBER (CM-1)

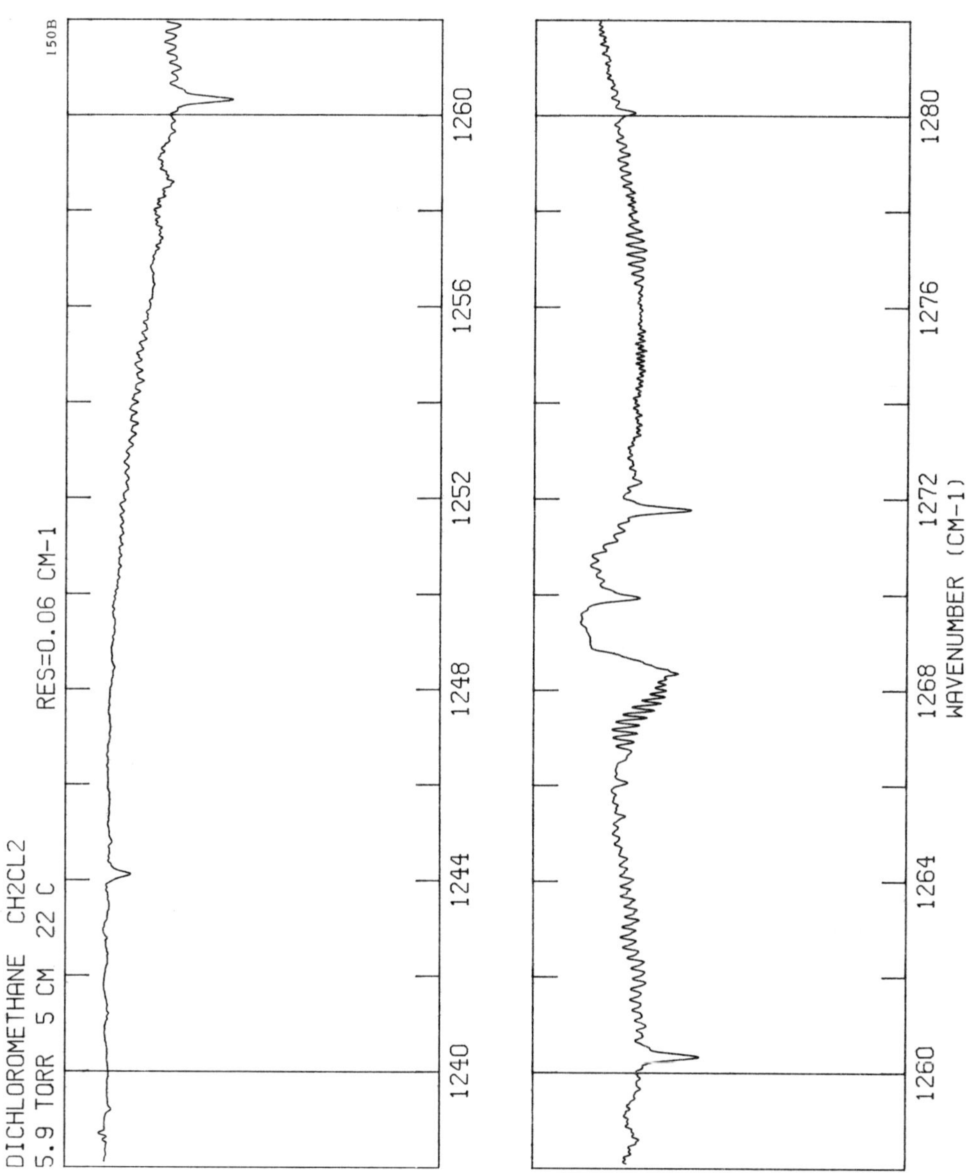
150B
DICHLOROMETHANE CH2CL2
5.9 TORR 5 CM 22 C
RES=0.06 CM-1
1240
1244
1248
1252
1256
1260
1264
1268
1272
1276
1280
WAVENUMBER (CM-1)

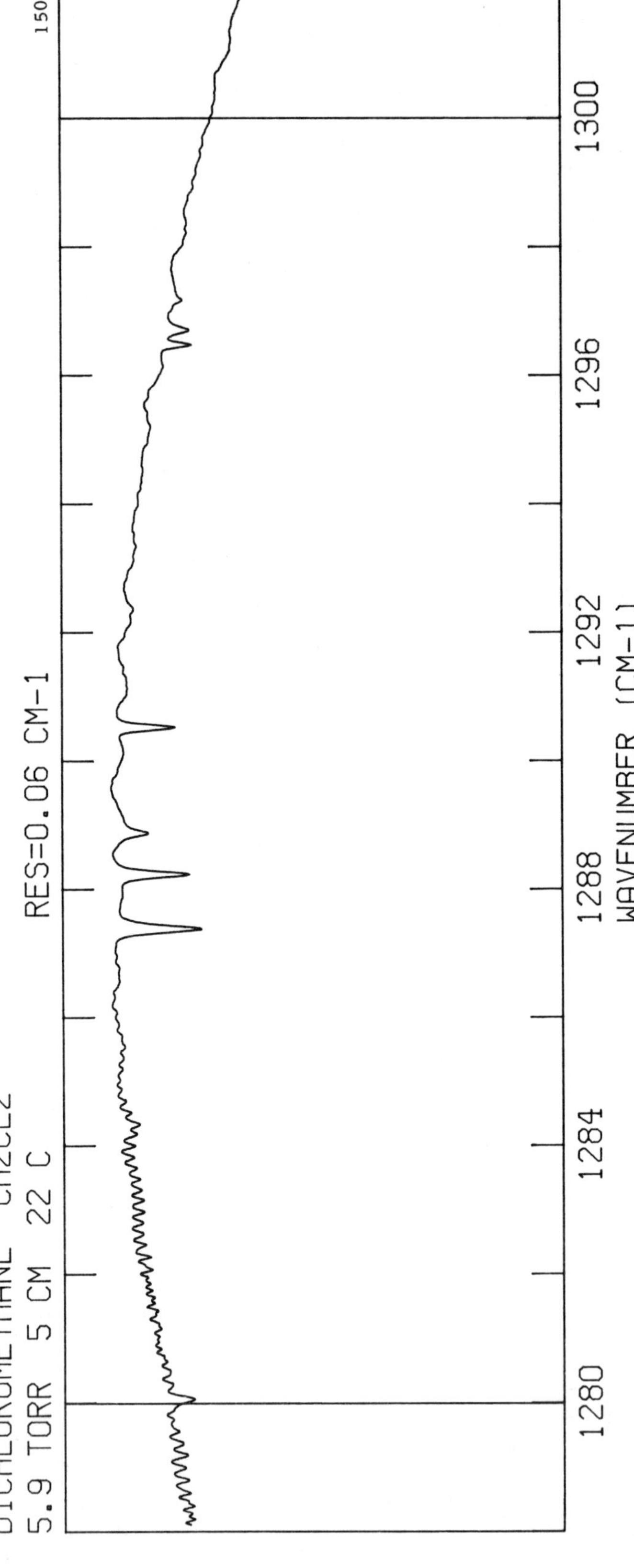
DICHLOROMETHANE CH2CL2
5.9 TORR 5 CM 22 C
RES=0.06 CM-1
150B
1280
1284
1288
1292
1296
1300
WAVENUMBER (CM-1)

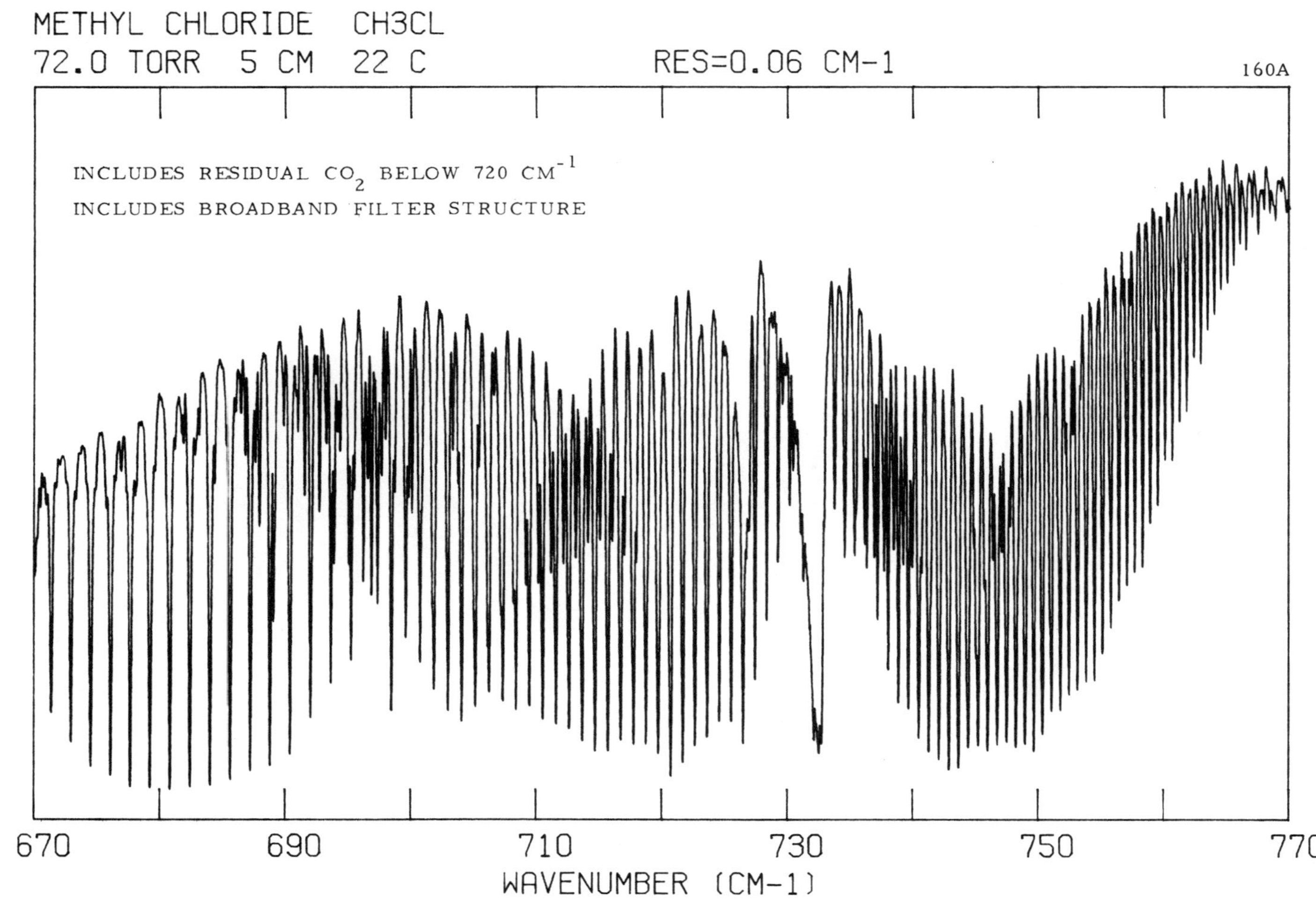

METHYL CHLORIDE CH3CL
72.0 TORR 5 CM 22 C
RES=0.06 CM-1
160A
INCLUDES RESIDUAL CO_2 BELOW 720 CM^{-1}
INCLUDES BROADBAND FILTER STRUCTURE
670
690
710
730
750
770
WAVENUMBER (CM-1)

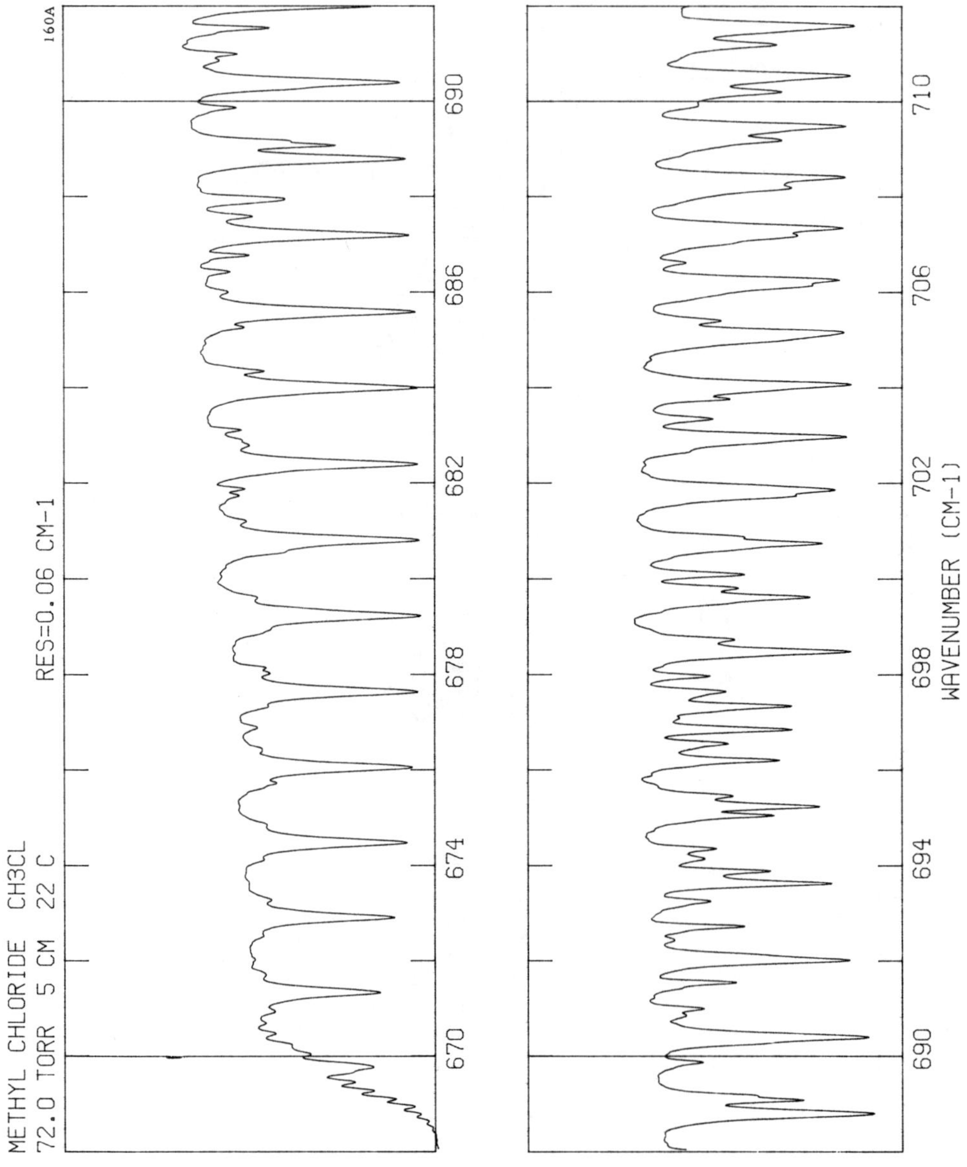
METHYL CHLORIDE CH3CL
72.0 TORR 5 CM 22 C
RES=0.06 CM-1
160A
670
674
678
682
686
690
694
698
702
706
710
WAVENUMBER (CM-1)

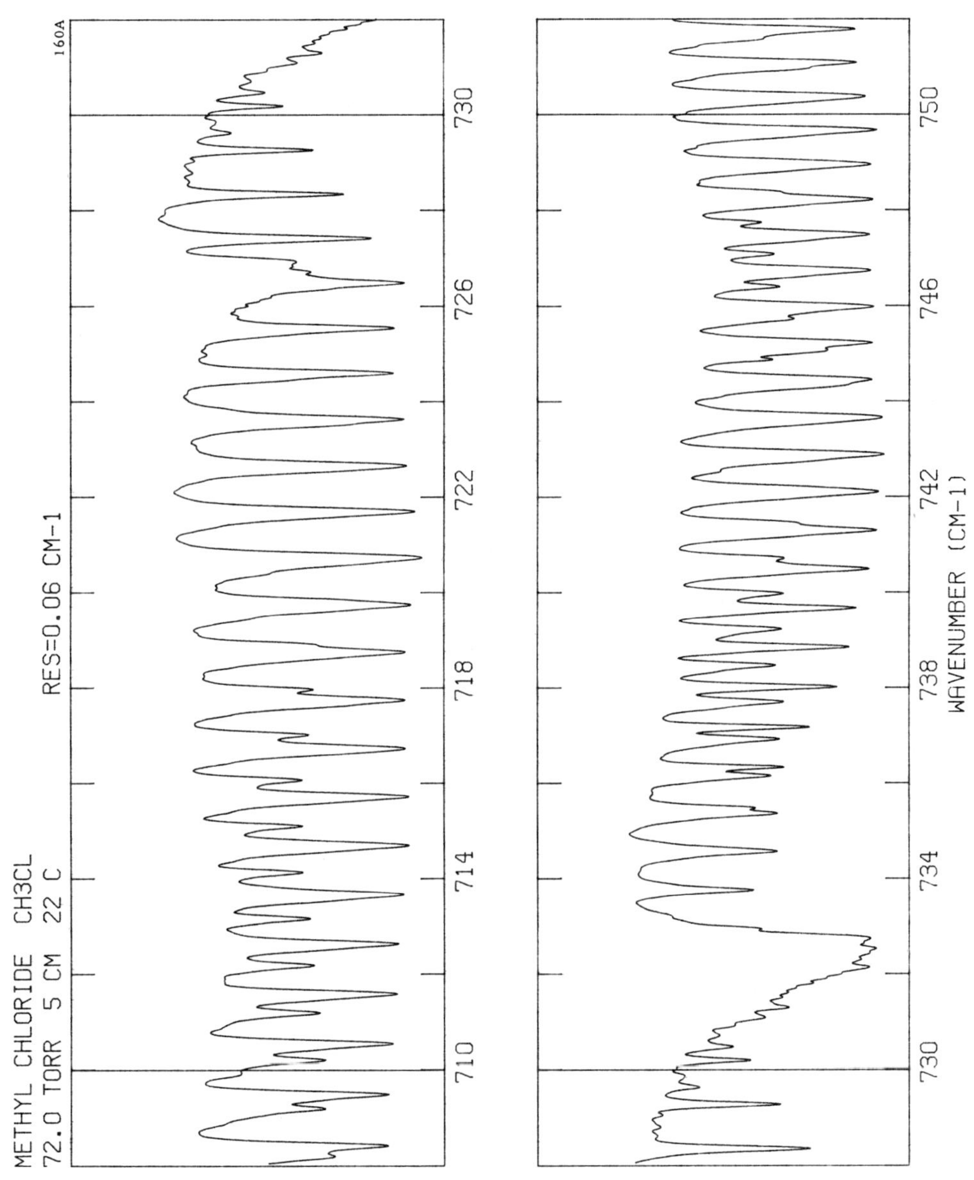

METHYL CHLORIDE CH3CL
72.0 TORR 5 CM 22 C
RES=0.06 CM-1
160A
710
714
718
722
726
730
730
734
738
742
746
750
WAVENUMBER (CM-1)

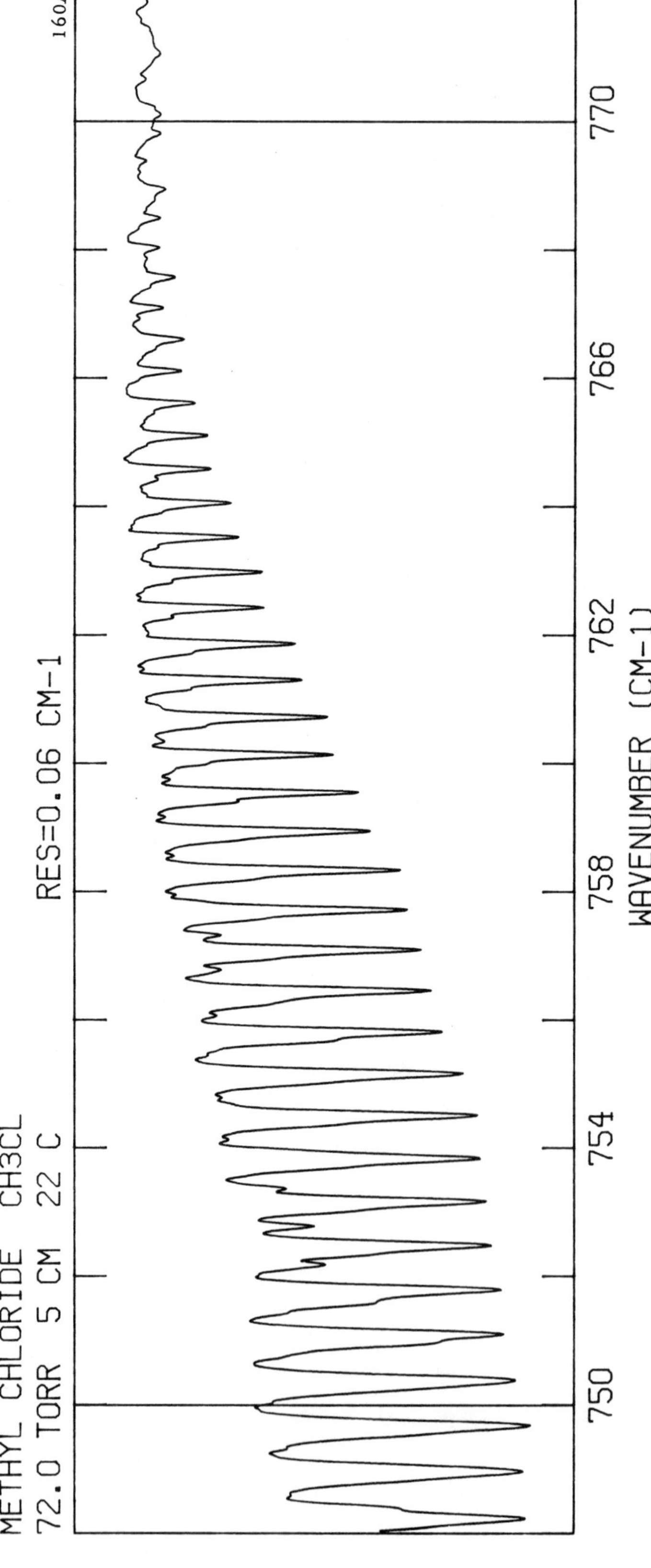
METHYL CHLORIDE CH3CL
72.0 TORR 5 CM 22 C
RES=0.06 CM-1
160A
750
754
758
762
766
770
WAVENUMBER (CM-1)

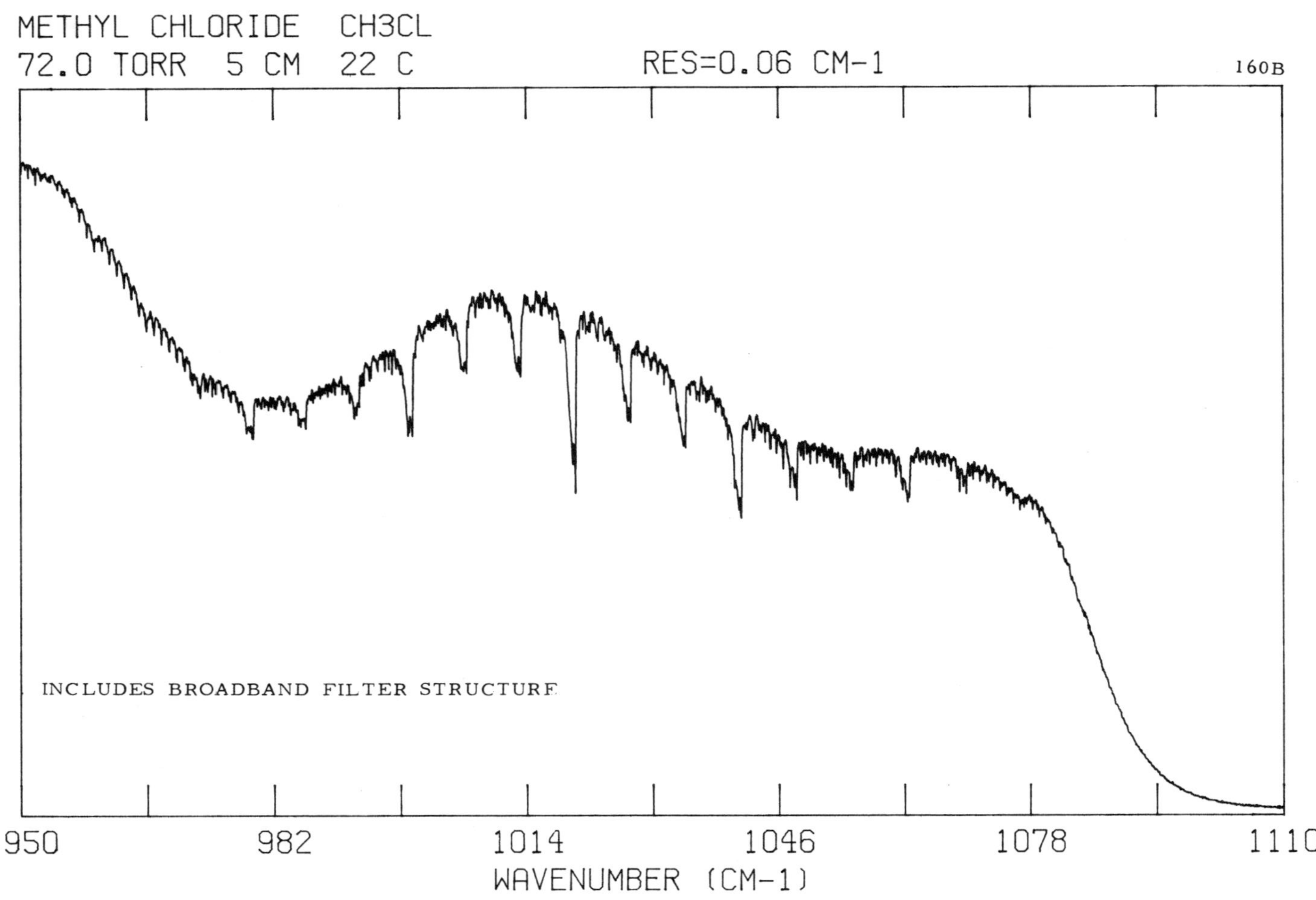
METHYL CHLORIDE CH3CL
72.0 TORR 5 CM 22 C
RES=0.06 CM-1
160B
INCLUDES BROADBAND FILTER STRUCTURF
950
982
1014
1046
1078
1110
WAVENUMBER (CM-1)

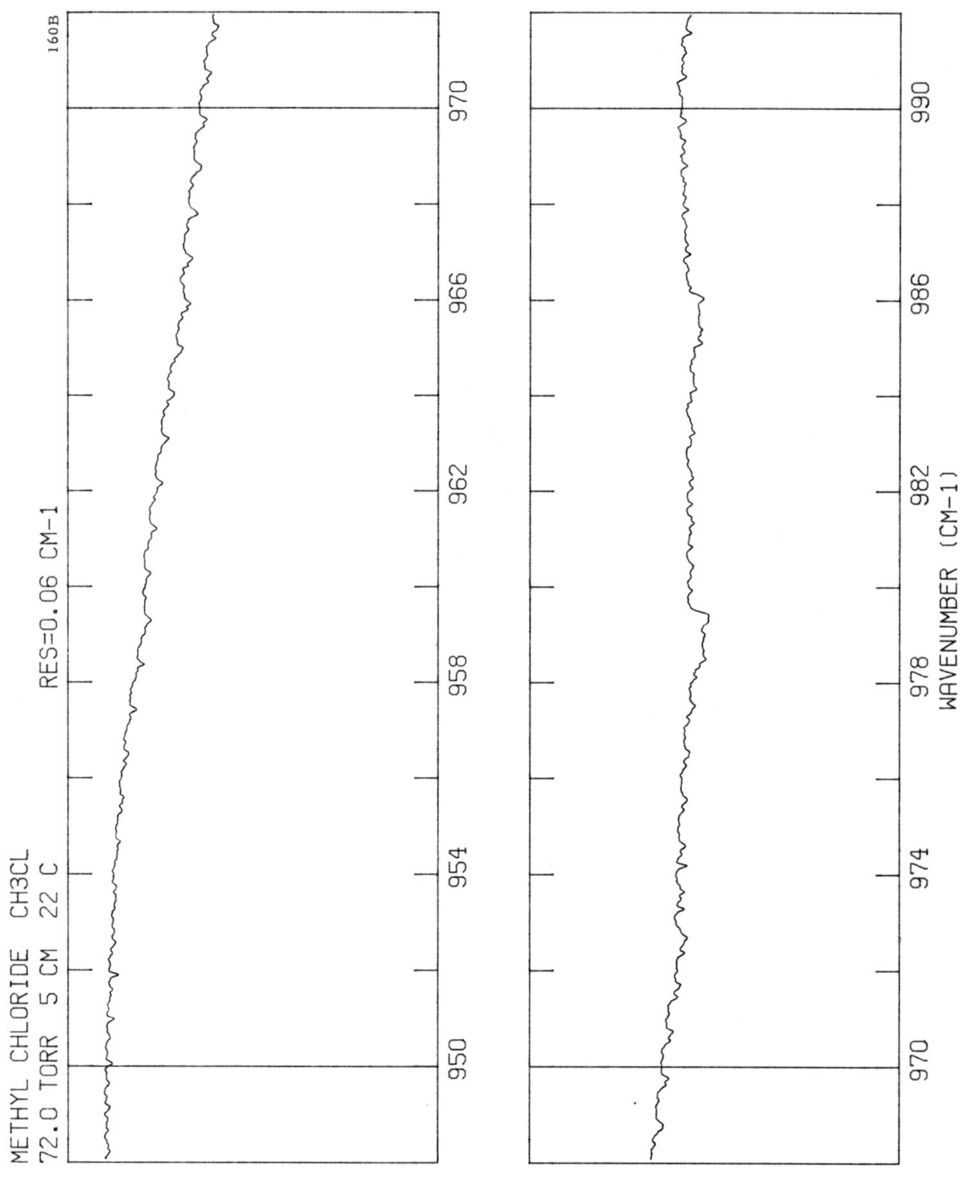
METHYL CHLORIDE CH3CL
72.0 TORR 5 CM 22 C
RES=0.06 CM-1
160B
950
954
958
962
966
970
974
978
982
986
990
WAVENUMBER (CM-1)

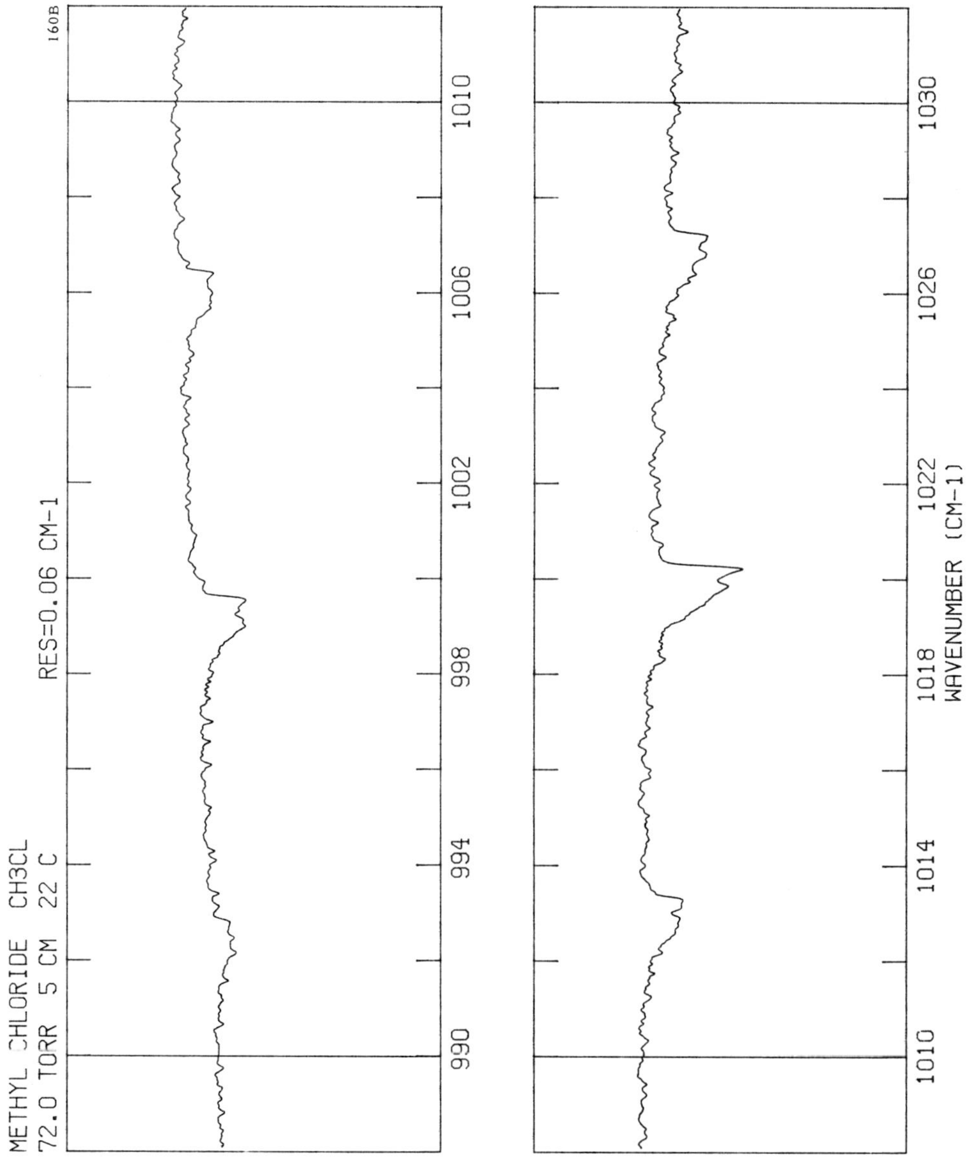
METHYL CHLORIDE CH3CL
72.0 TORR 5 CM 22 C
RES=0.06 CM-1
160B
990
994
998
1002
1006
1010
1010
1014
1018
1022
1026
1030
WAVENUMBER (CM-1)

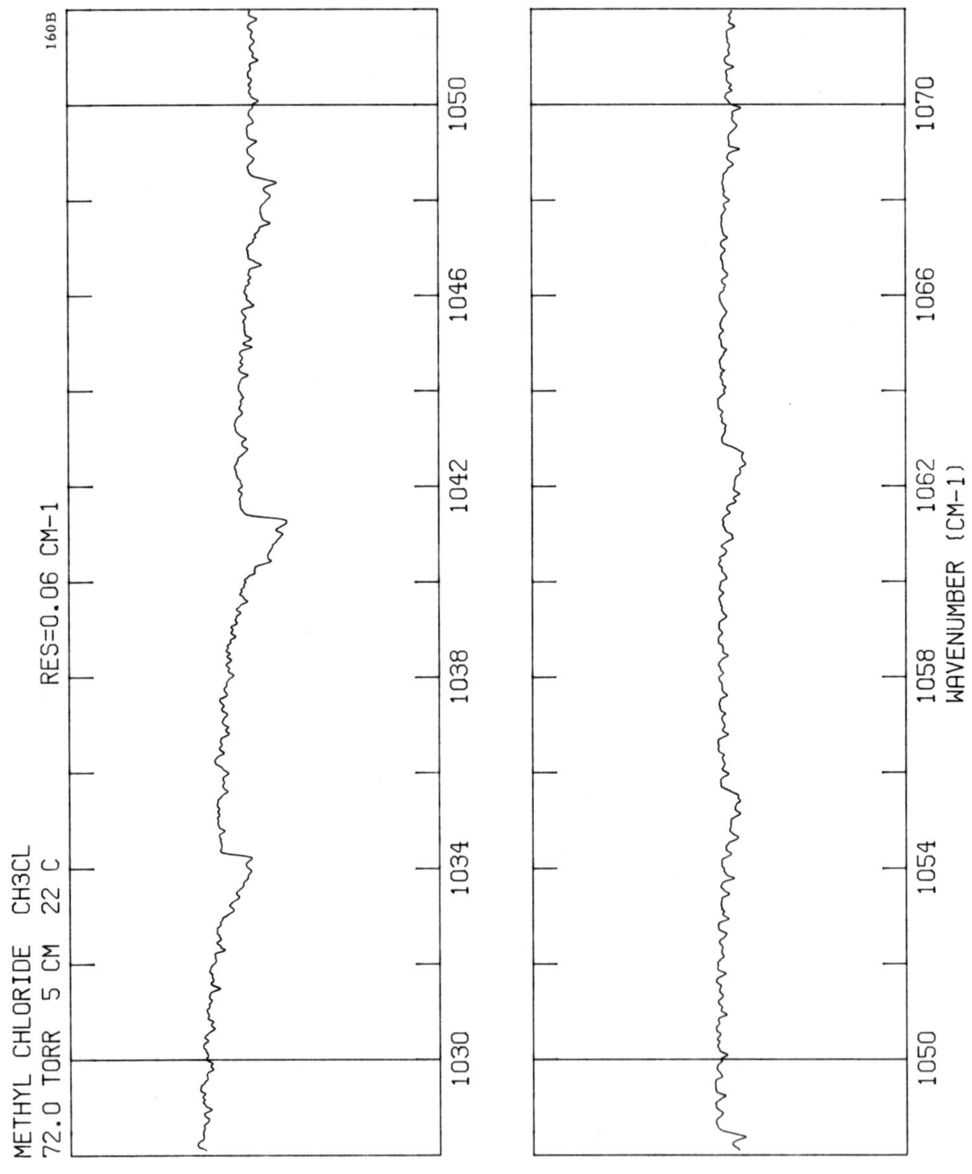
160B
METHYL CHLORIDE CH3CL
72.0 TORR 5 CM 22 C
RES=0.06 CM-1
1030
1034
1038
1042
1046
1050
1050
1054
1058
1062
1066
1070
WAVENUMBER (CM-1)

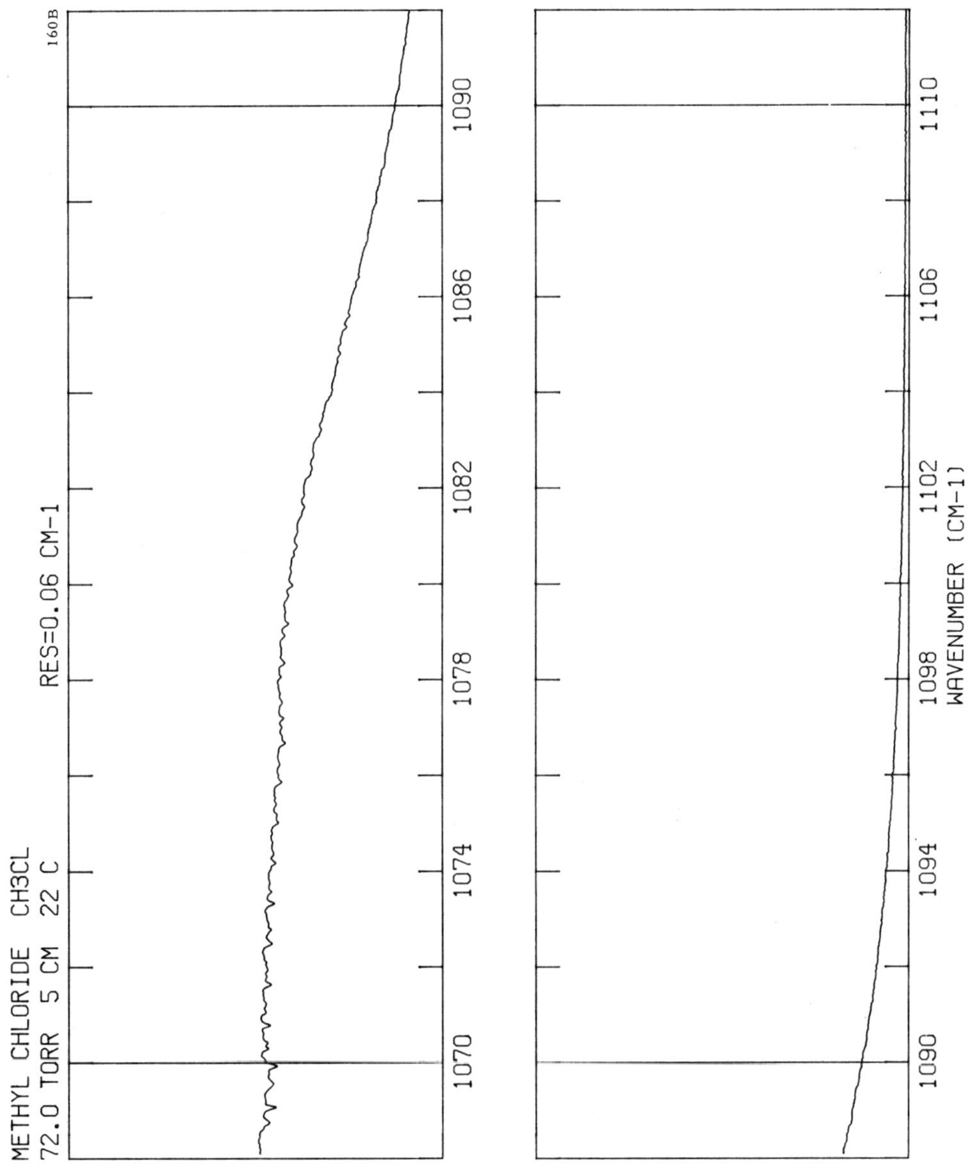

METHYL CHLORIDE CH3CL
72.0 TORR 5 CM 22 C
RES=0.06 CM-1
160B
1070
1074
1078
1082
1086
1090
1090
1094
1098
1102
1106
1110
WAVENUMBER (CM-1)

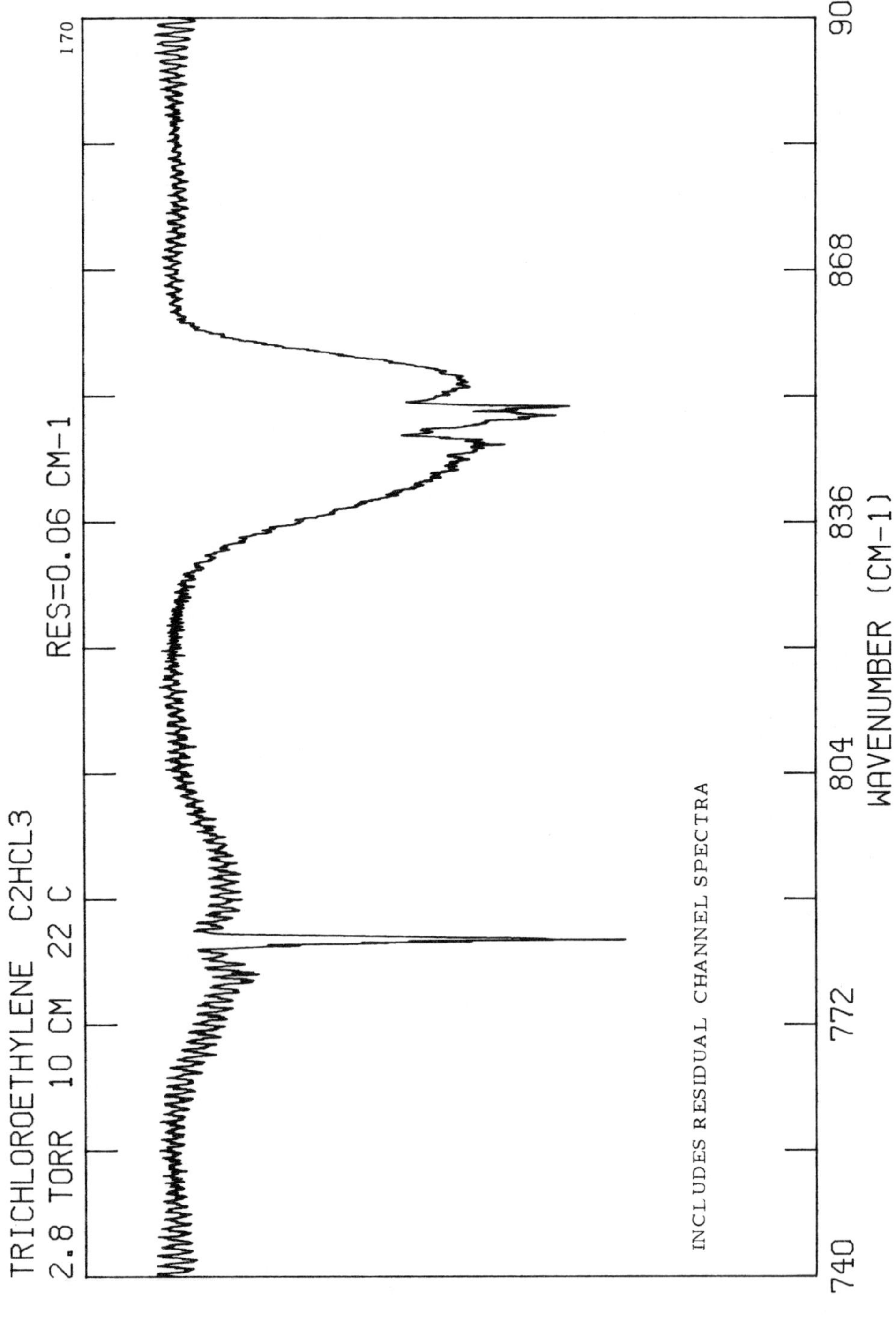

TRICHLOROETHYLENE C2HCL3
2.8 TORR 10 CM 22 C
RES=0.06 CM-1
170
INCLUDES RESIDUAL CHANNEL SPECTRA
740
772
804
836
868
900
WAVENUMBER (CM-1)

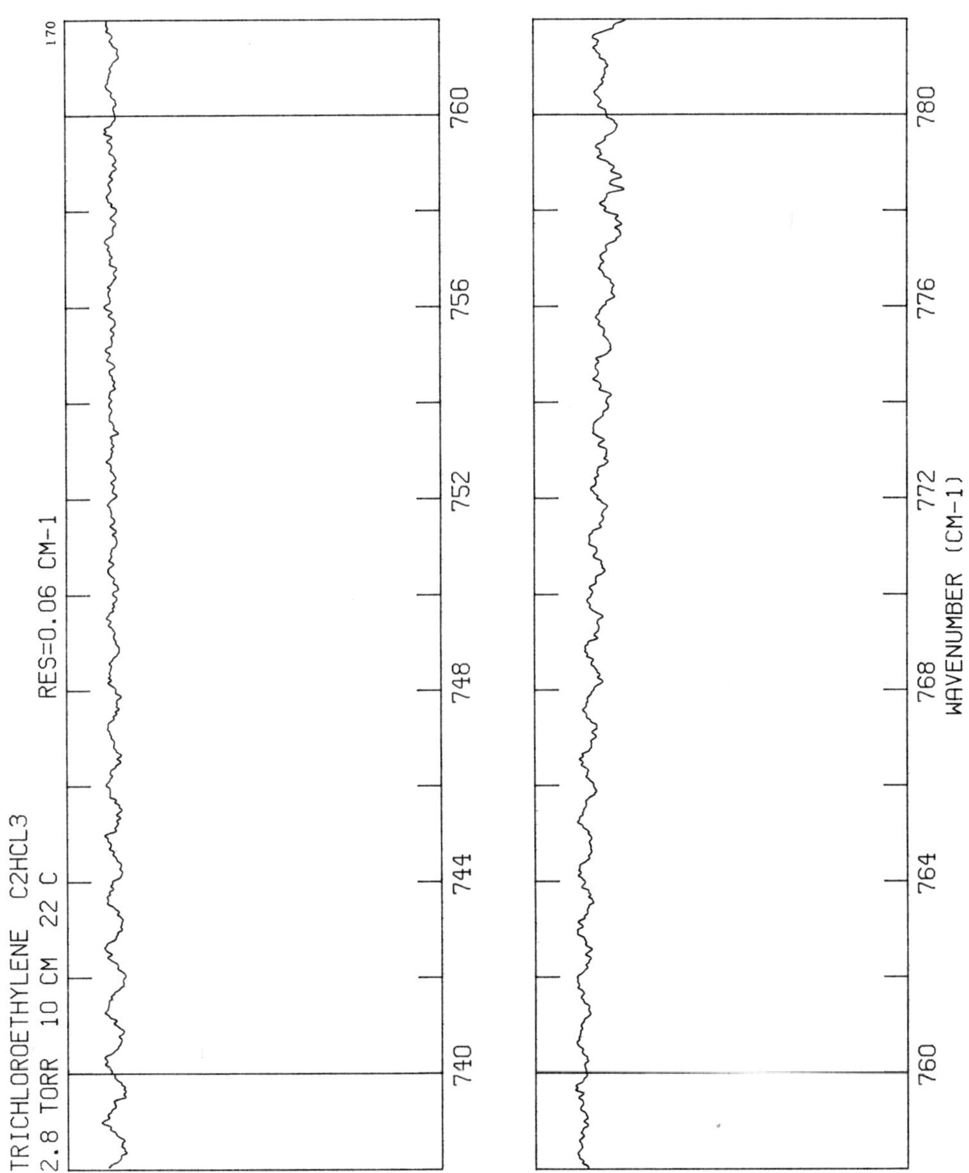

TRICHLOROETHYLENE C2HCL3
2.8 TORR 10 CM 22 C
RES=0.06 CM-1
170
740
744
748
752
756
760
760
764
768
772
776
780
WAVENUMBER (CM-1)

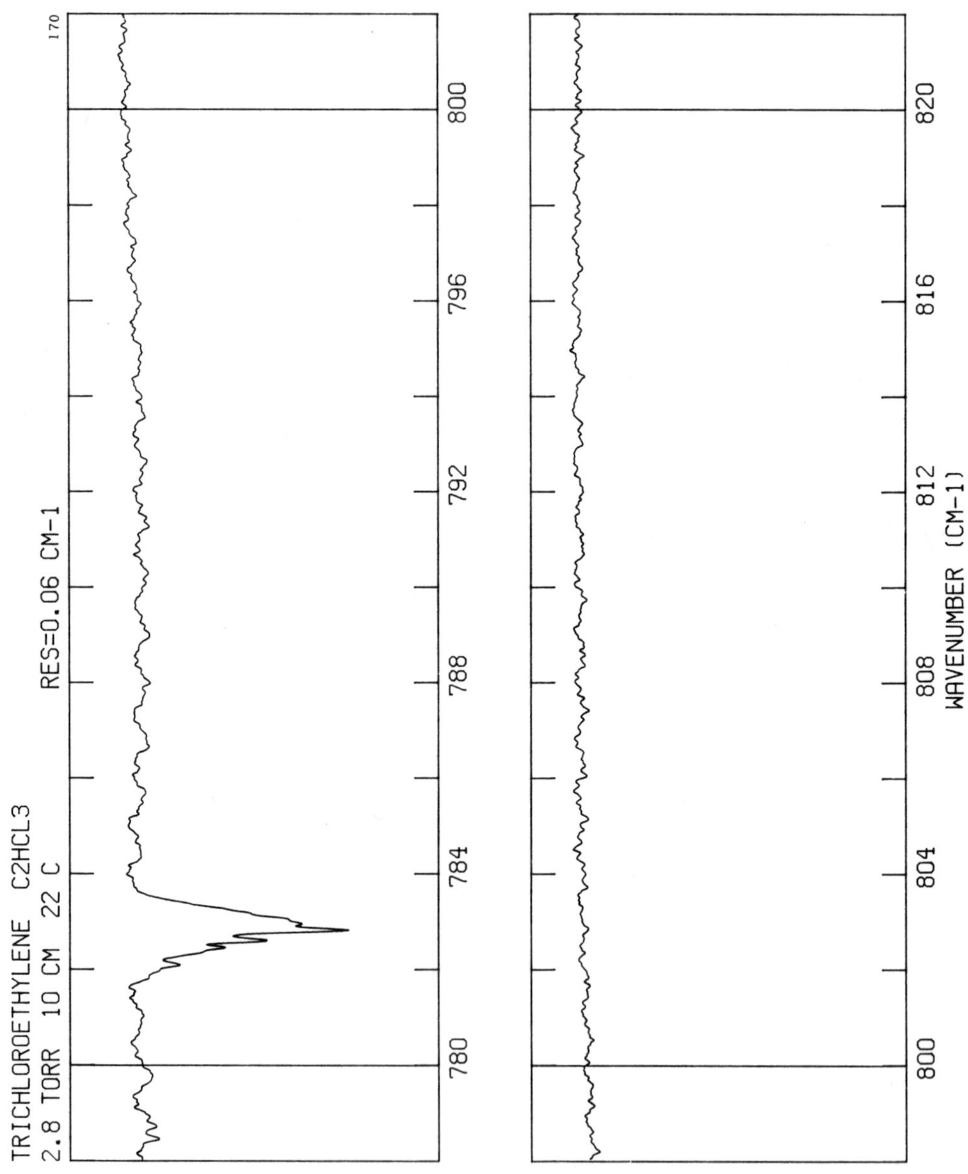

TRICHLOROETHYLENE C2HCL3
2.8 TORR 10 CM 22 C
RES=0.06 CM-1
170
780
784
788
792
796
800
804
808
812
816
820
WAVENUMBER (CM-1)

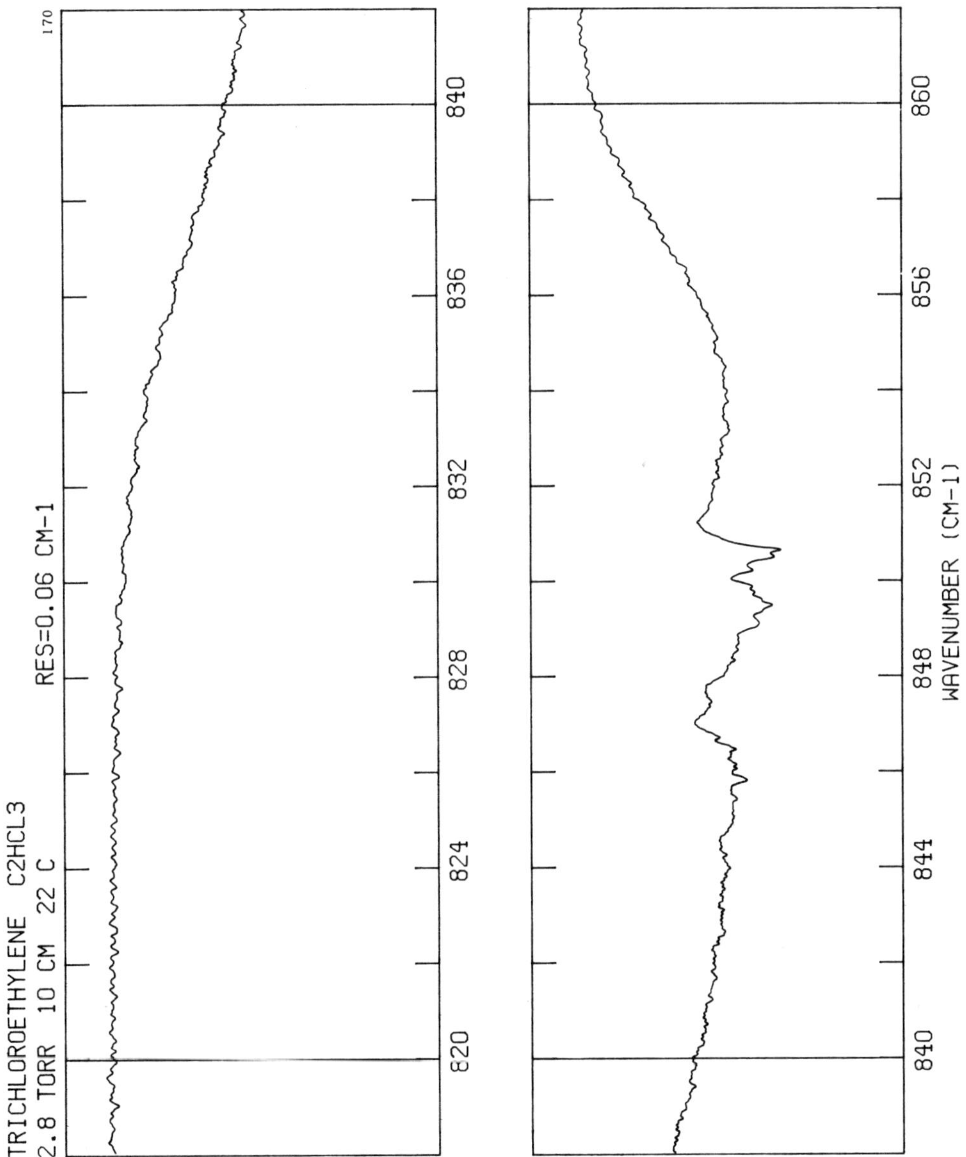

170
TRICHLOROETHYLENE C2HCL3
2.8 TORR 10 CM 22 C
RES=0.06 CM-1
820
824
828
832
836
840
840
844
848
852
856
860
WAVENUMBER (CM-1)

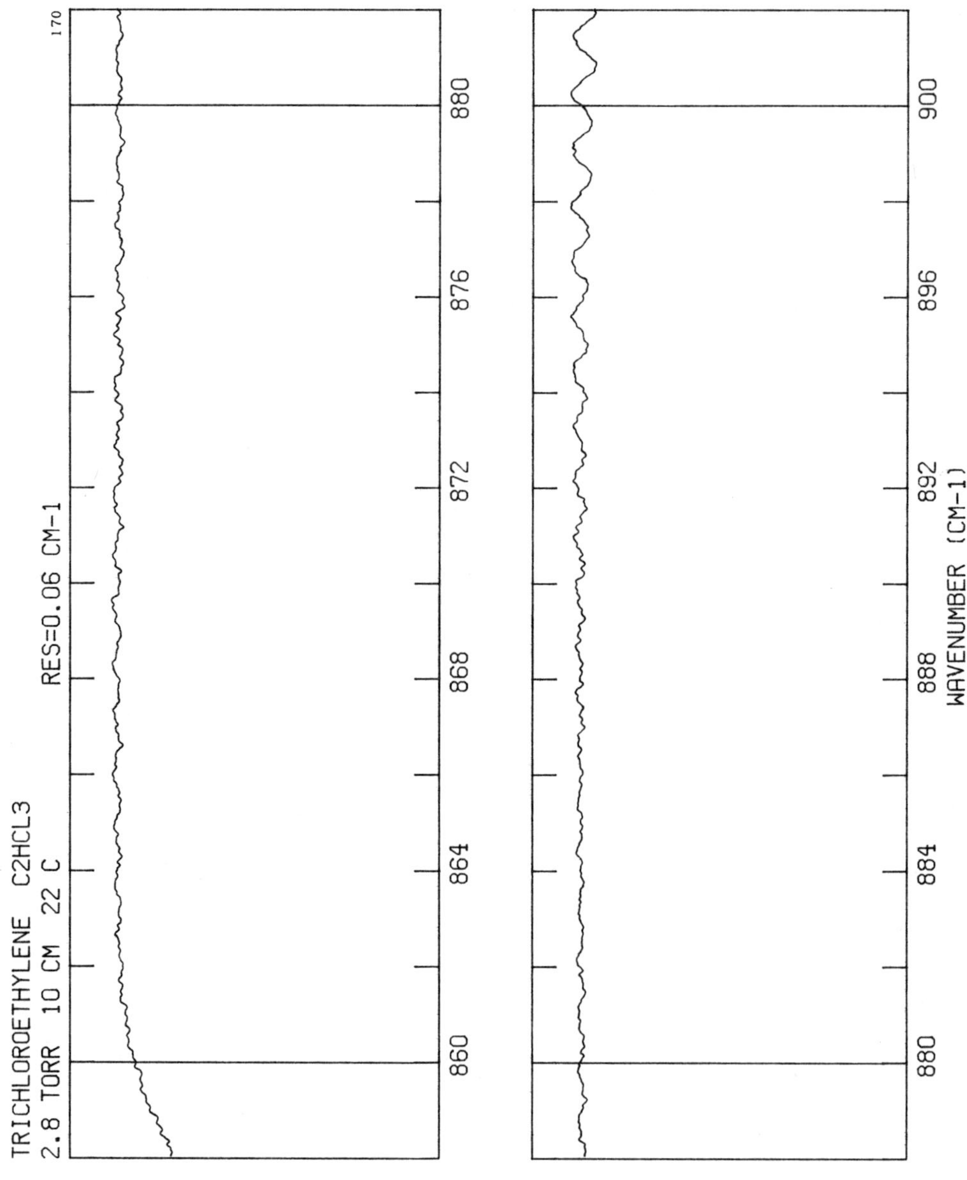
170
TRICHLOROETHYLENE C2HCL3
2.8 TORR 10 CM 22 C
RES=0.06 CM-1
860
864
868
872
876
880
880
884
888
892
896
900
WAVENUMBER (CM-1)

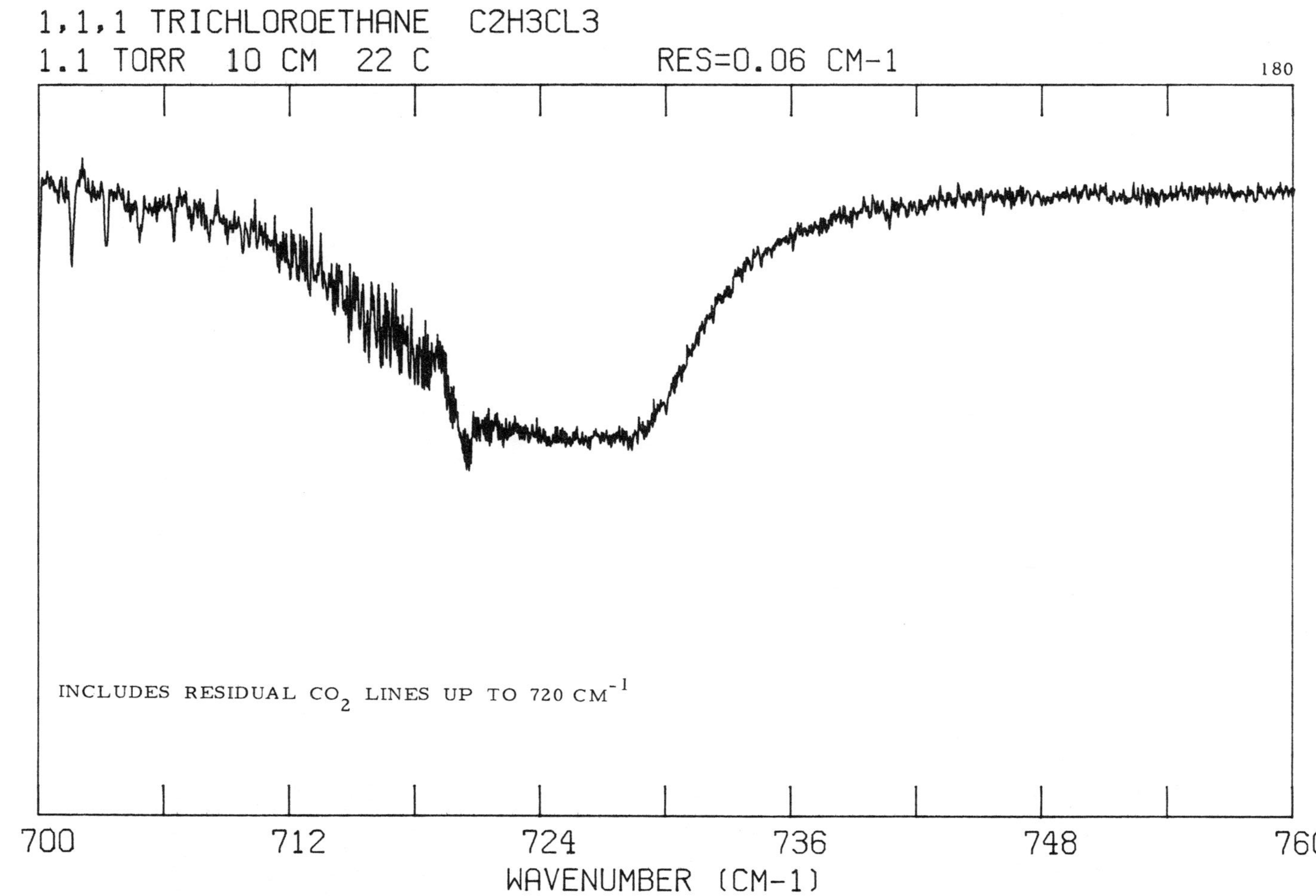
1,1,1 TRICHLOROETHANE C2H3CL3
1.1 TORR 10 CM 22 C RES=0.06 CM-1
180
INCLUDES RESIDUAL CO2 LINES UP TO 720 CM-1
700
712
724
736
748
760
WAVENUMBER (CM-1)

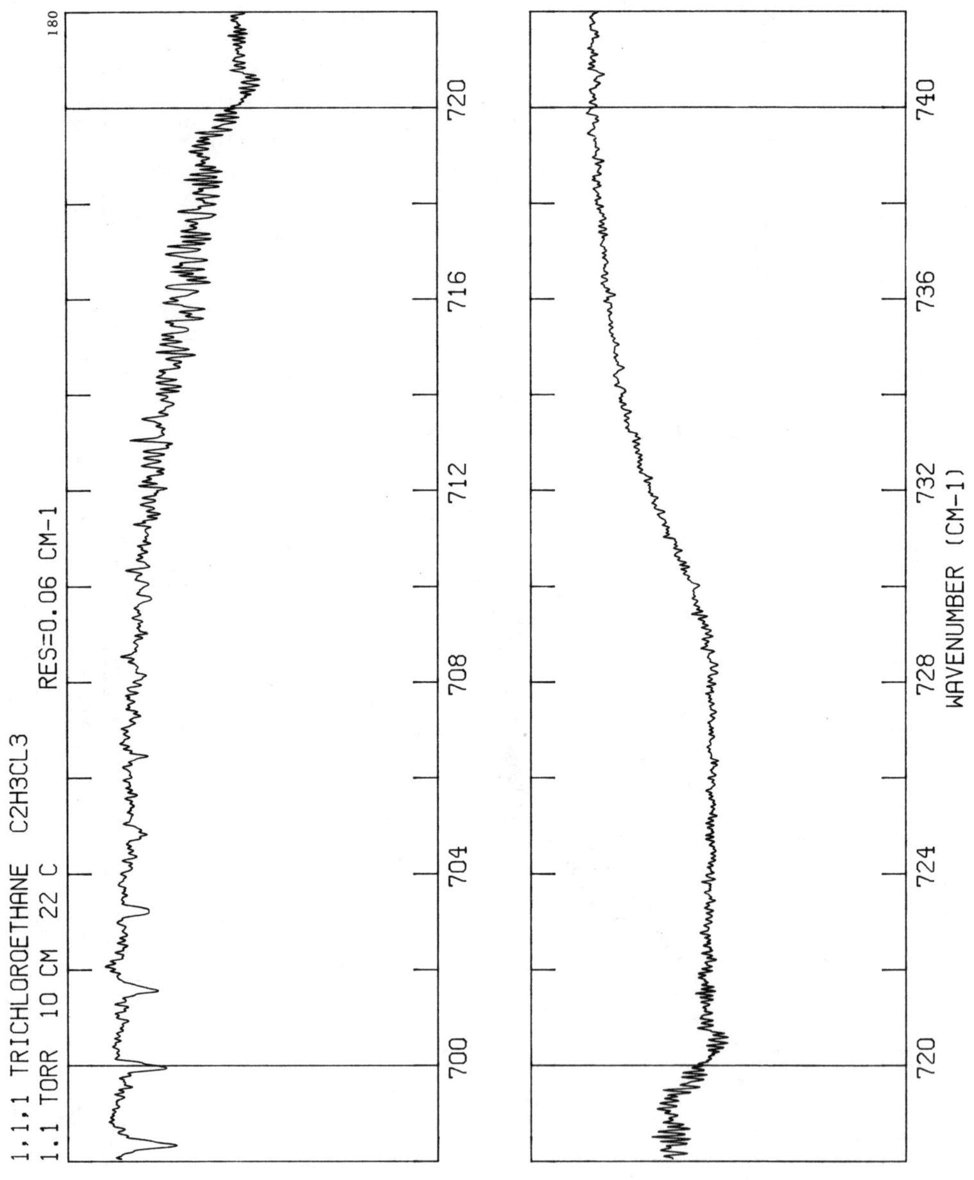
1,1,1 TRICHLOROETHANE C2H3CL3
1.1 TORR 10 CM 22 C
RES=0.06 CM-1
180
700
704
708
712
716
720
720
724
728
732
736
740
WAVENUMBER (CM-1)

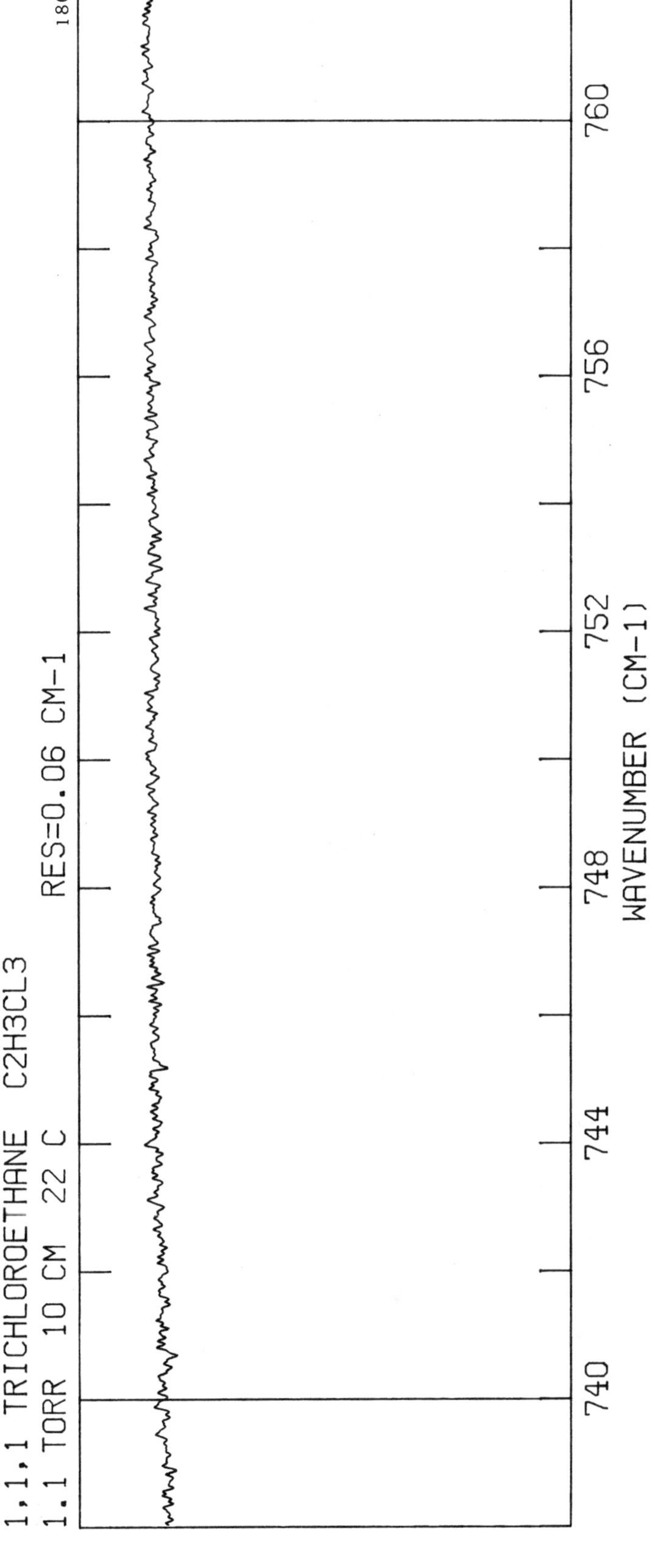
1,1,1 TRICHLOROETHANE C2H3CL3
1.1 TORR 10 CM 22 C
RES=0.06 CM-1
180
740
744
748
752
756
760
WAVENUMBER (CM-1)

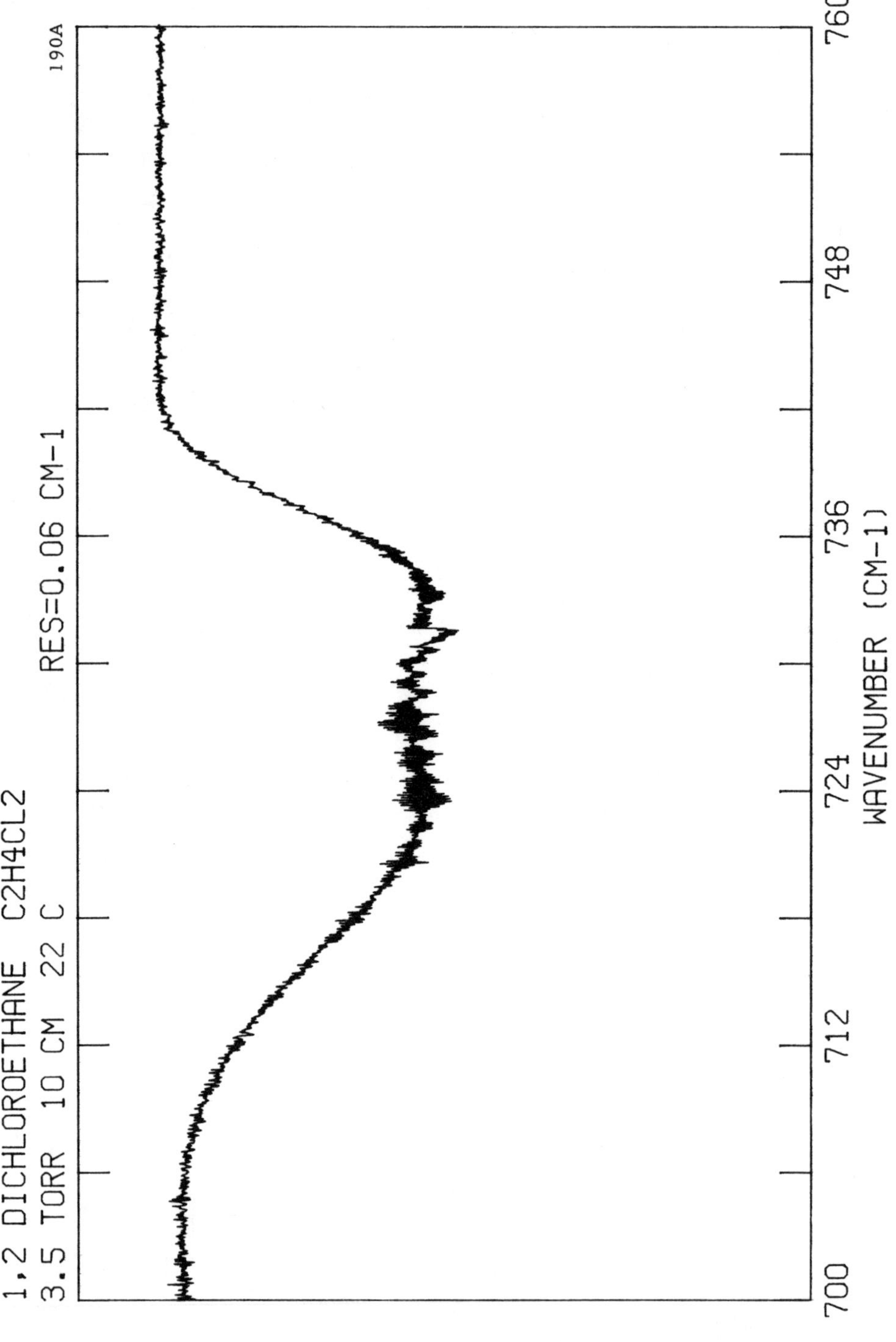
1,2 DICHLOROETHANE C2H4CL2
3.5 TORR 10 CM 22 C
RES=0.06 CM-1
190A
700
712
724
736
748
760
WAVENUMBER (CM-1)

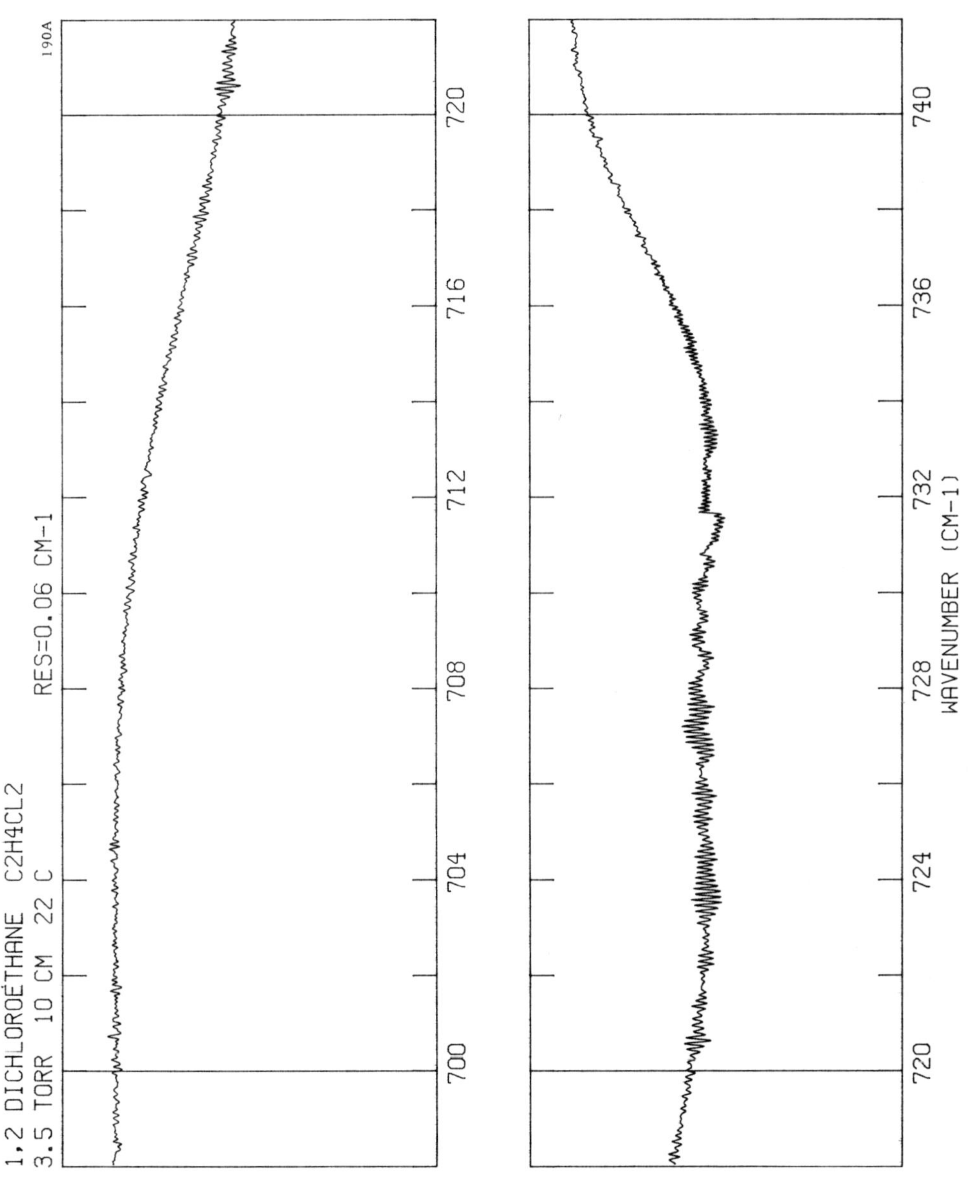
1,2 DICHLOROÉTHANE C2H4CL2
3.5 TORR 10 CM 22 C
RES=0.06 CM-1
190A
700
704
708
712
716
720
720
724
728
732
736
740
WAVENUMBER (CM-1)

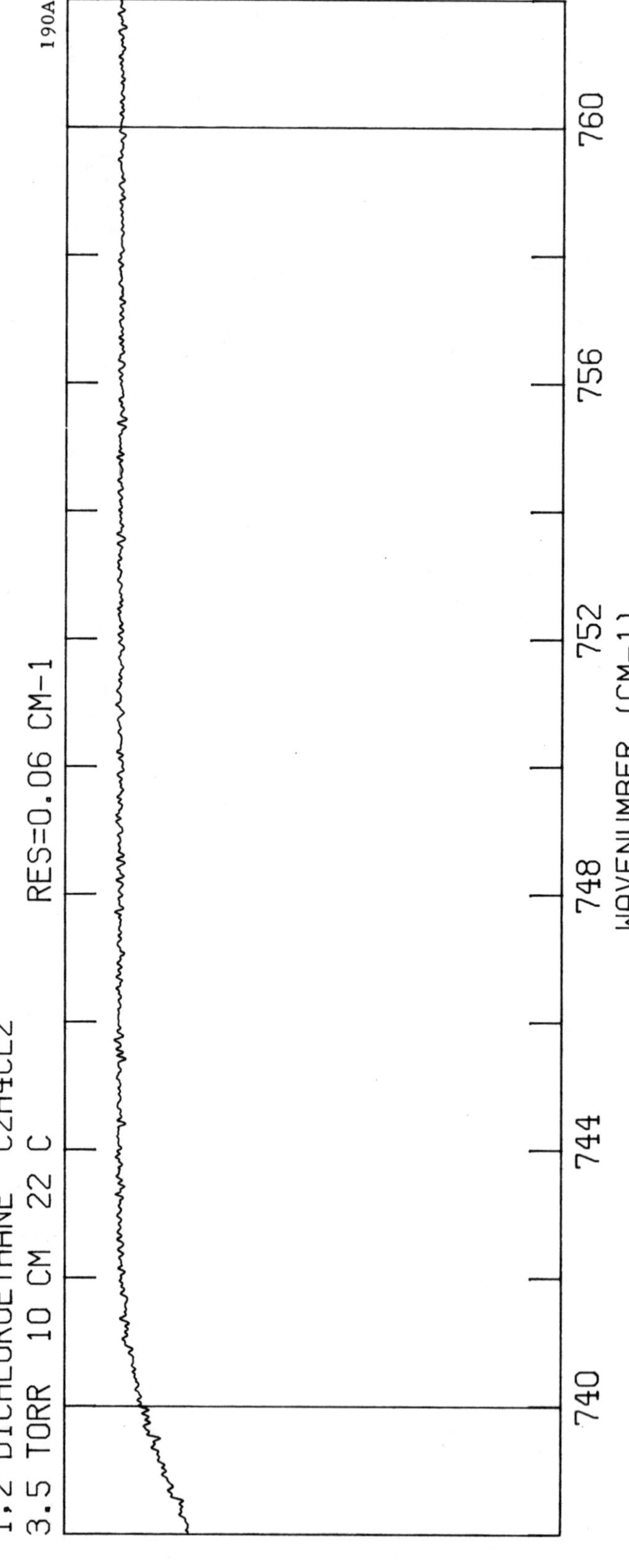
190A
1,2 DICHLOROETHANE C2H4CL2
3.5 TORR 10 CM 22 C
RES=0.06 CM-1
740
744
748
752
756
760
WAVENUMBER (CM-1)

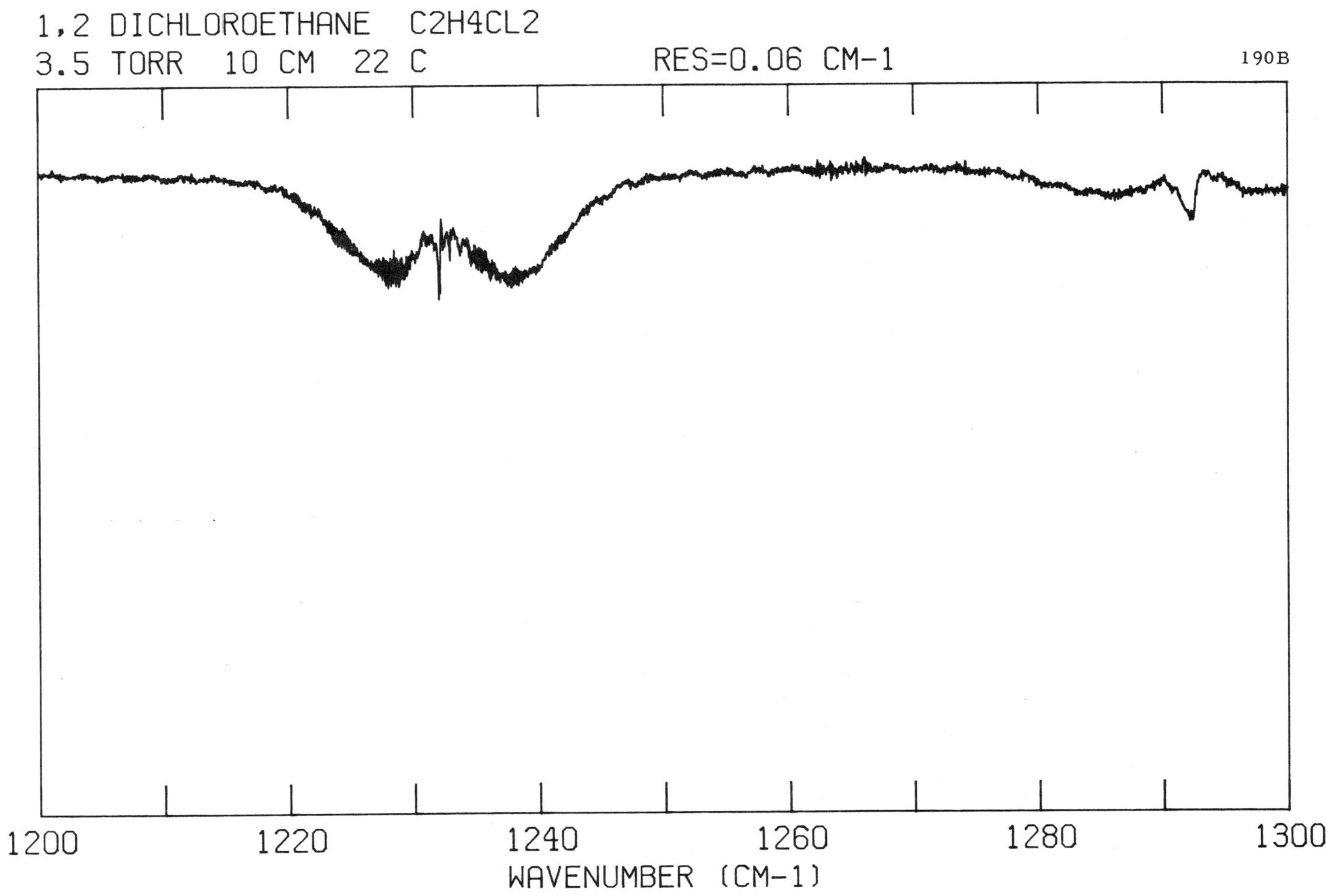

1,2 DICHLOROETHANE C2H4CL2
3.5 TORR 10 CM 22 C
RES=0.06 CM-1
190B
1200
1220
1240
1260
1280
1300
WAVENUMBER (CM-1)

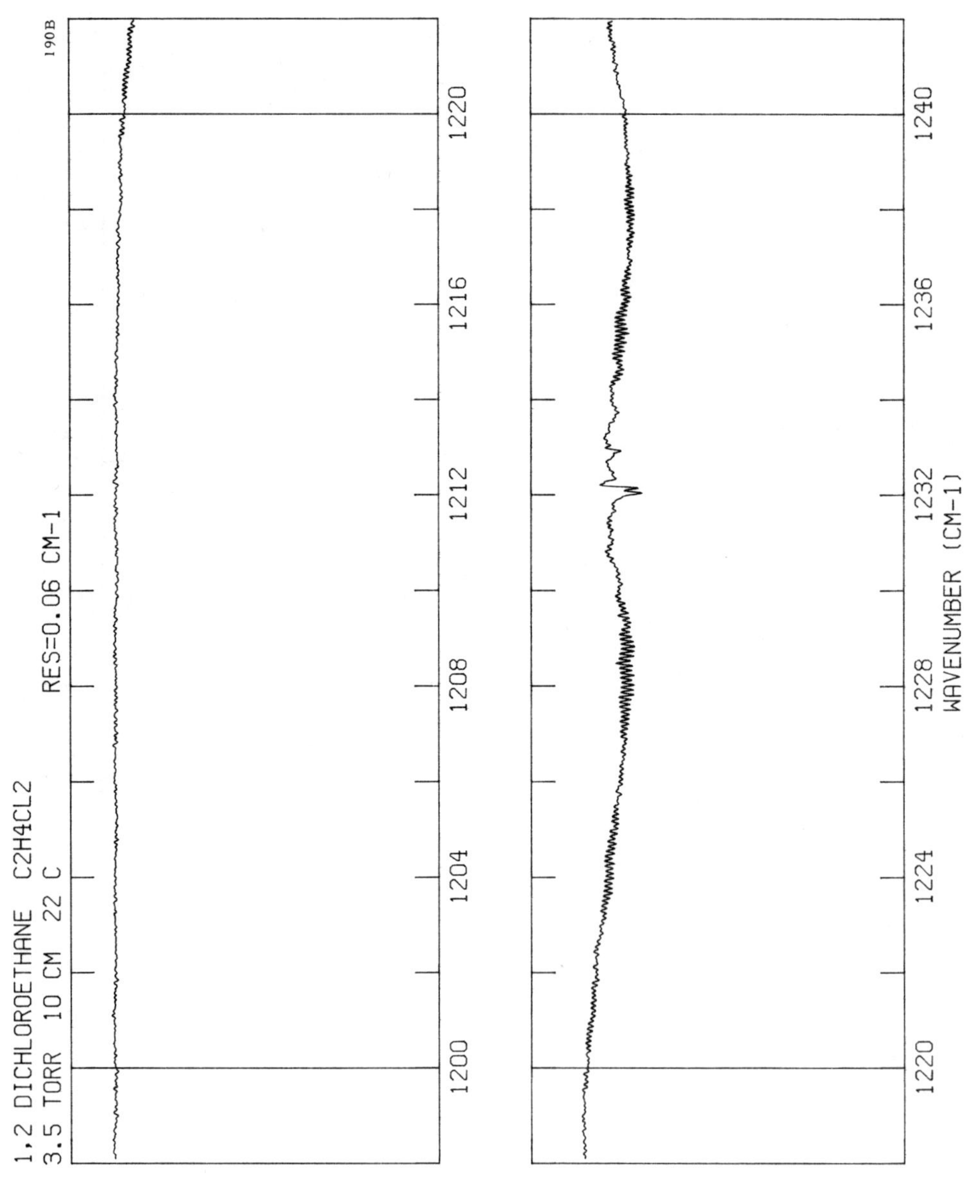
1,2 DICHLOROETHANE C2H4CL2
3.5 TORR 10 CM 22 C
RES=0.06 CM-1
190B
1200
1204
1208
1212
1216
1220
1220
1224
1228
1232
1236
1240
WAVENUMBER (CM-1)

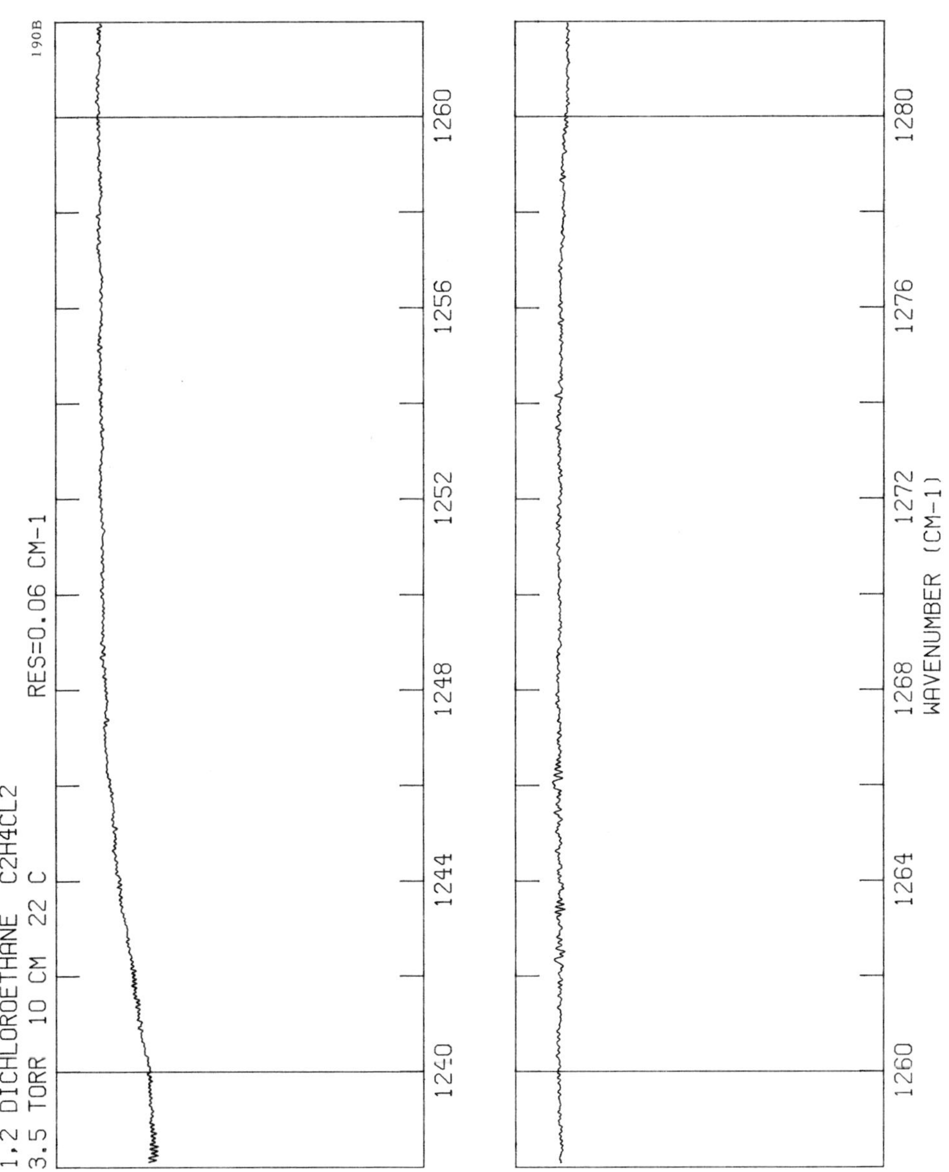

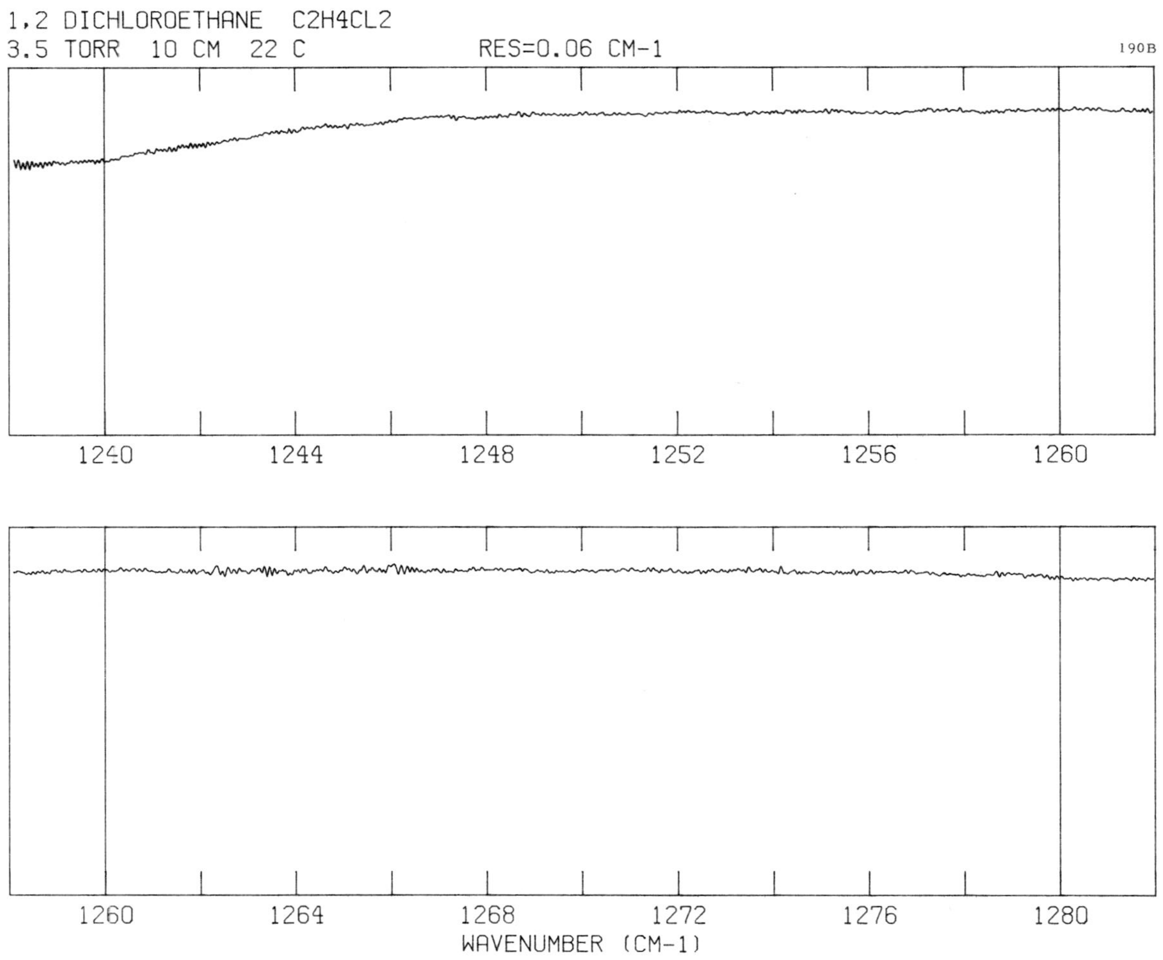

190B
1,2 DICHLOROETHANE C2H4CL2
3.5 TORR 10 CM 22 C
RES=0.06 CM-1
1240
1244
1248
1252
1256
1260
1264
1268
1272
1276
1280
WAVENUMBER (CM-1)

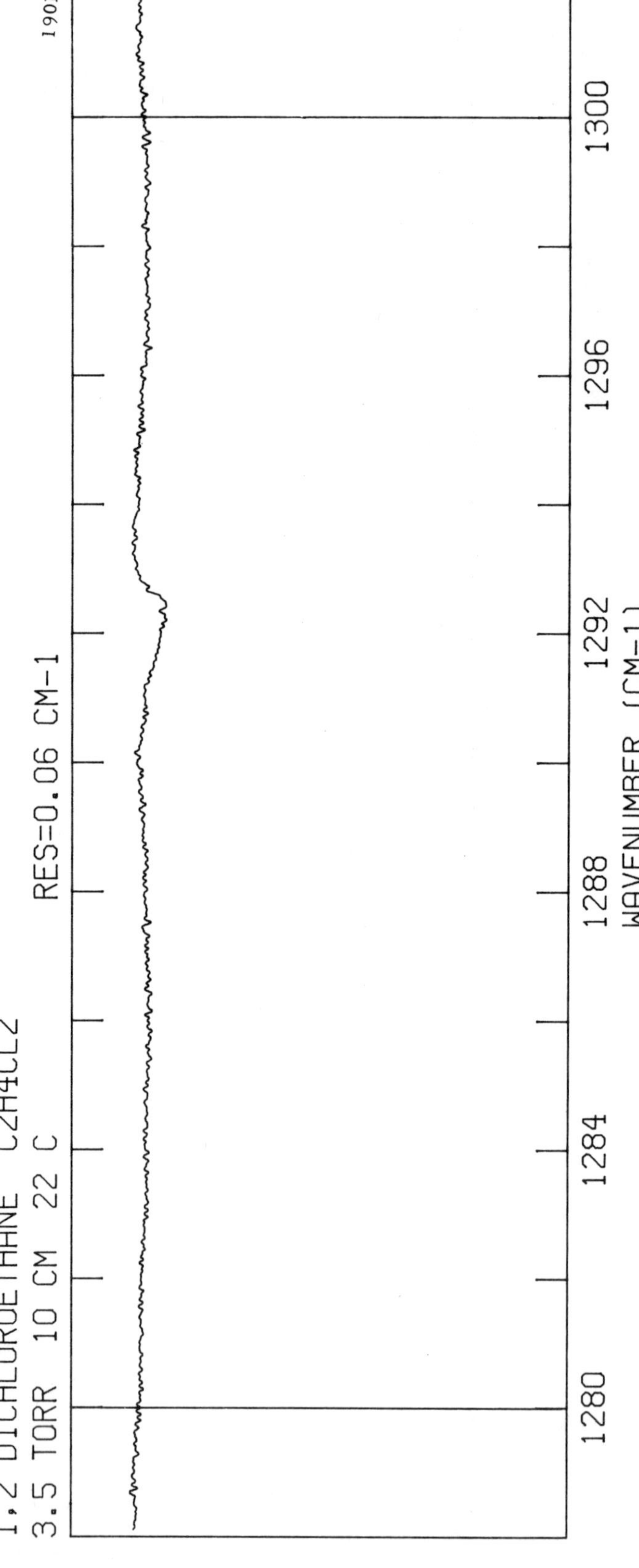
1,2 DICHLOROETHANE C2H4CL2
3.5 TORR 10 CM 22 C
RES=0.06 CM-1
190B
1280
1284
1288
1292
1296
1300
WAVENUMBER (CM-1)

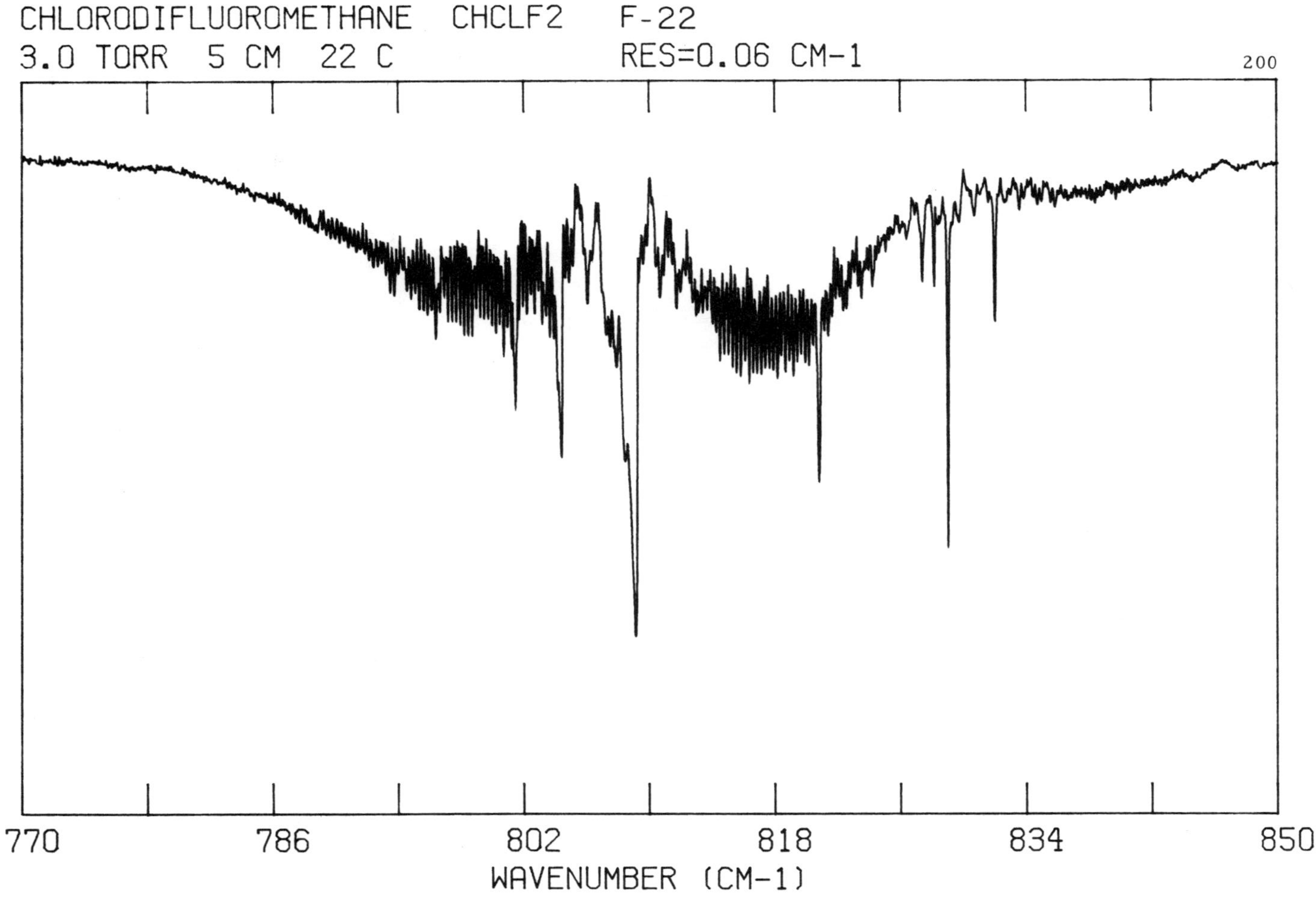
CHLORODIFLUOROMETHANE CHCLF2 F-22
3.0 TORR 5 CM 22 C RES=0.06 CM-1
200
770
786
802
818
834
850
WAVENUMBER (CM-1)

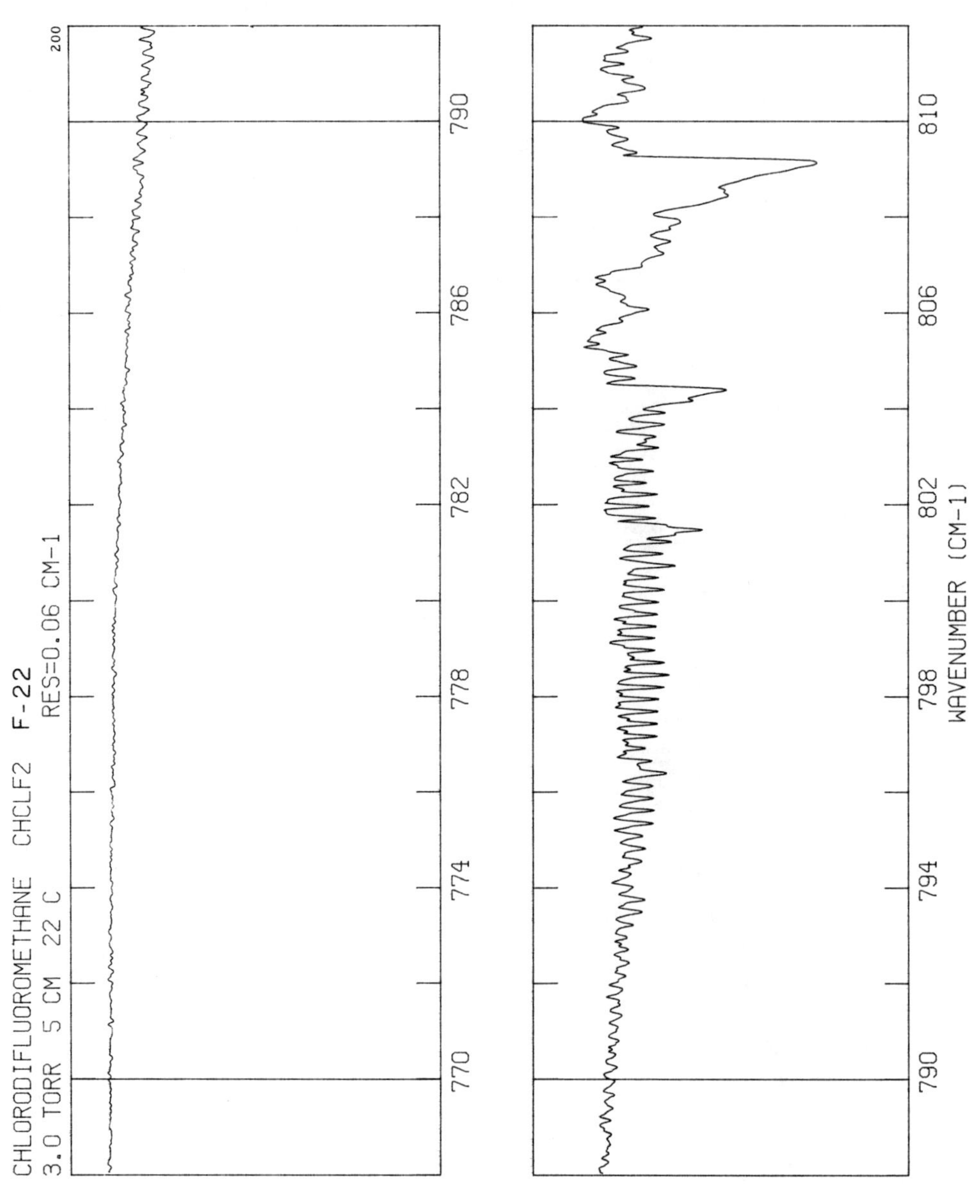
CHLORODIFLUOROMETHANE CHCLF2 F-22
3.0 TORR 5 CM 22 C RES=0.06 CM-1
200
770
774
778
782
786
790
790
794
798
802
806
810
WAVENUMBER (CM-1)

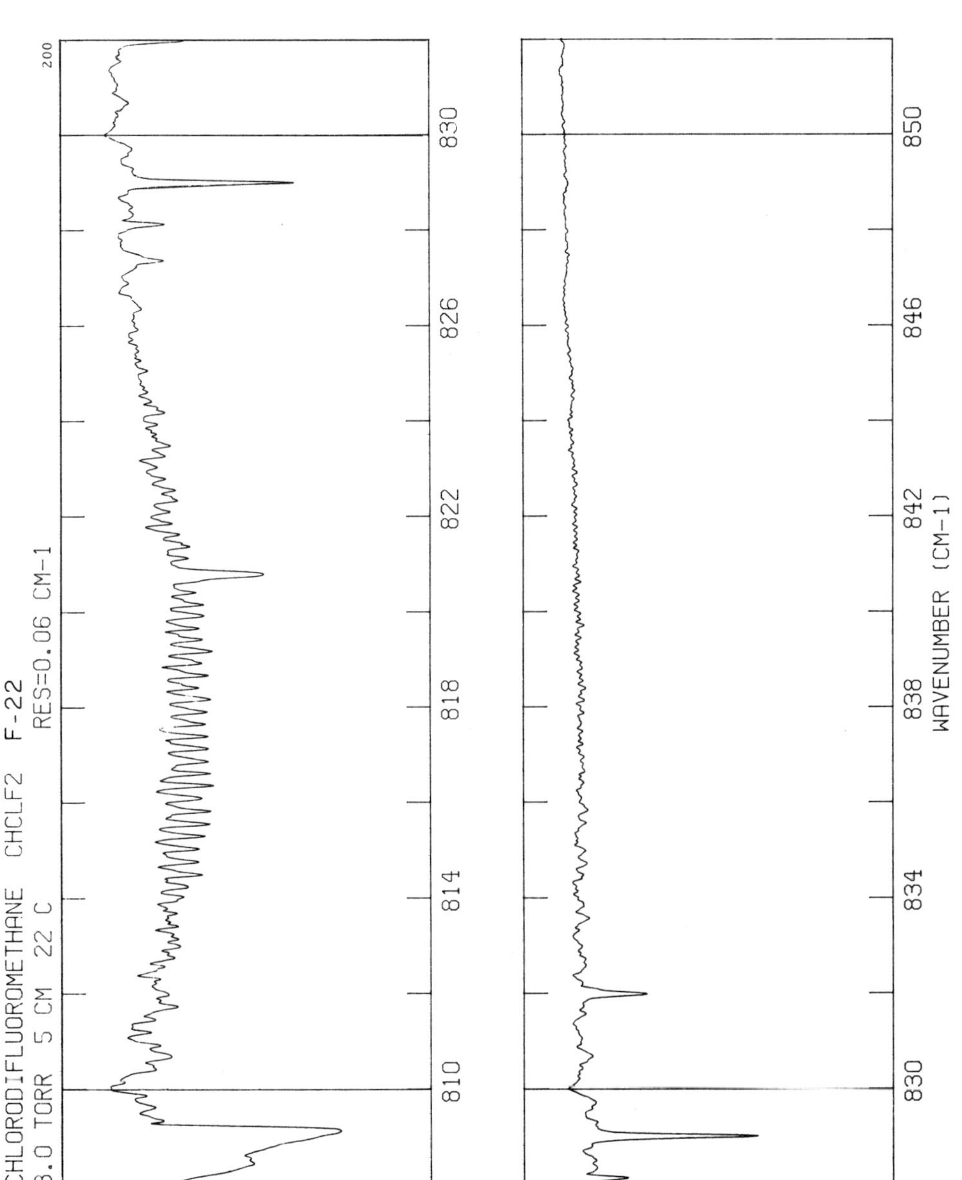
CHLORODIFLUOROMETHANE CHCLF2 F-22
3.0 TORR 5 CM 22 C RES=0.06 CM-1
200
810
814
818
822
826
830
830
834
838
842
846
850
WAVENUMBER (CM-1)

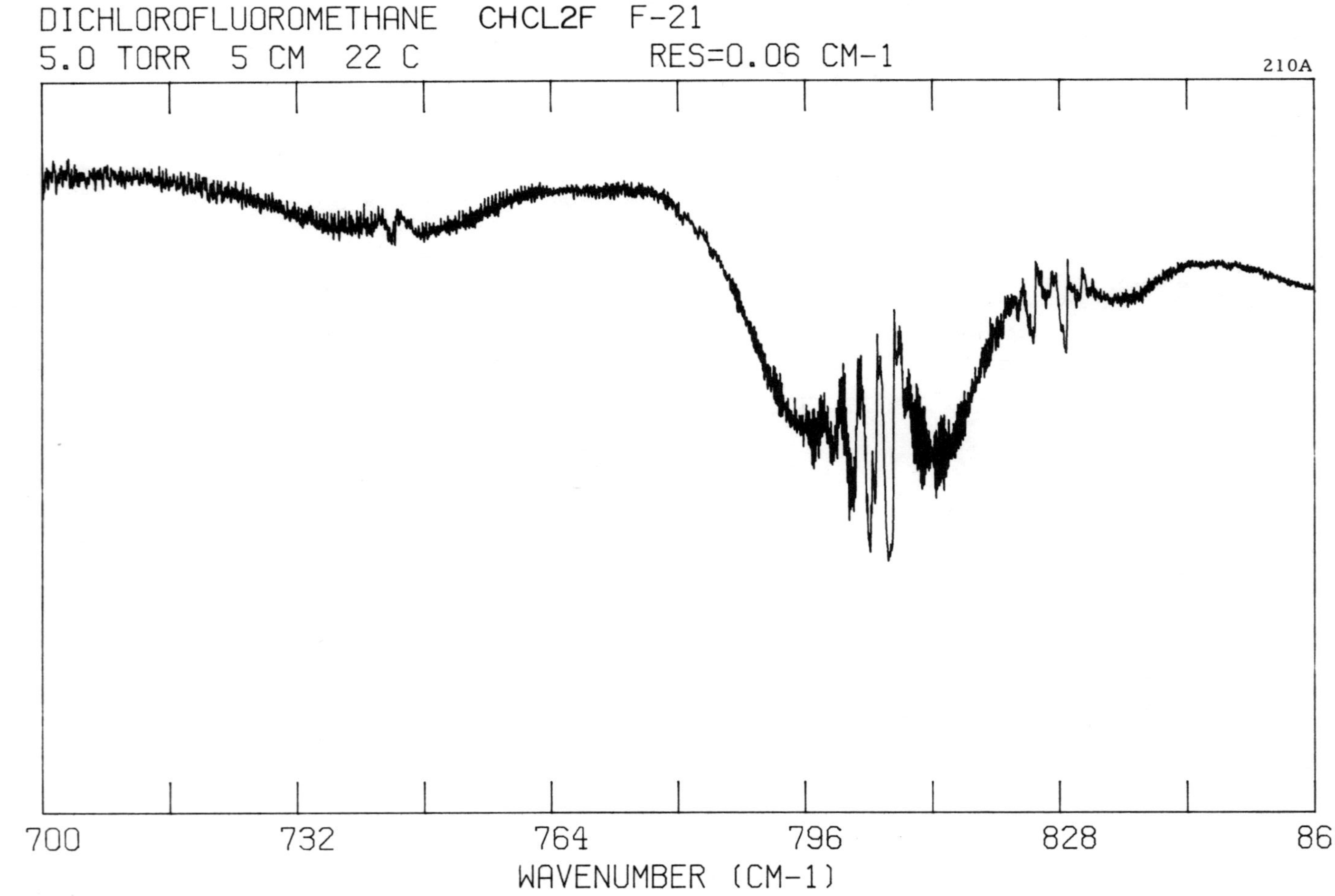
DICHLOROFLUOROMETHANE CHCL2F F-21
5.0 TORR 5 CM 22 C RES=0.06 CM-1
210A
700
732
764
796
828
860
WAVENUMBER (CM-1)

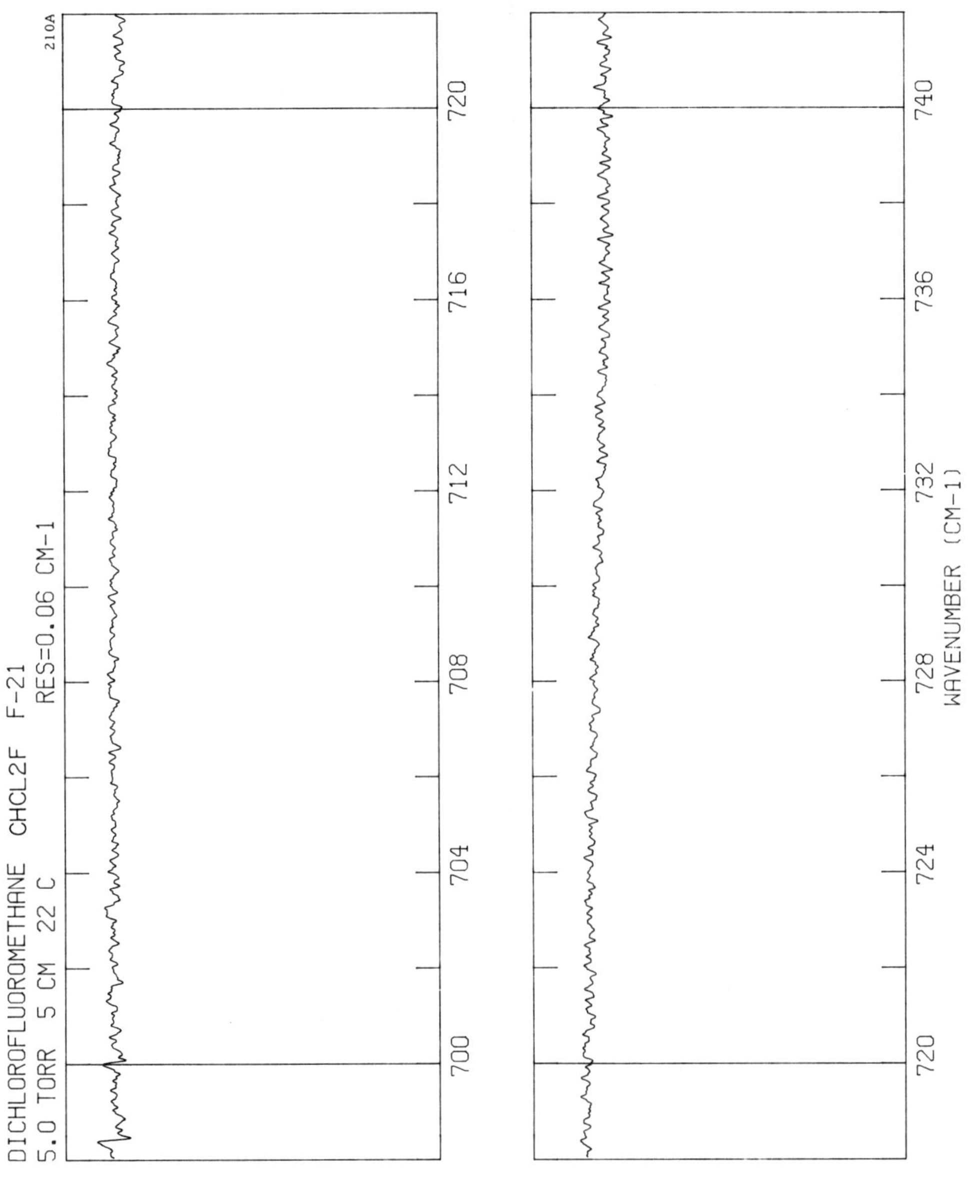

DICHLOROFLUOROMETHANE CHCL2F F-21
5.0 TORR 5 CM 22 C
RES=0.06 CM-1
210A
700
704
708
712
716
720
720
724
728
732
736
740
WAVENUMBER (CM-1)

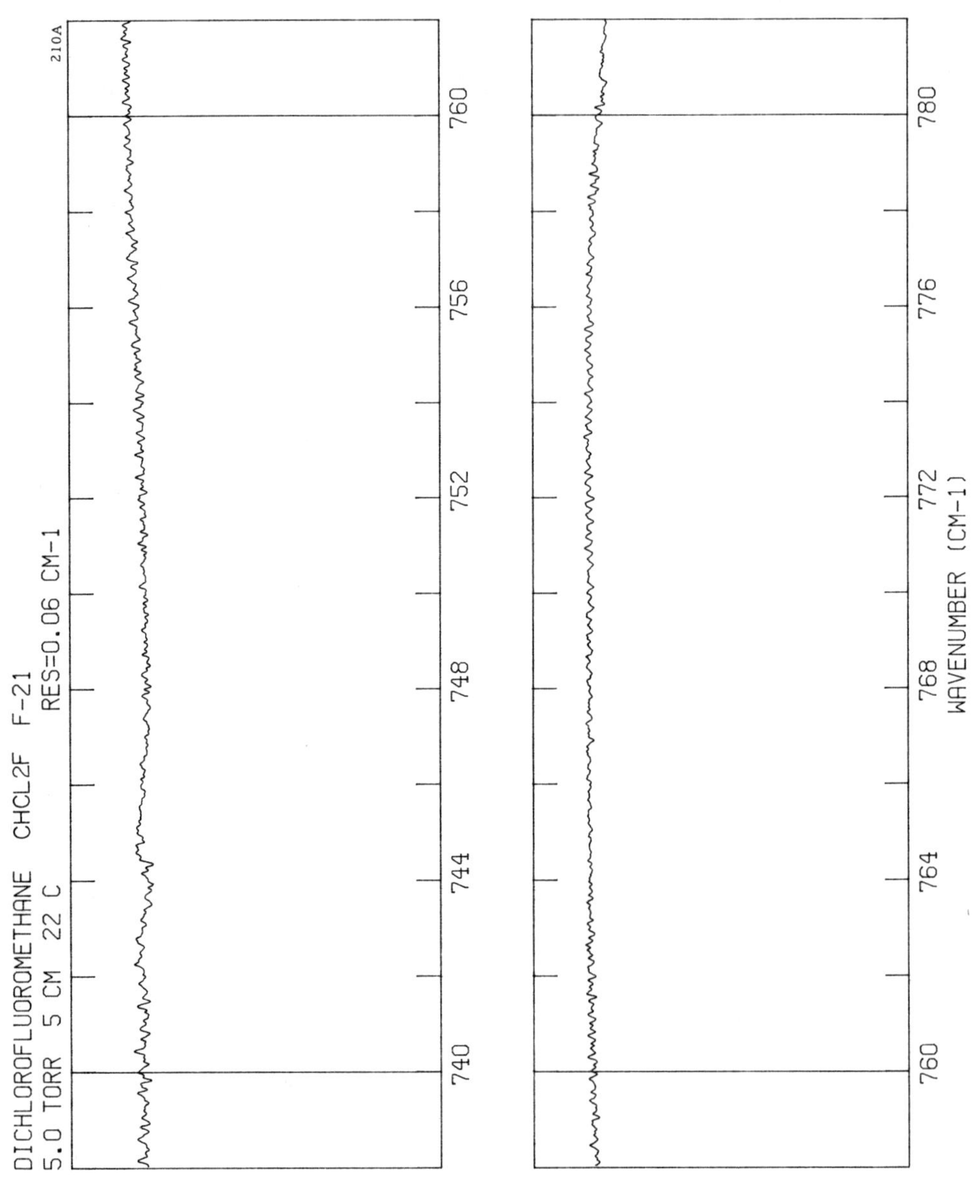
DICHLOROFLUOROMETHANE CHCL2F F-21
5.0 TORR 5 CM 22 C RES=0.06 CM-1
210A
740
744
748
752
756
760
760
764
768
772
776
780
WAVENUMBER (CM-1)

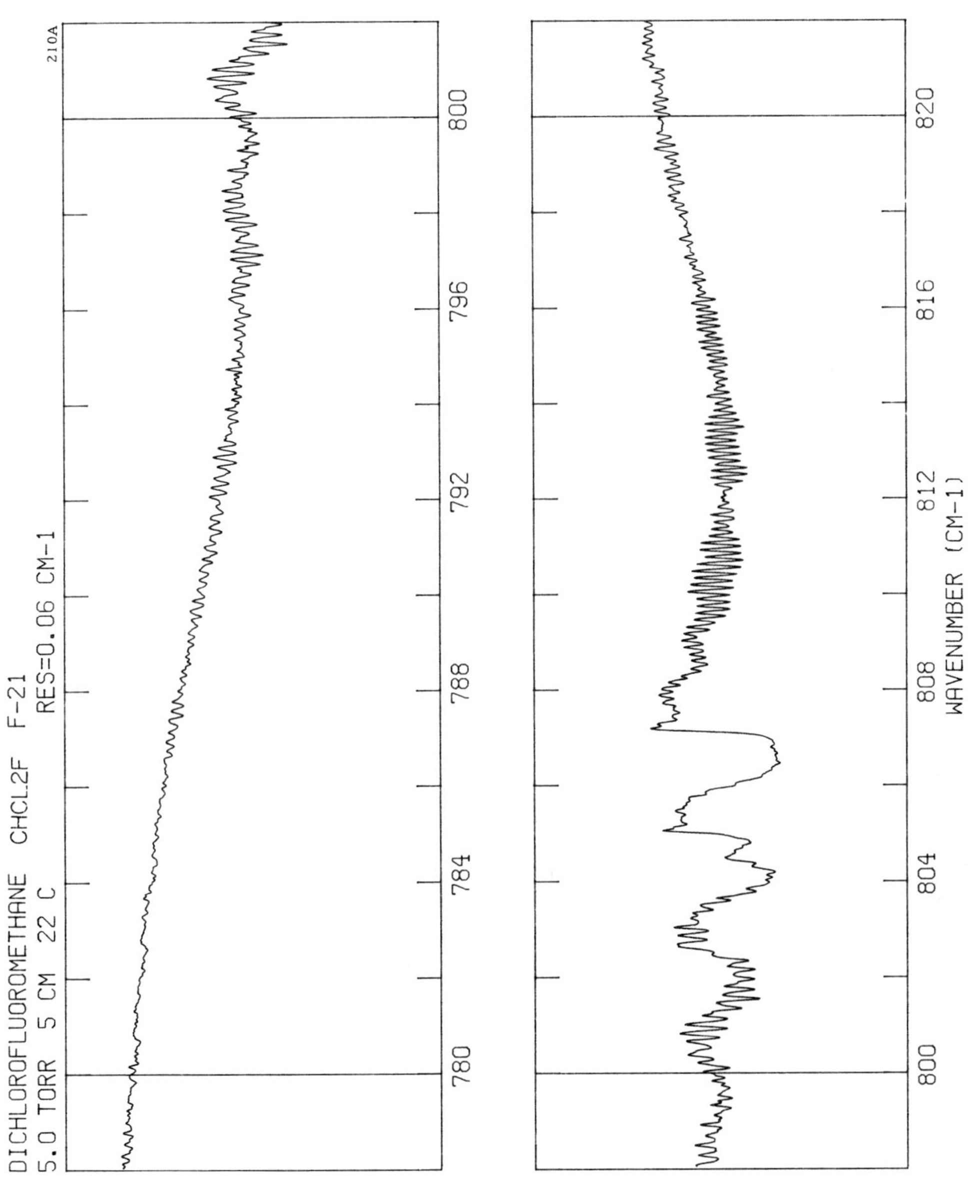
DICHLOROFLUOROMETHANE CHCL2F F-21
5.0 TORR 5 CM 22 C RES=0.06 CM-1
210A
780
784
788
792
796
800
800
804
808
812
816
820
WAVENUMBER (CM-1)

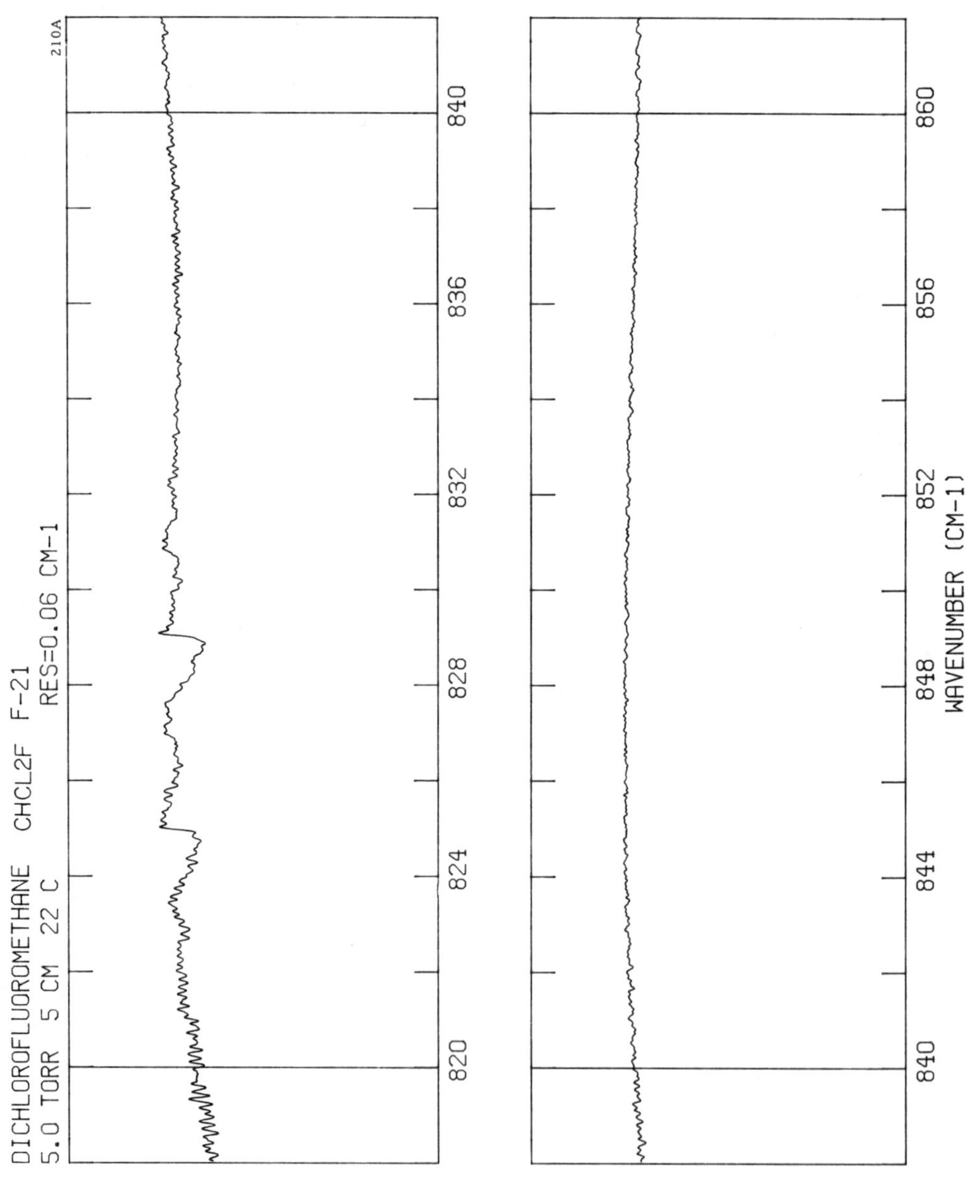

210A
DICHLOROFLUOROMETHANE CHCL2F F-21
5.0 TORR 5 CM 22 C RES=0.06 CM-1
820
824
828
832
836
840
840
844
848
852
856
860
WAVENUMBER (CM-1)

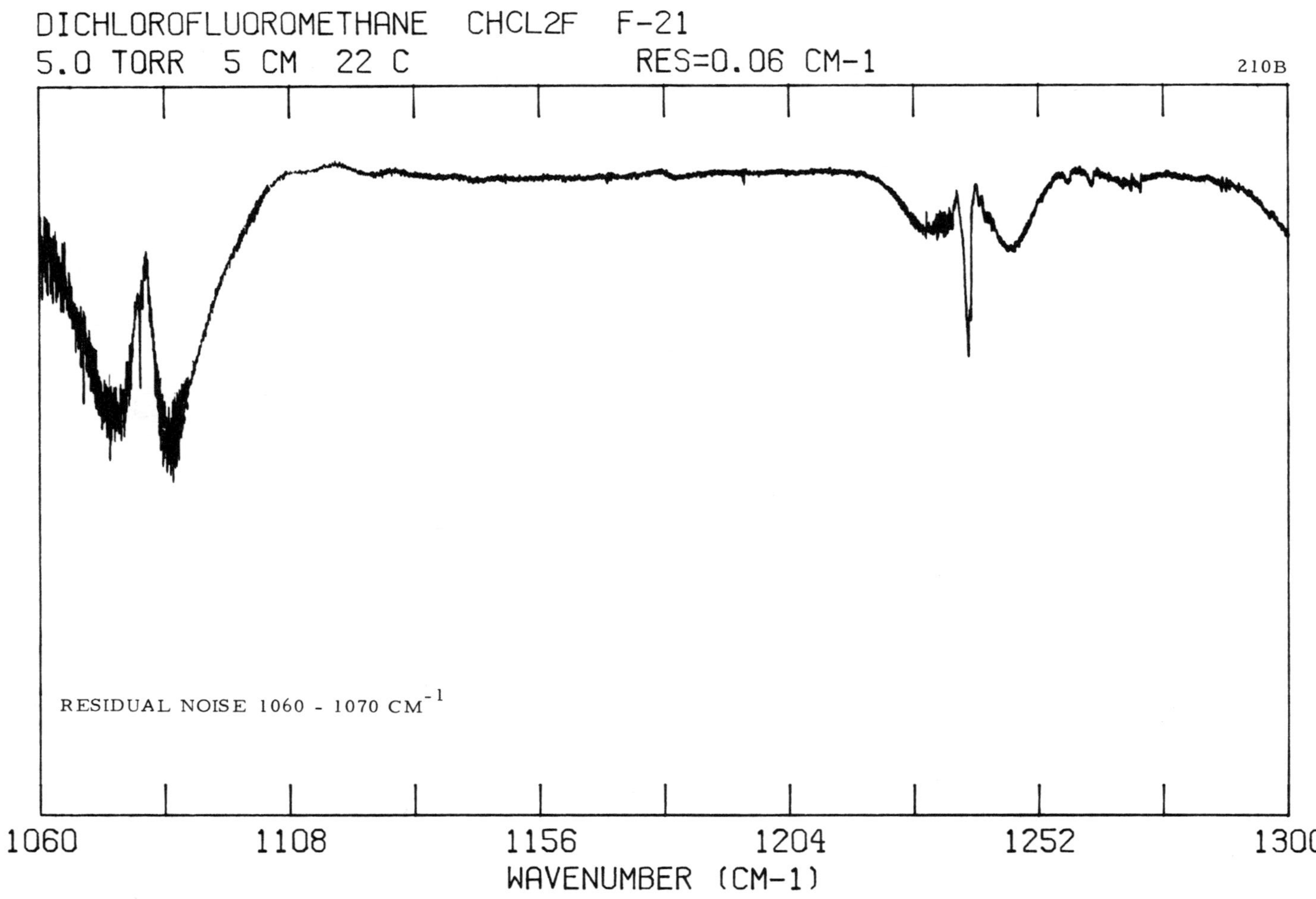

DICHLOROFLUOROMETHANE CHCL2F F-21
5.0 TORR 5 CM 22 C RES=0.06 CM-1
210B
RESIDUAL NOISE 1060 - 1070 CM-1
1060
1108
1156
1204
1252
1300
WAVENUMBER (CM-1)

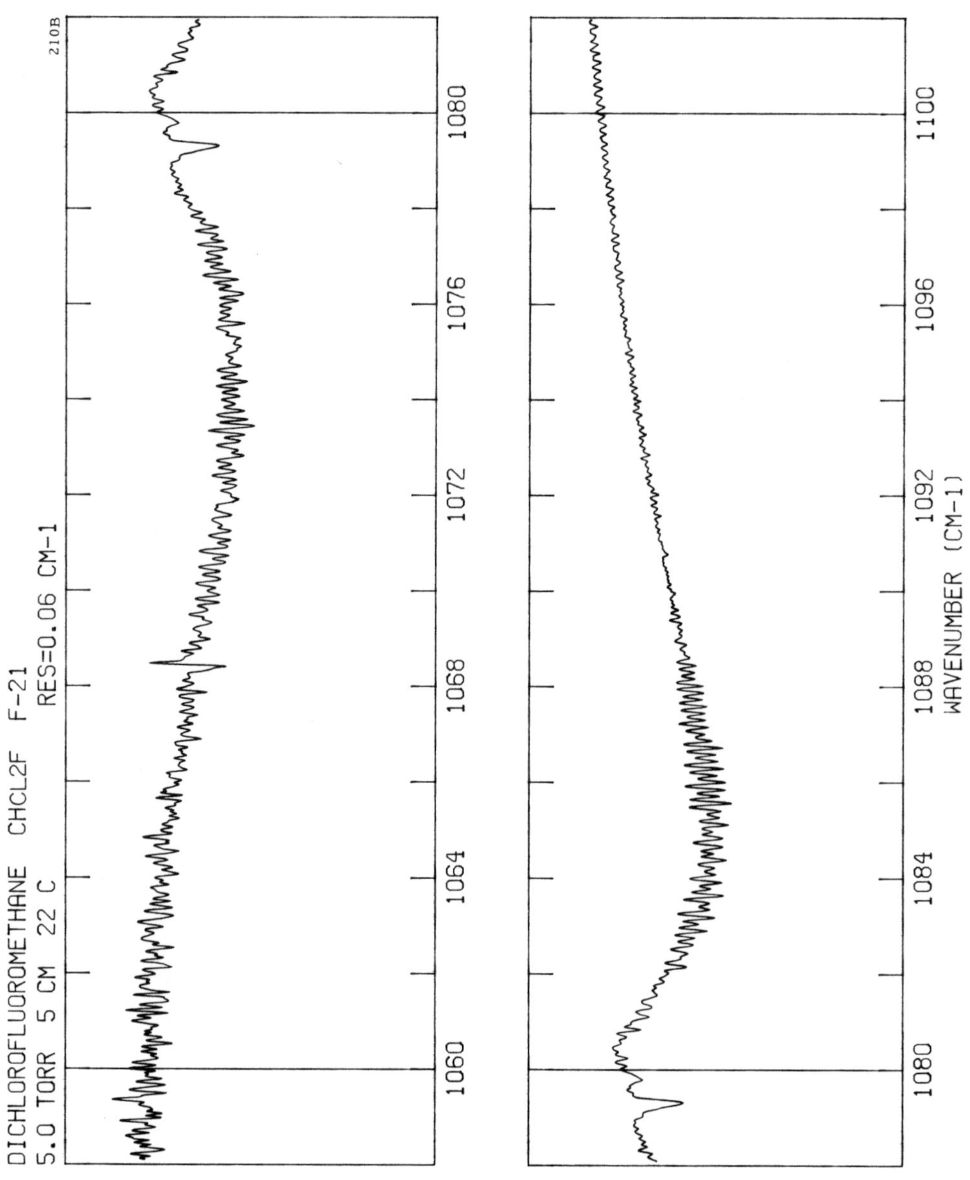
DICHLOROFLUOROMETHANE CHCL2F F-21
5.0 TORR 5 CM 22 C RES=0.06 CM-1
210B
1060
1064
1068
1072
1076
1080
1084
1088
1092
1096
1100
WAVENUMBER (CM-1)

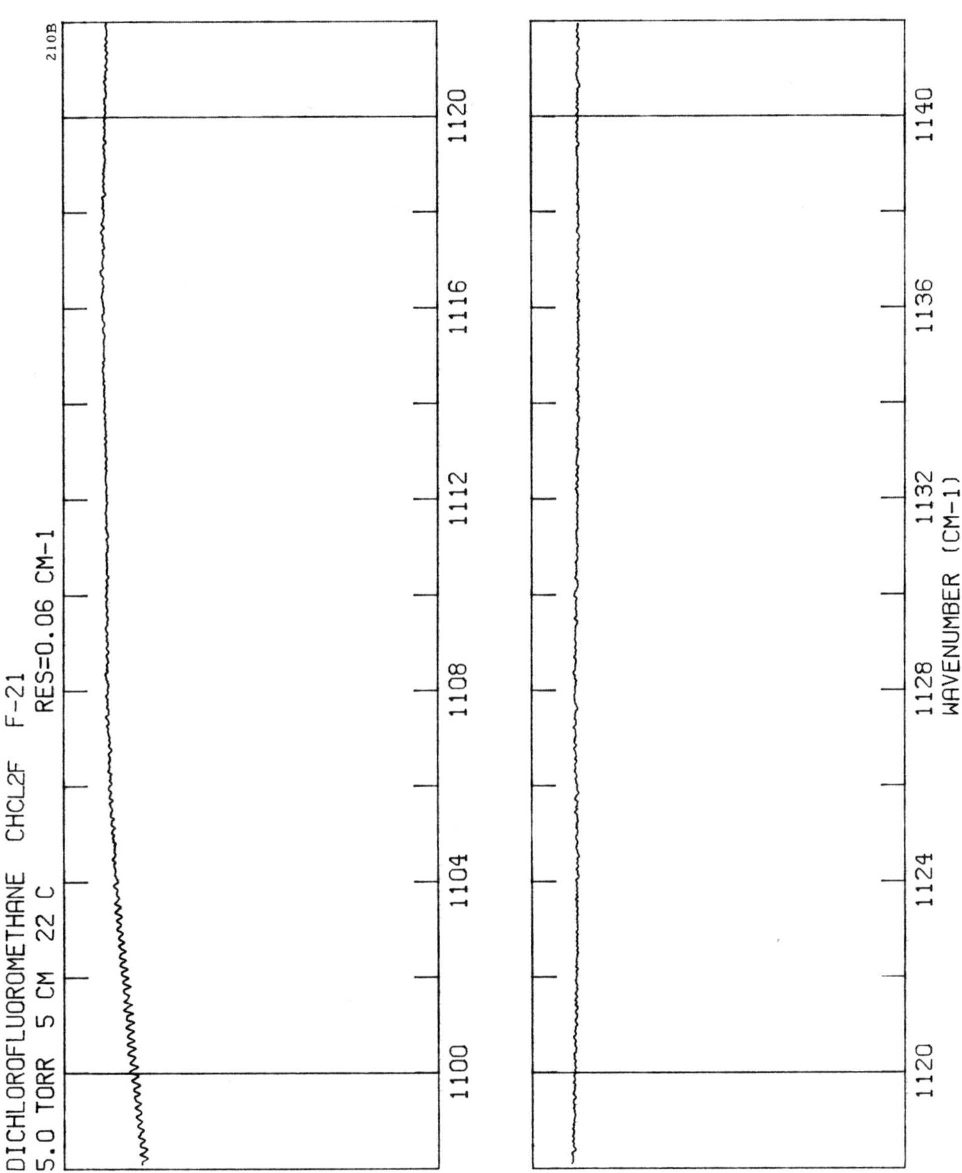
DICHLOROFLUOROMETHANE CHCL2F F-21
5.0 TORR 5 CM 22 C RES=0.06 CM-1
210B
1100
1104
1108
1112
1116
1120
1124
1128
1132
1136
1140
WAVENUMBER (CM-1)

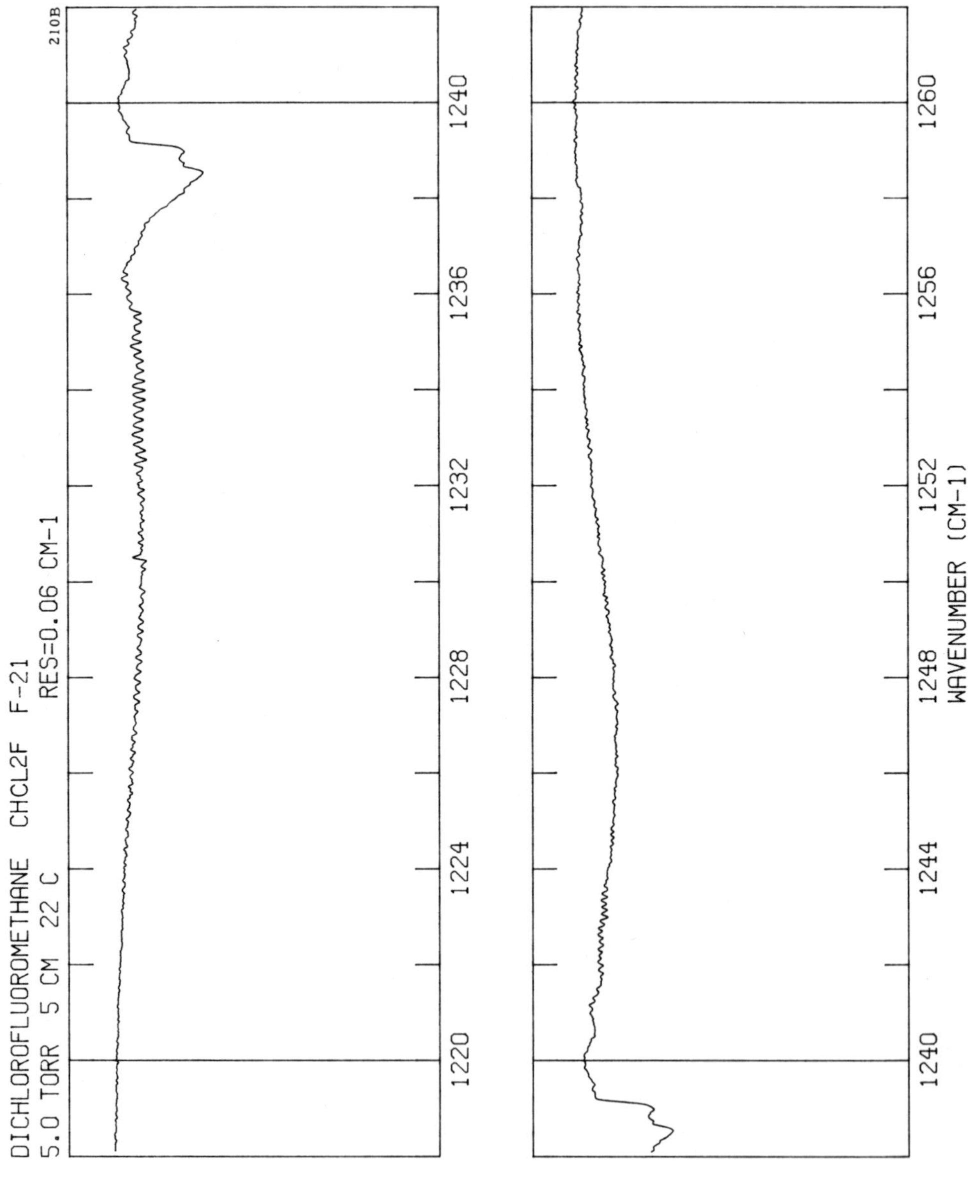
DICHLOROFLUOROMETHANE CHCL2F F-21
5.0 TORR 5 CM 22 C RES=0.06 CM-1
210B
1220
1224
1228
1232
1236
1240
1240
1244
1248
1252
1256
1260
WAVENUMBER (CM-1)

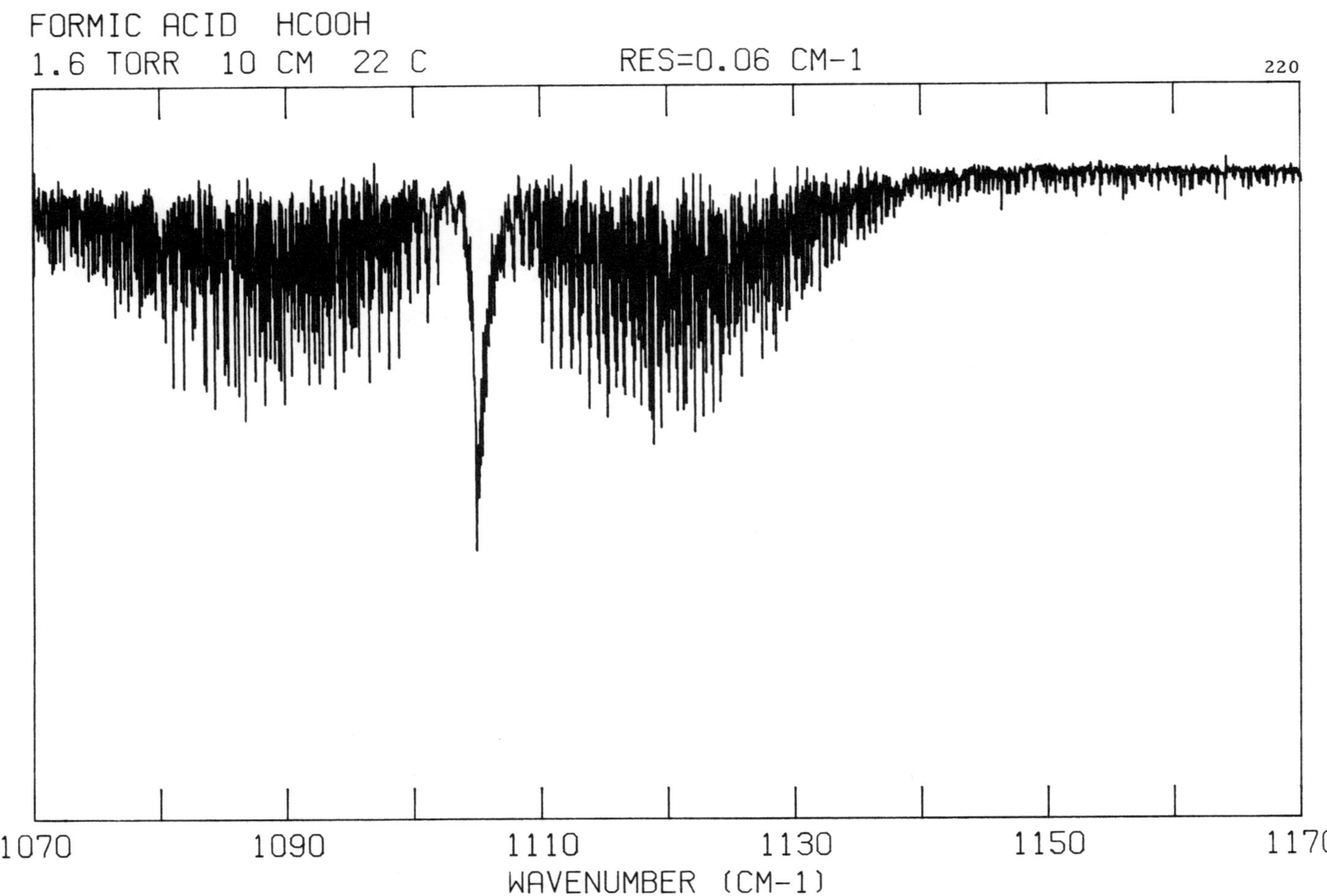
FORMIC ACID HCOOH
1.6 TORR 10 CM 22 C
RES=0.06 CM-1
220
1070
1090
1110
1130
1150
1170
WAVENUMBER (CM-1)

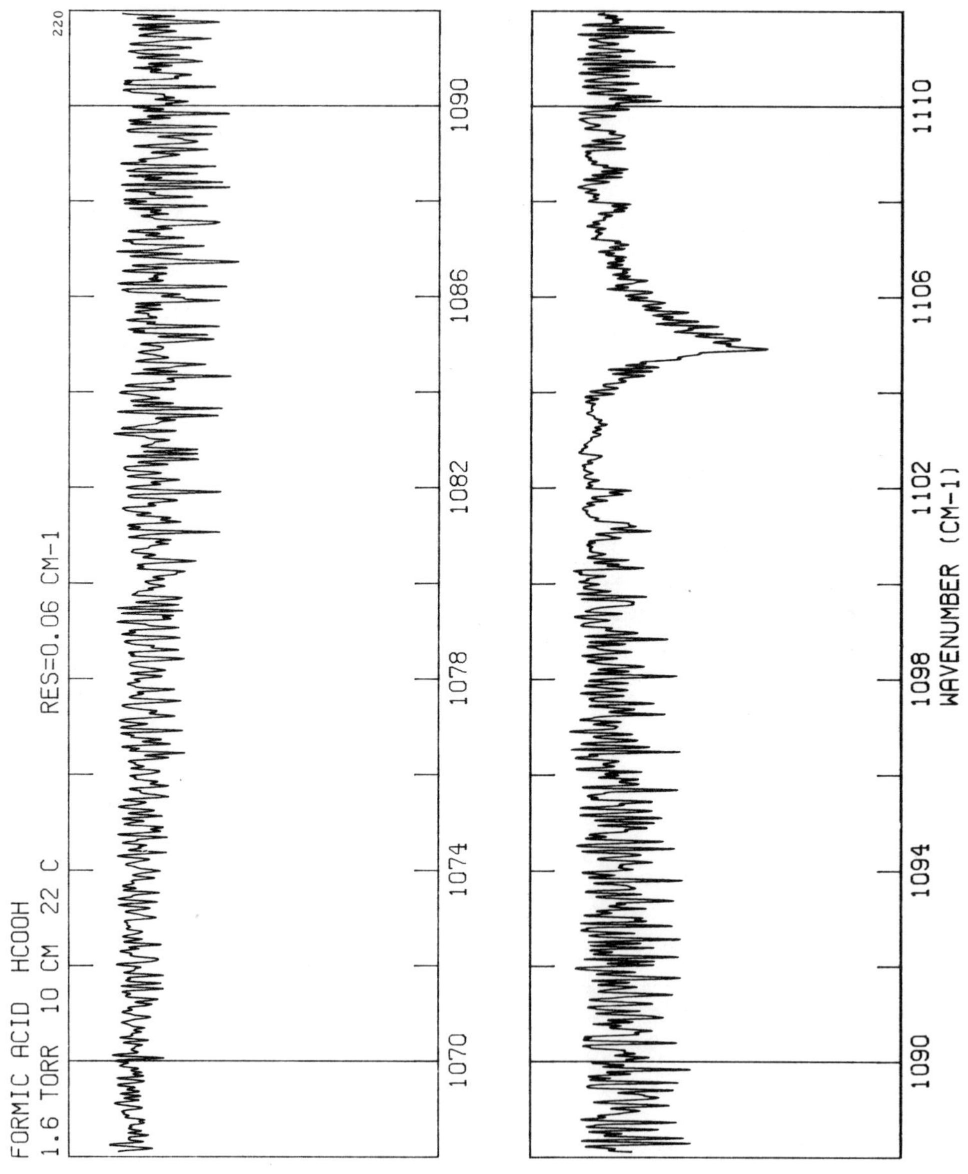

FORMIC ACID HCOOH
1.6 TORR 10 CM 22 C
RES=0.06 CM-1
220
1070
1074
1078
1082
1086
1090
1094
1098
1102
1106
1110
WAVENUMBER (CM-1)

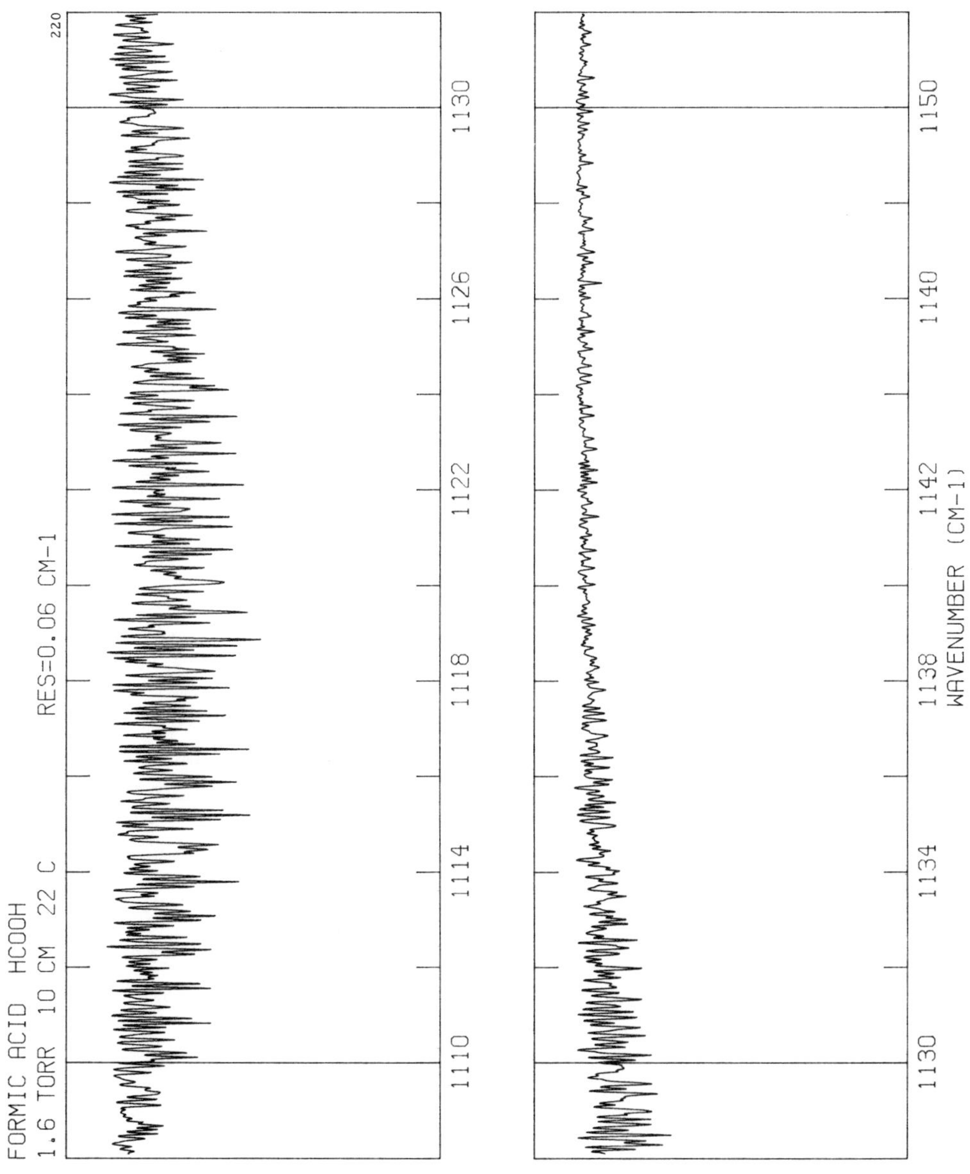
FORMIC ACID HCOOH
1.6 TORR 10 CM 22 C
RES=0.06 CM-1
220
1110
1114
1118
1122
1126
1130
1130
1134
1138
1142
1140
1150
WAVENUMBER (CM-1)

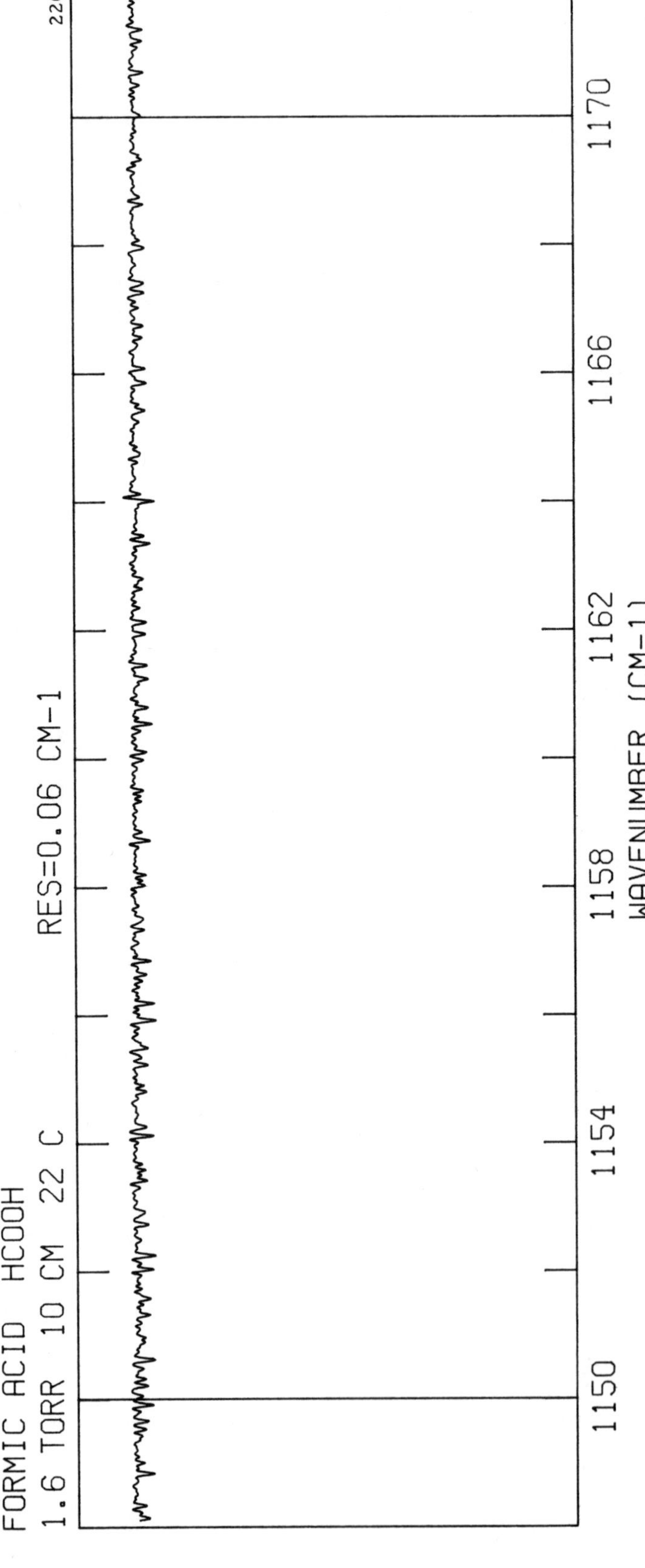
FORMIC ACID HCOOH
1.6 TORR 10 CM 22 C
RES=0.06 CM-1
220
1150
1154
1158
1162
1166
1170
WAVENUMBER (CM-1)

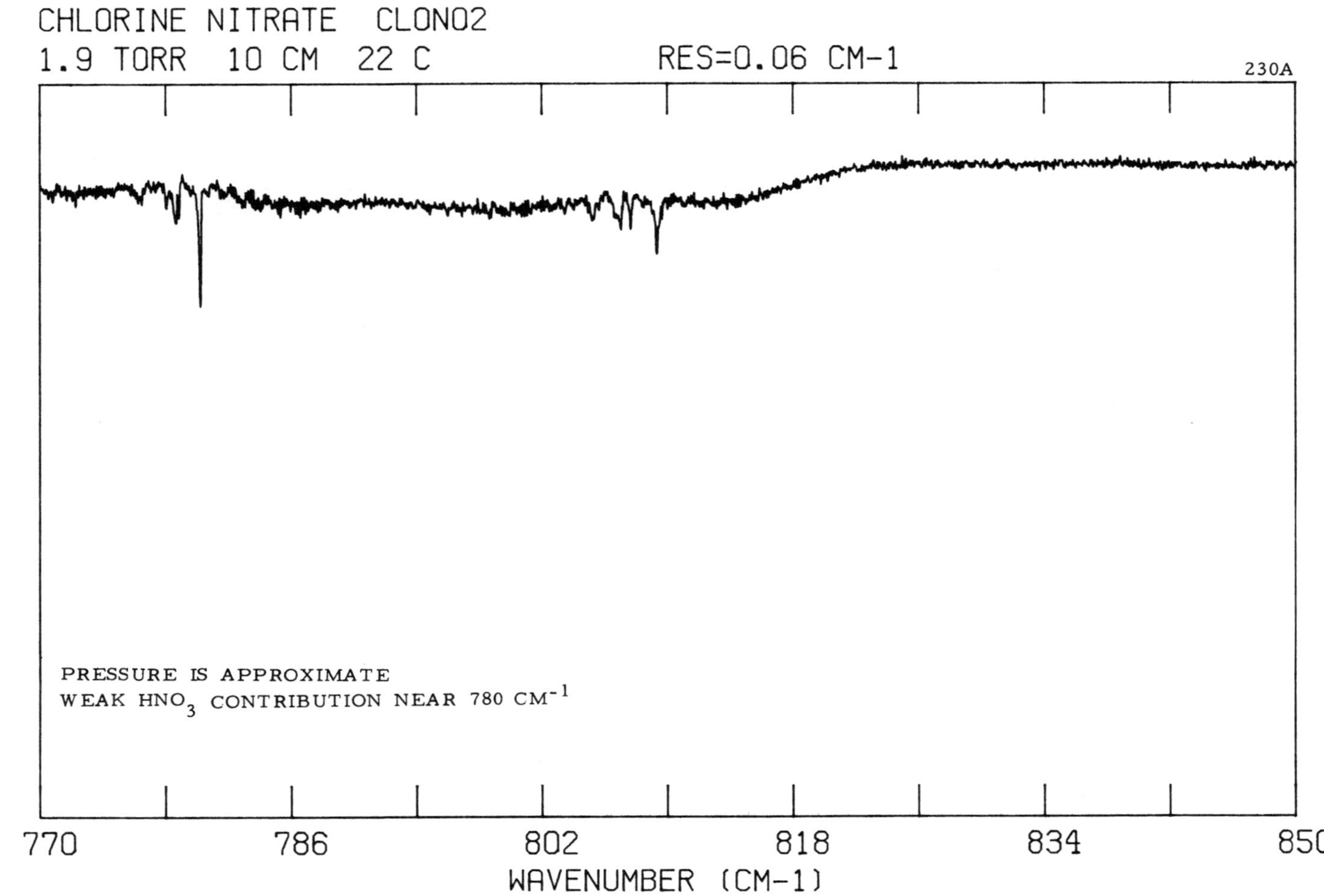
CHLORINE NITRATE CLONO2
1.9 TORR 10 CM 22 C
RES=0.06 CM-1
230A
PRESSURE IS APPROXIMATE
WEAK HNO3 CONTRIBUTION NEAR 780 CM-1
770
786
802
818
834
850
WAVENUMBER (CM-1)

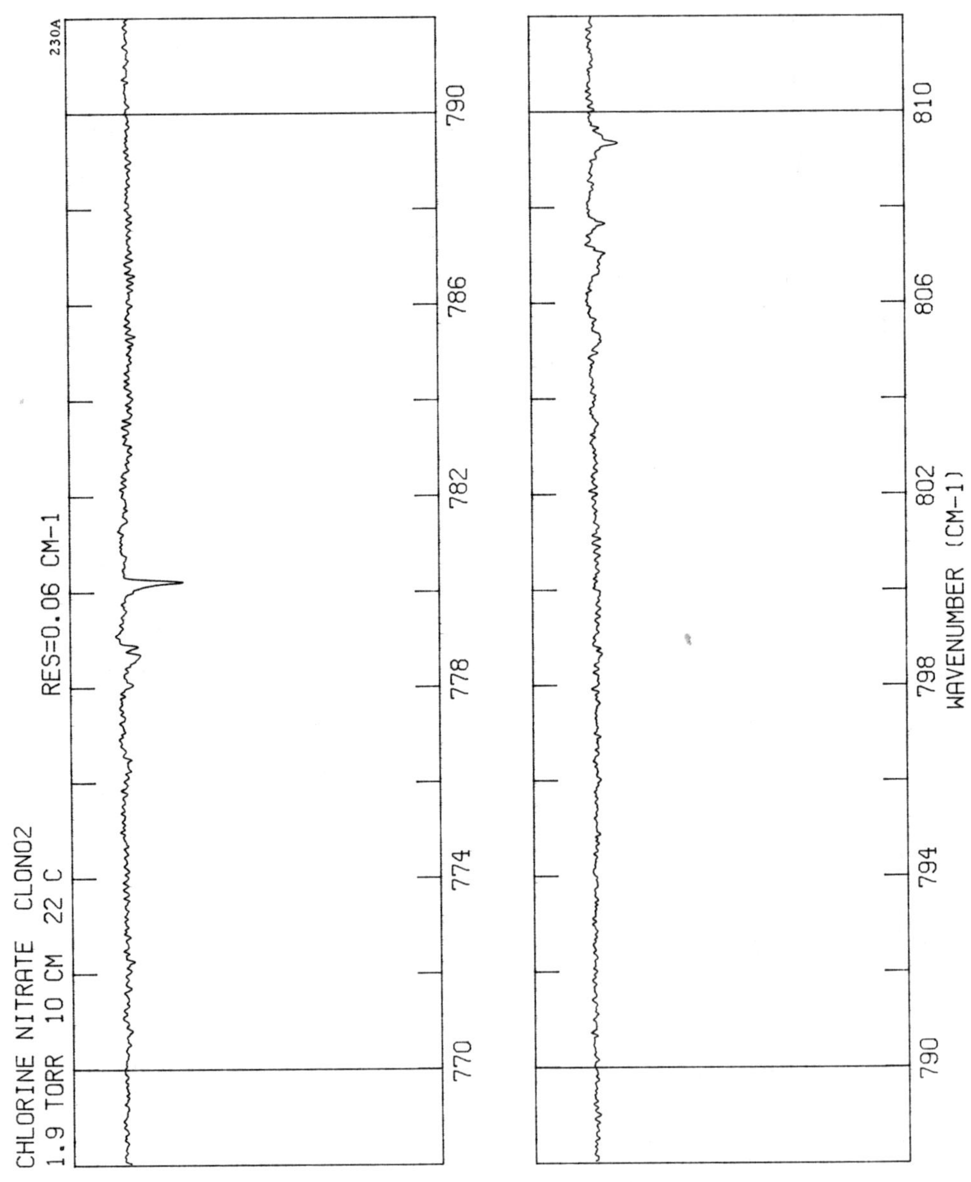
CHLORINE NITRATE CLONO2
1.9 TORR 10 CM 22 C
RES=0.06 CM-1
230A
770
774
778
782
786
790
790
794
798
802
806
810
WAVENUMBER (CM-1)

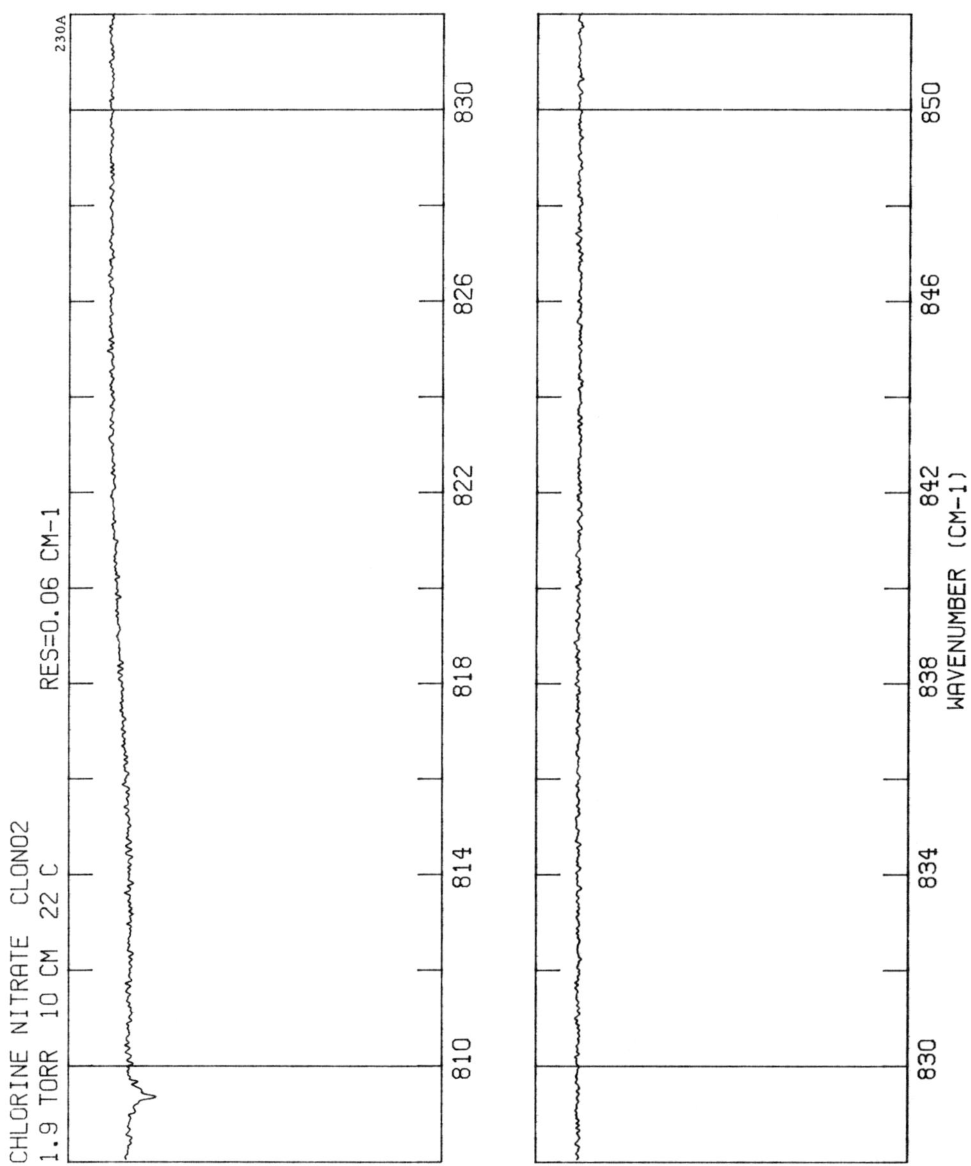
CHLORINE NITRATE CLONO2
1.9 TORR 10 CM 22 C
RES=0.06 CM-1
230A
810
814
818
822
826
830
830
834
838
842
846
850
WAVENUMBER (CM-1)

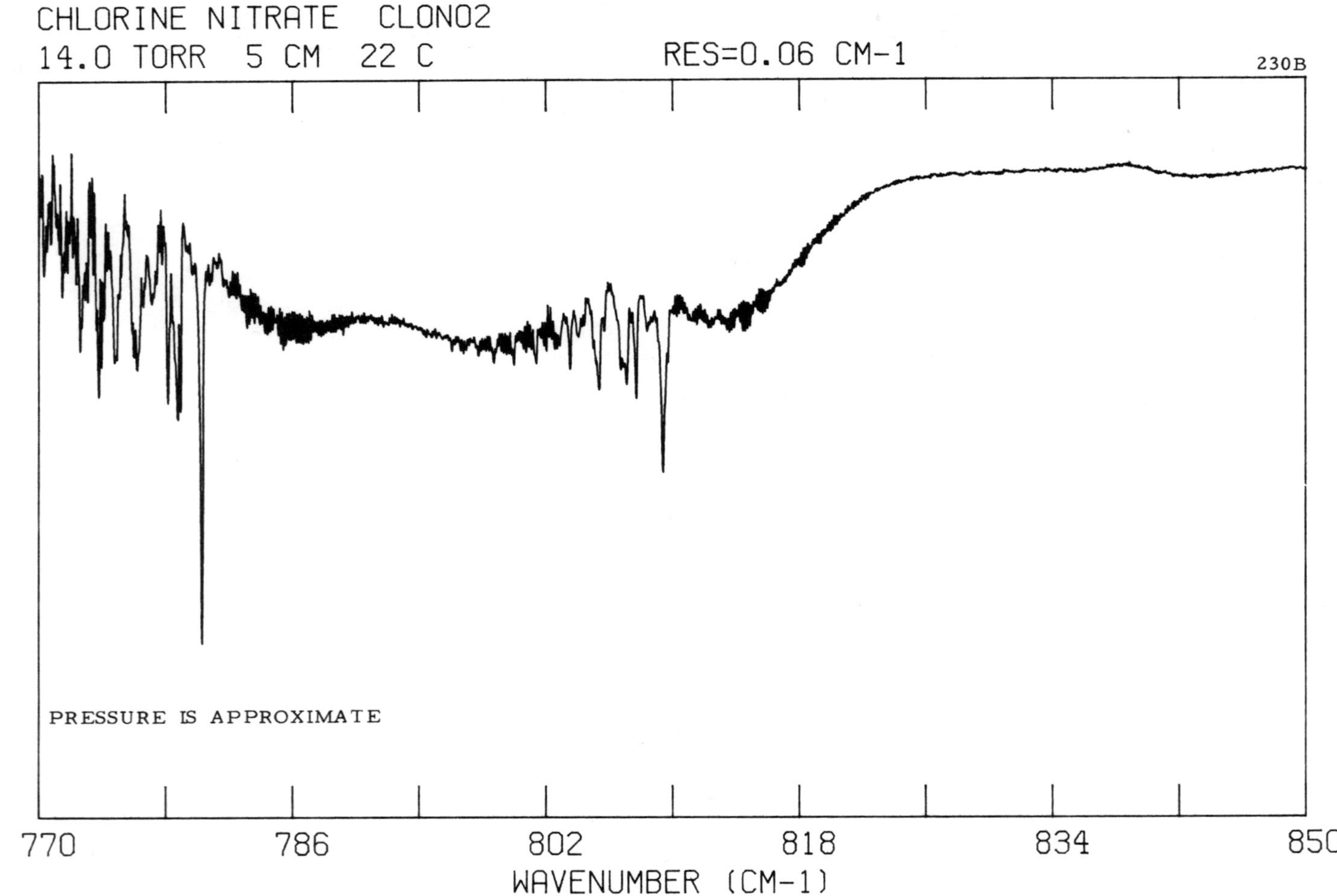
CHLORINE NITRATE CLONO2
14.0 TORR 5 CM 22 C
RES=0.06 CM-1
230B
PRESSURE IS APPROXIMATE
770
786
802
818
834
850
WAVENUMBER (CM-1)

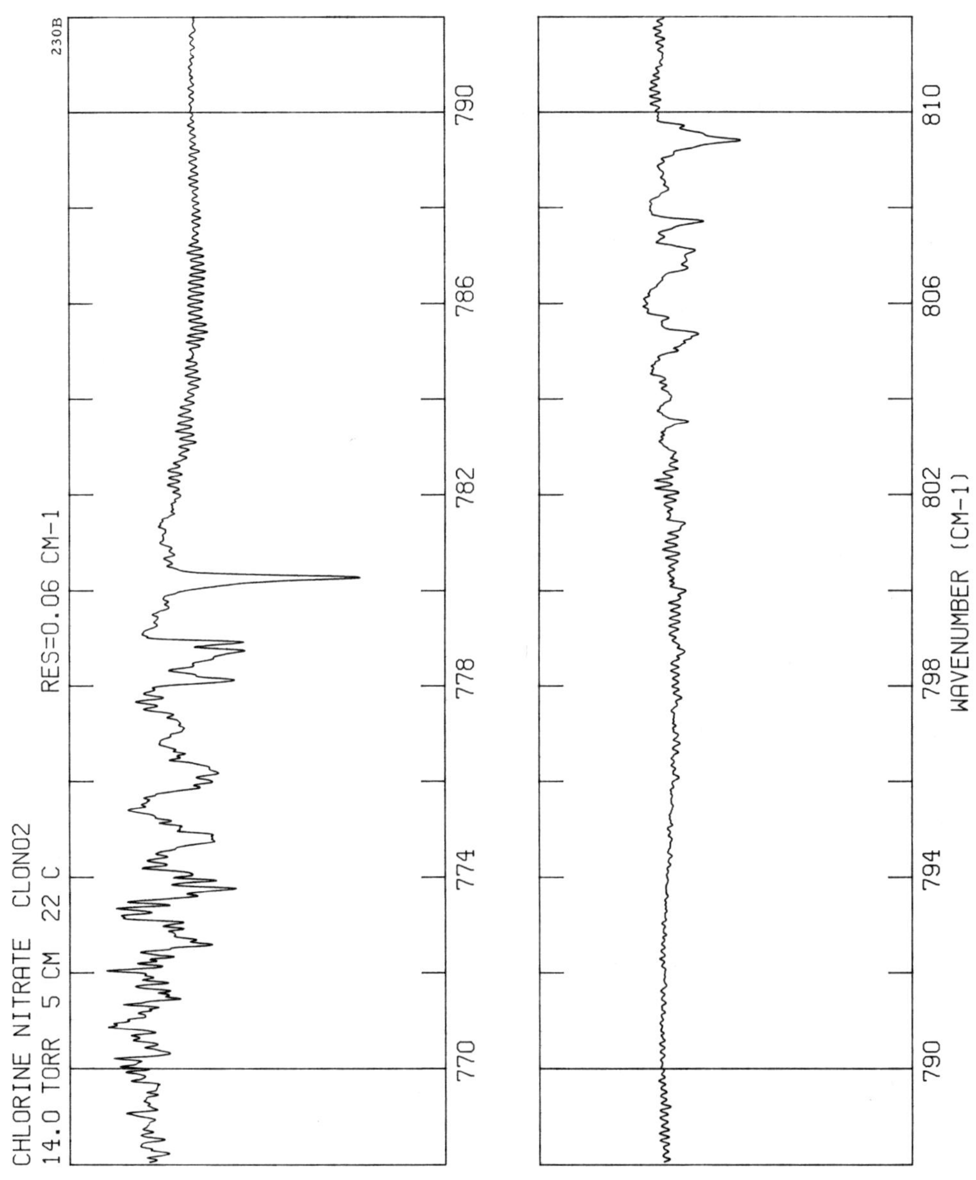
230B
CHLORINE NITRATE CLONO2
14.0 TORR 5 CM 22 C
RES=0.06 CM-1
770
774
778
782
786
790
790
794
798
802
806
810
WAVENUMBER (CM-1)

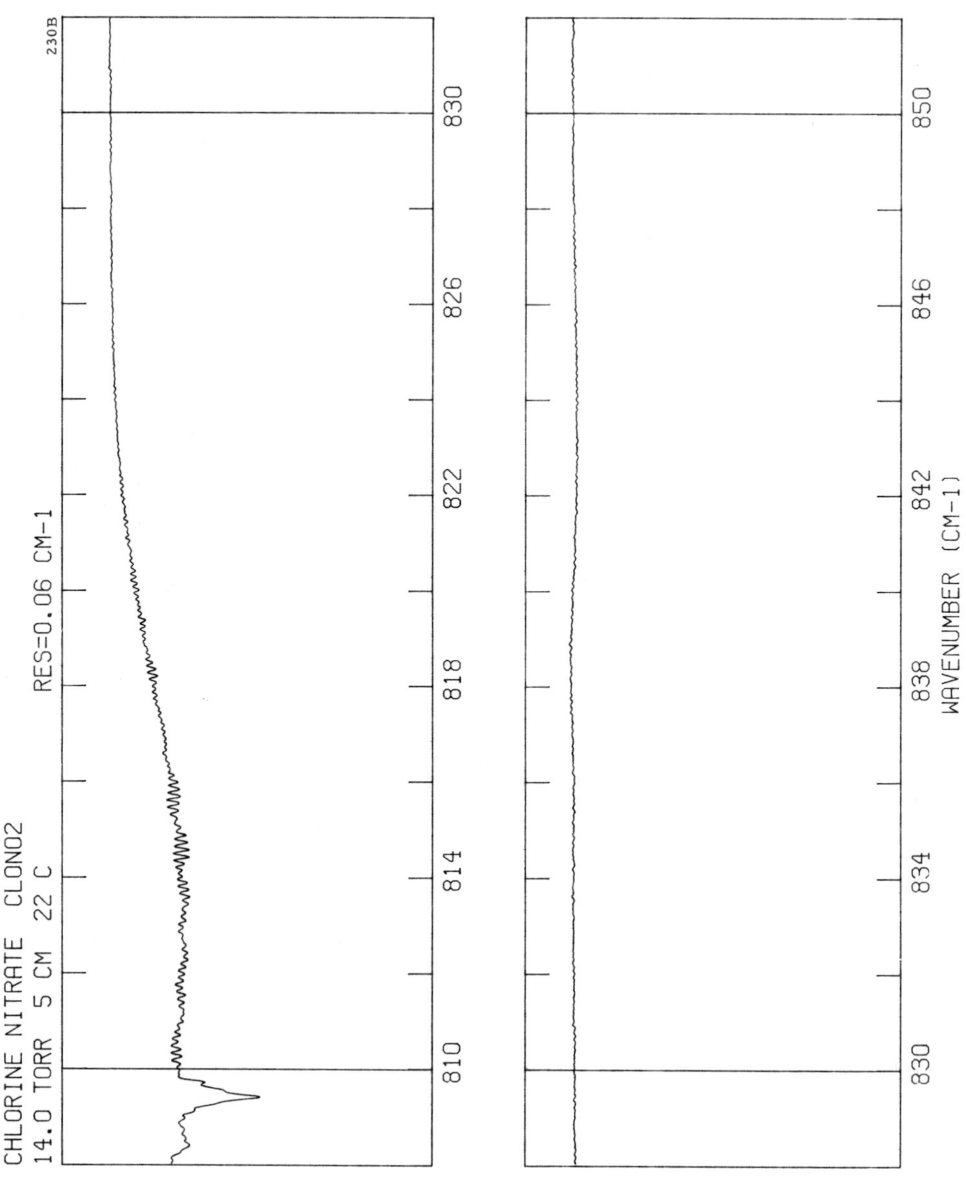
230B
CHLORINE NITRATE CLONO2
14.0 TORR 5 CM 22 C
RES=0.06 CM-1
810
814
818
822
826
830
830
834
838
842
846
850
WAVENUMBER (CM-1)

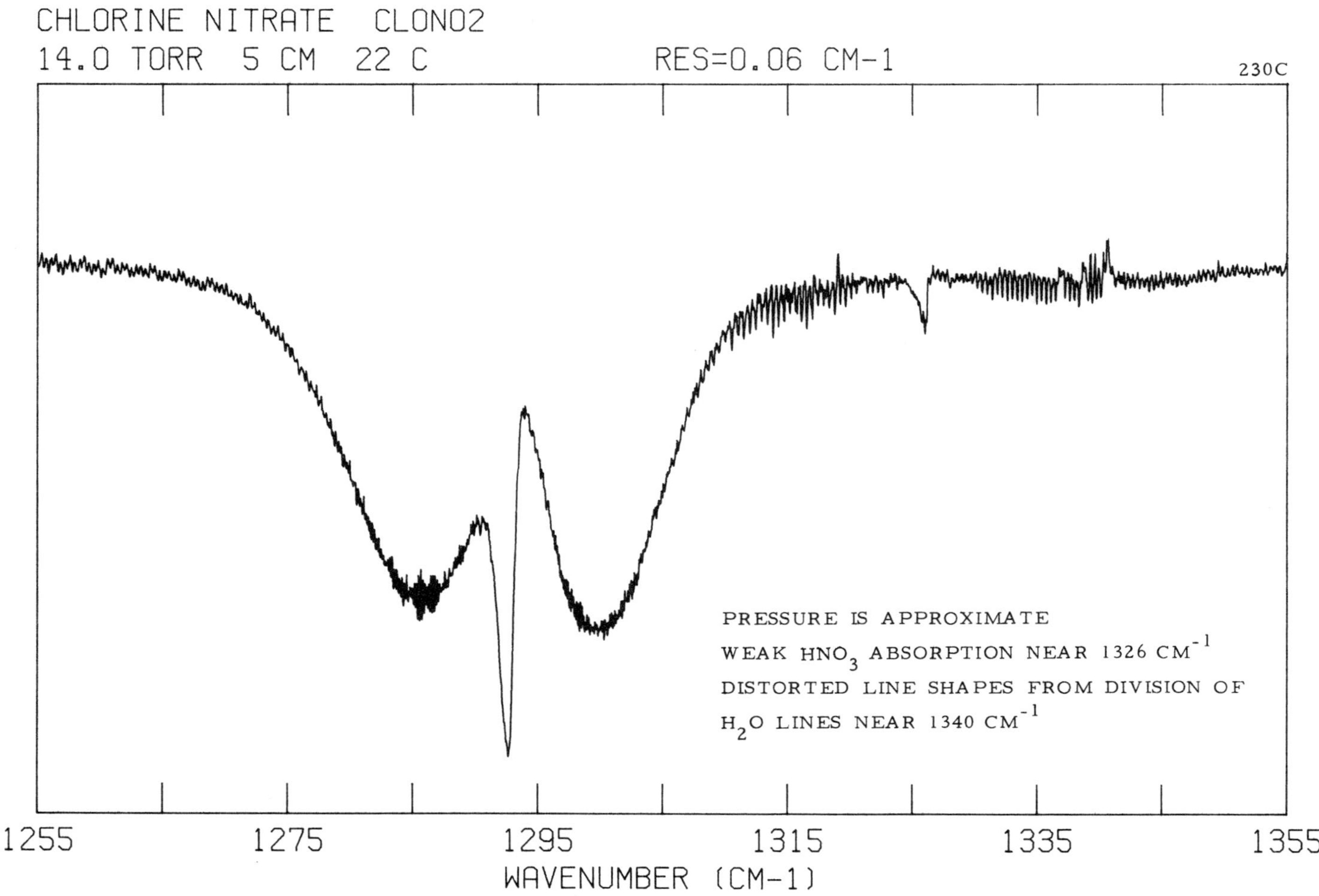
CHLORINE NITRATE CLONO2
14.0 TORR 5 CM 22 C
RES=0.06 CM-1
230C
PRESSURE IS APPROXIMATE
WEAK HNO_3 ABSORPTION NEAR 1326 CM^{-1}
DISTORTED LINE SHAPES FROM DIVISION OF
H_2O LINES NEAR 1340 CM^{-1}
1255
1275
1295
1315
1335
1355
WAVENUMBER (CM-1)

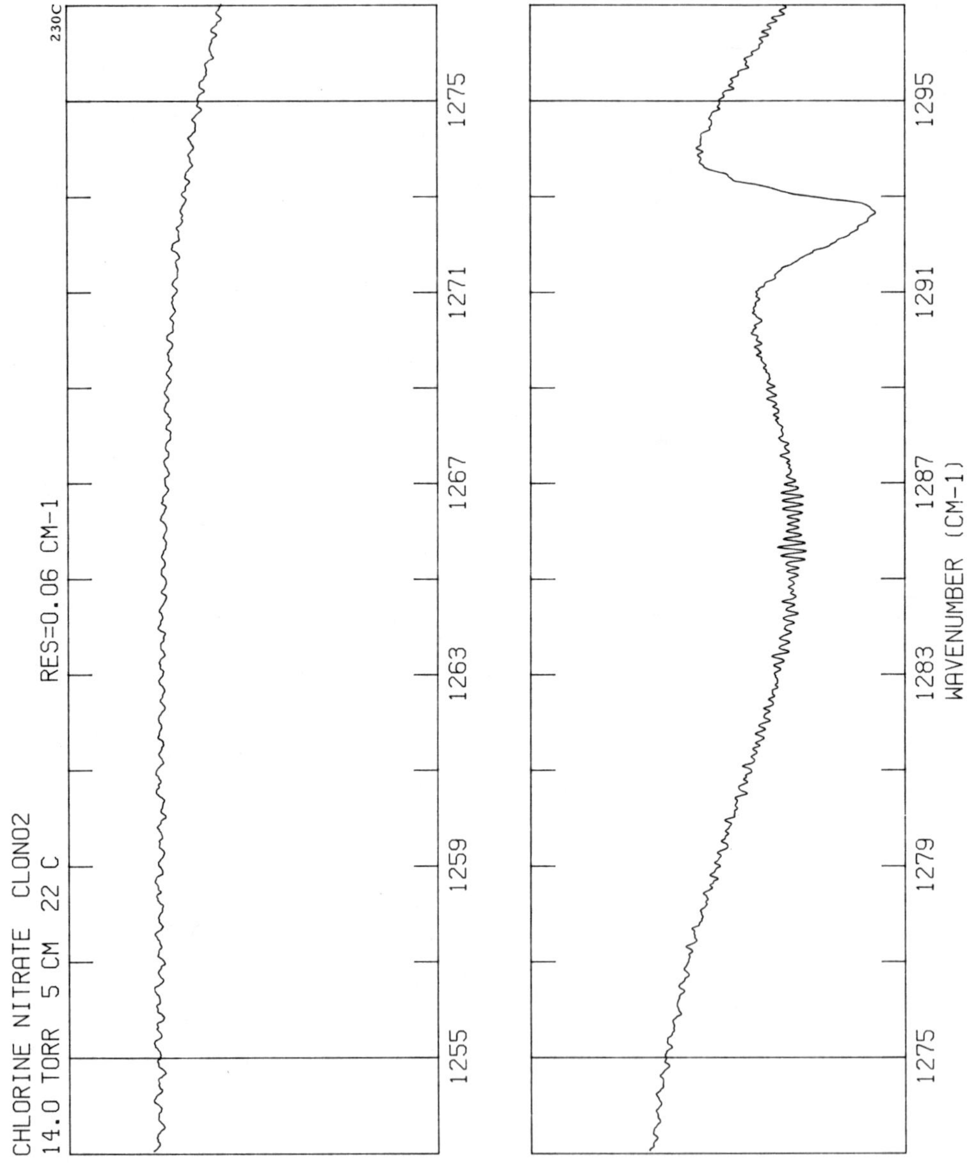
CHLORINE NITRATE CLONO2
14.0 TORR 5 CM 22 C
RES=0.06 CM-1
230C
1255
1259
1263
1267
1271
1275
1275
1279
1283
1287
1291
1295
WAVENUMBER (CM-1)

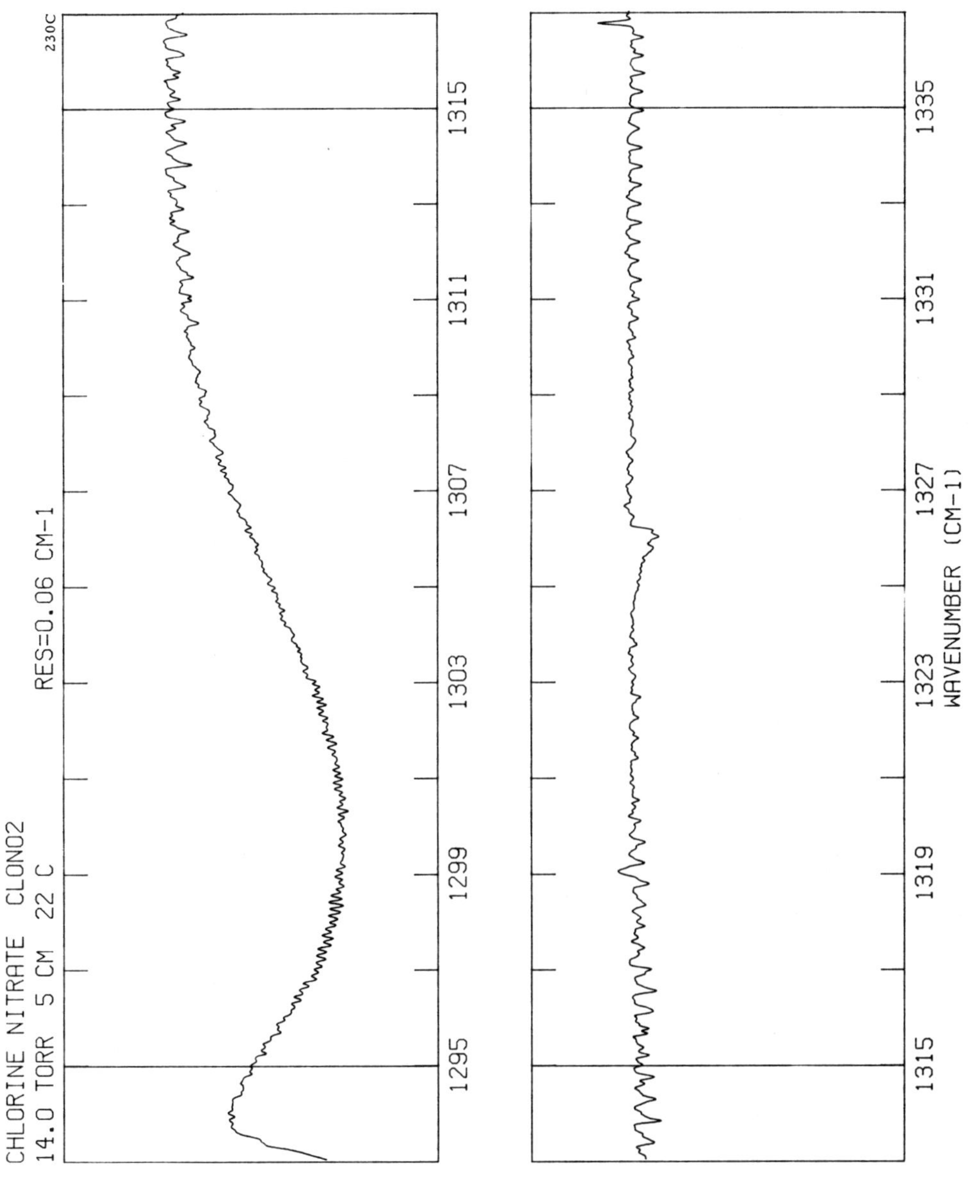
CHLORINE NITRATE CLONO2
14.0 TORR 5 CM 22 C
RES=0.06 CM-1
230C
1295
1299
1303
1307
1311
1315
1315
1319
1323
1327
1331
1335
WAVENUMBER (CM-1)

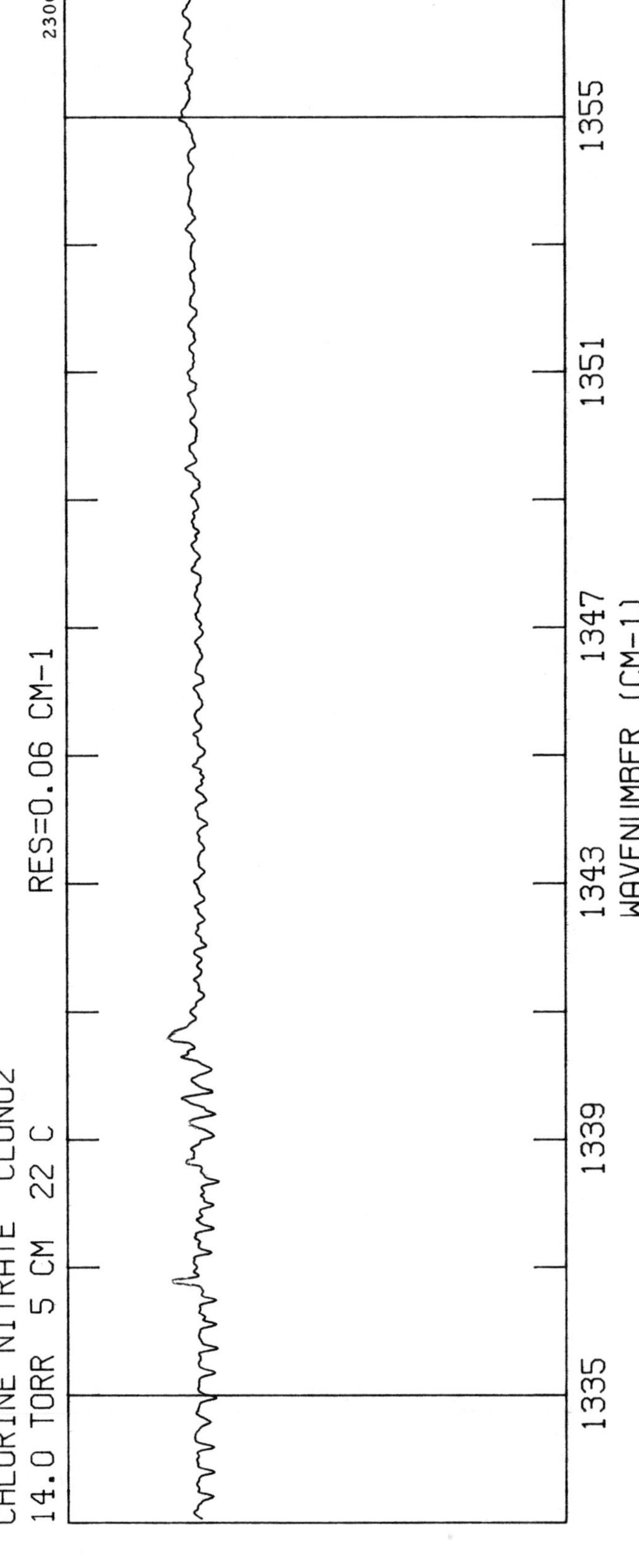
CHLORINE NITRATE CLONO2
14.0 TORR 5 CM 22 C
RES=0.06 CM-1
230C
1335
1339
1343
1347
1351
1355
WAVENUMBER (CM-1)

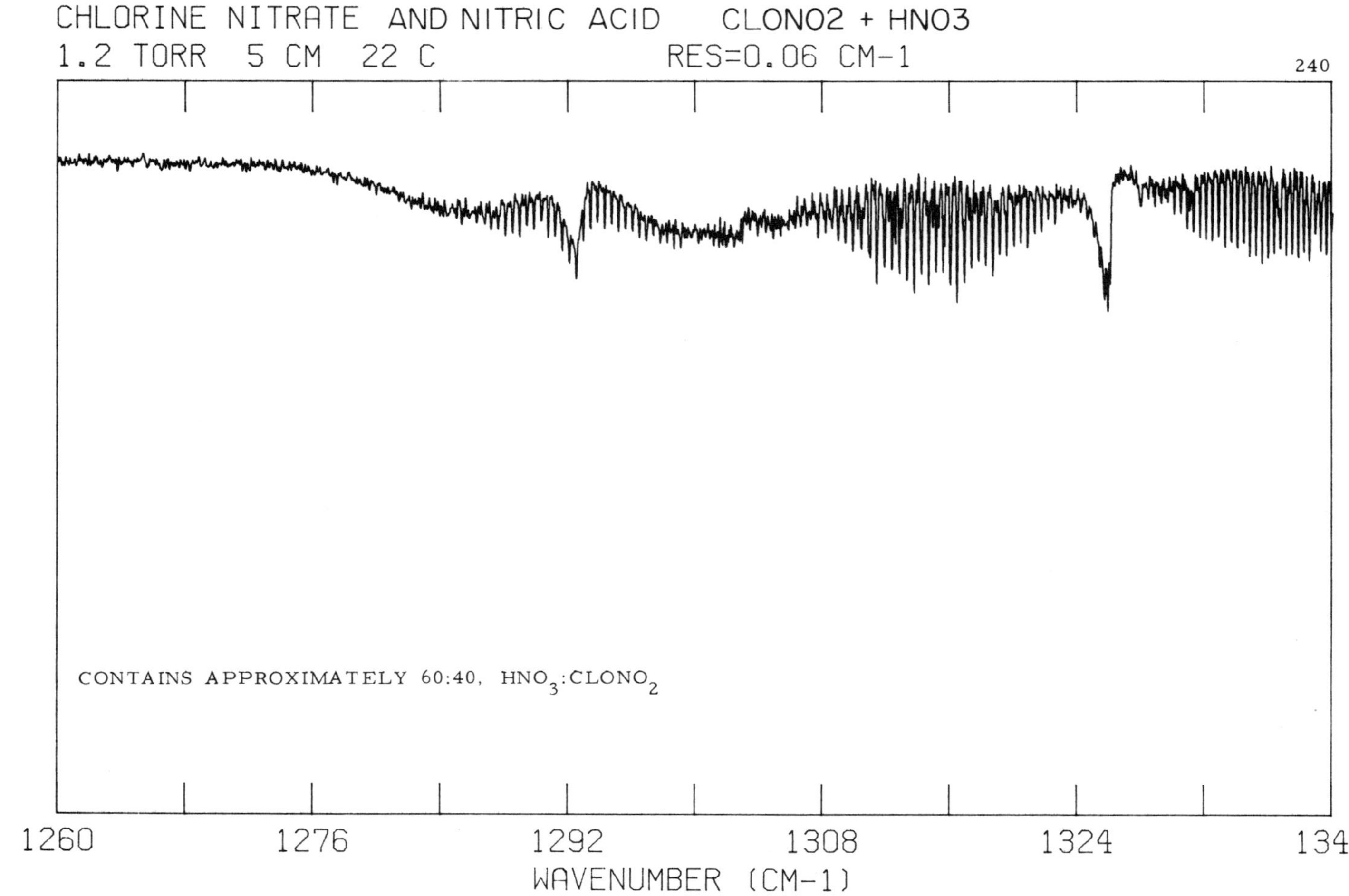
CHLORINE NITRATE AND NITRIC ACID CLONO2 + HNO3
1.2 TORR 5 CM 22 C RES=0.06 CM-1
240
CONTAINS APPROXIMATELY 60:40, HNO3:CLONO2
1260
1276
1292
1308
1324
1340
WAVENUMBER (CM-1)

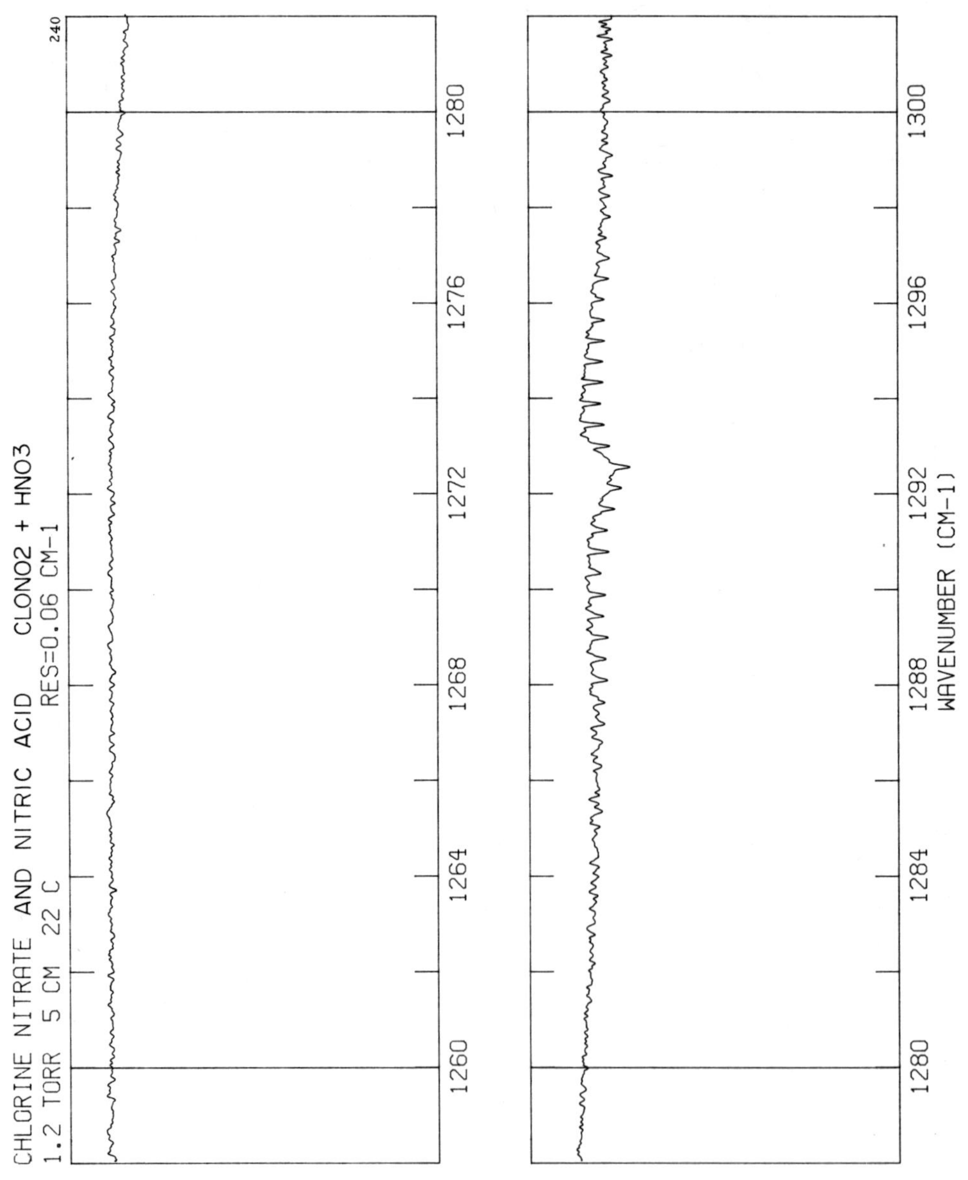
CHLORINE NITRATE AND NITRIC ACID CLONO2 + HNO3
1.2 TORR 5 CM 22 C RES=0.06 CM-1
240
1260
1264
1268
1272
1276
1280
1284
1288
1292
1296
1300
WAVENUMBER (CM-1)

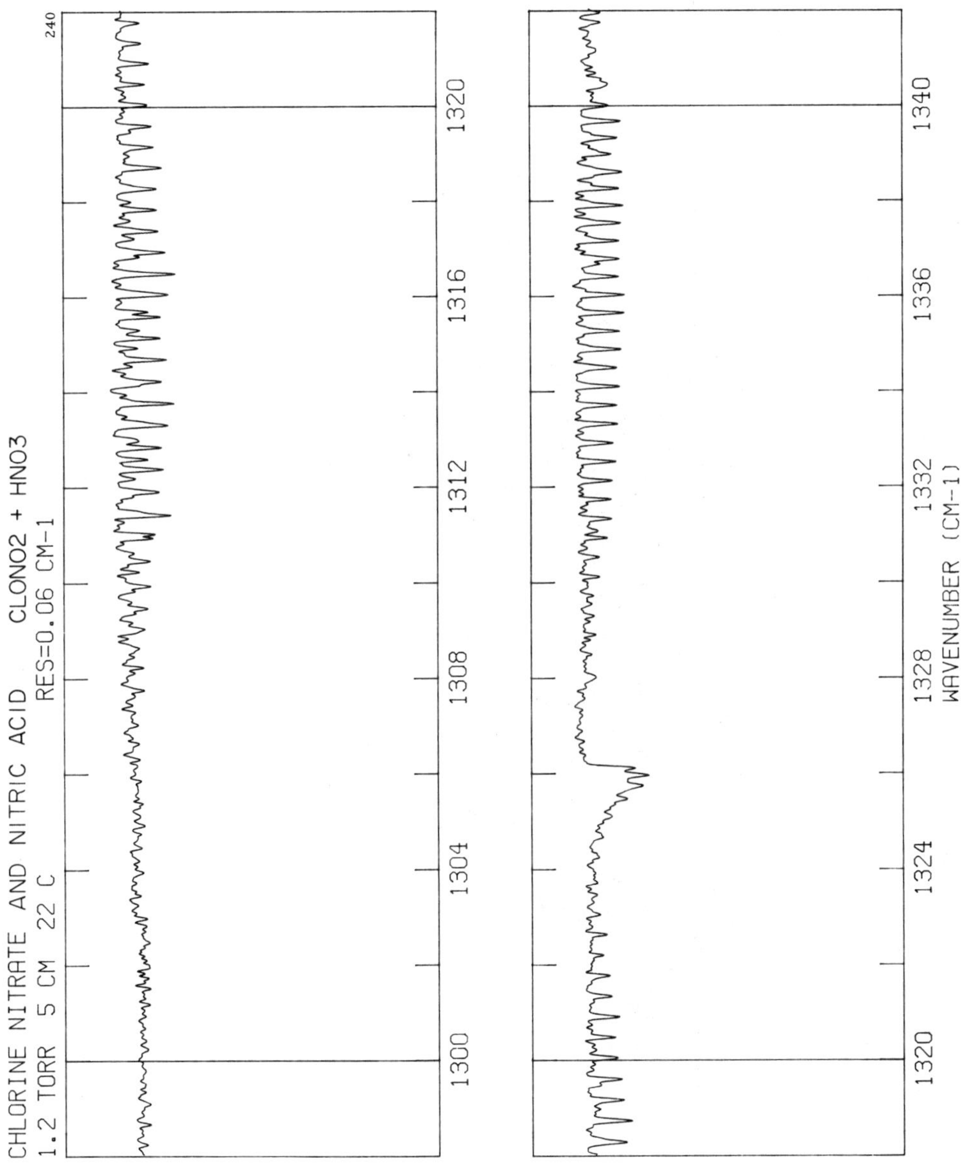

CHLORINE NITRATE AND NITRIC ACID CLONO2 + HNO3
1.2 TORR 5 CM 22 C RES=0.06 CM-1
240
1300
1304
1308
1312
1316
1320
1320
1324
1328
1332
1336
1340
WAVENUMBER (CM-1)

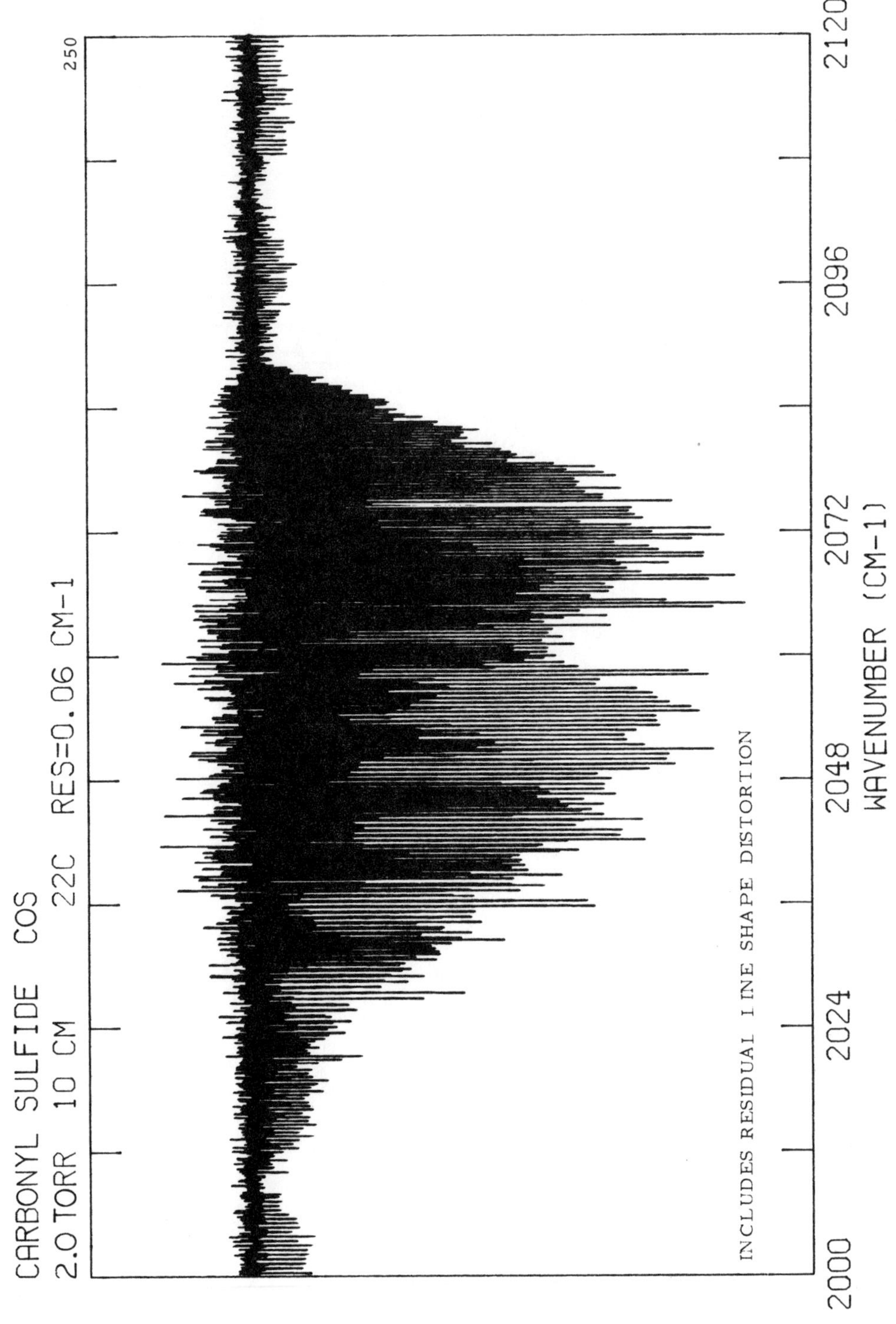
CARBONYL SULFIDE COS
2.0 TORR 10 CM 22C RES=0.06 CM-1
250
INCLUDES RESIDUAL I INE SHAPE DISTORTION
2000
2024
2048
2072
2096
2120
WAVENUMBER (CM-1)

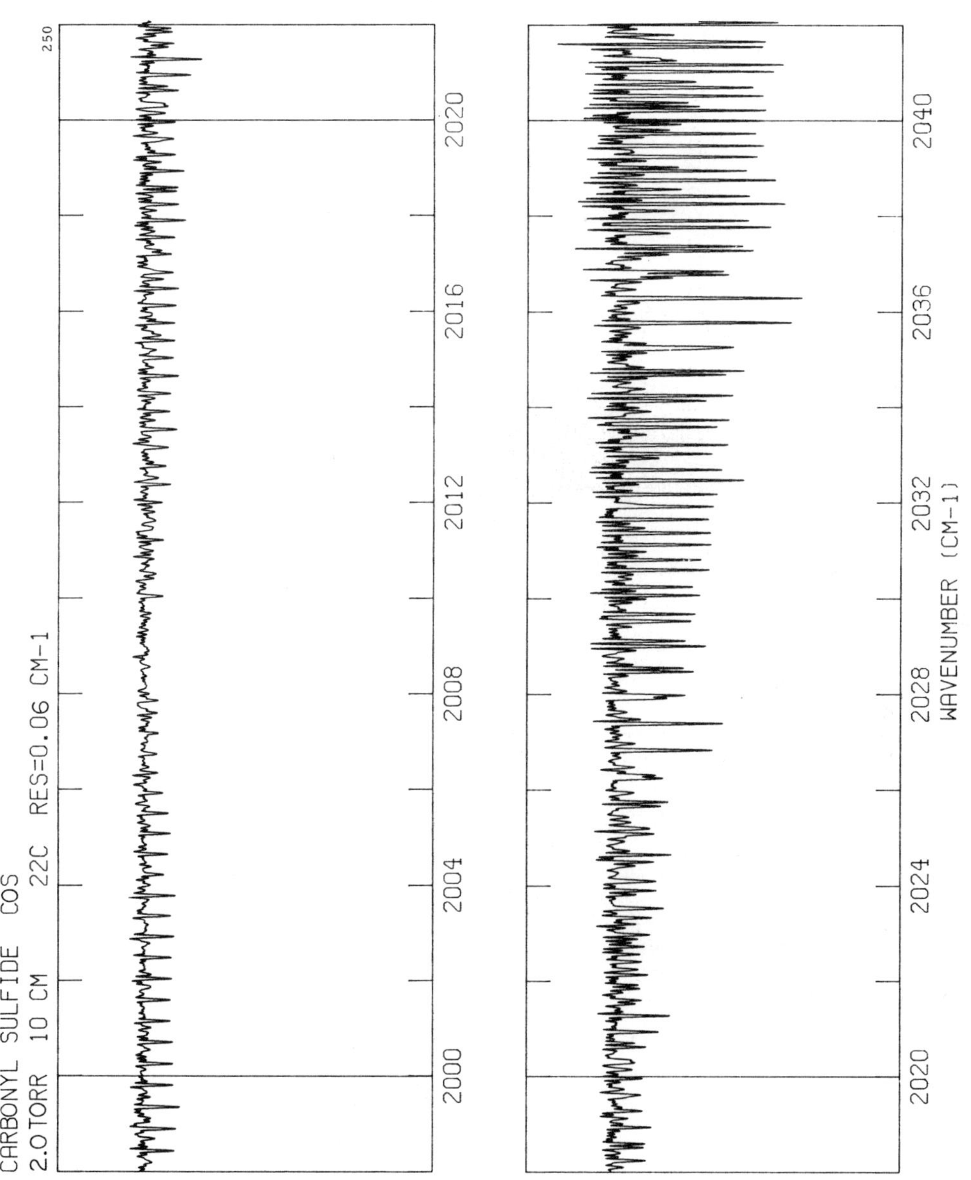

CARBONYL SULFIDE COS
2.0 TORR 10 CM 22C RES=0.06 CM-1
250
2000
2004
2008
2012
2016
2020
2024
2028
2032
2036
2040
WAVENUMBER (CM-1)

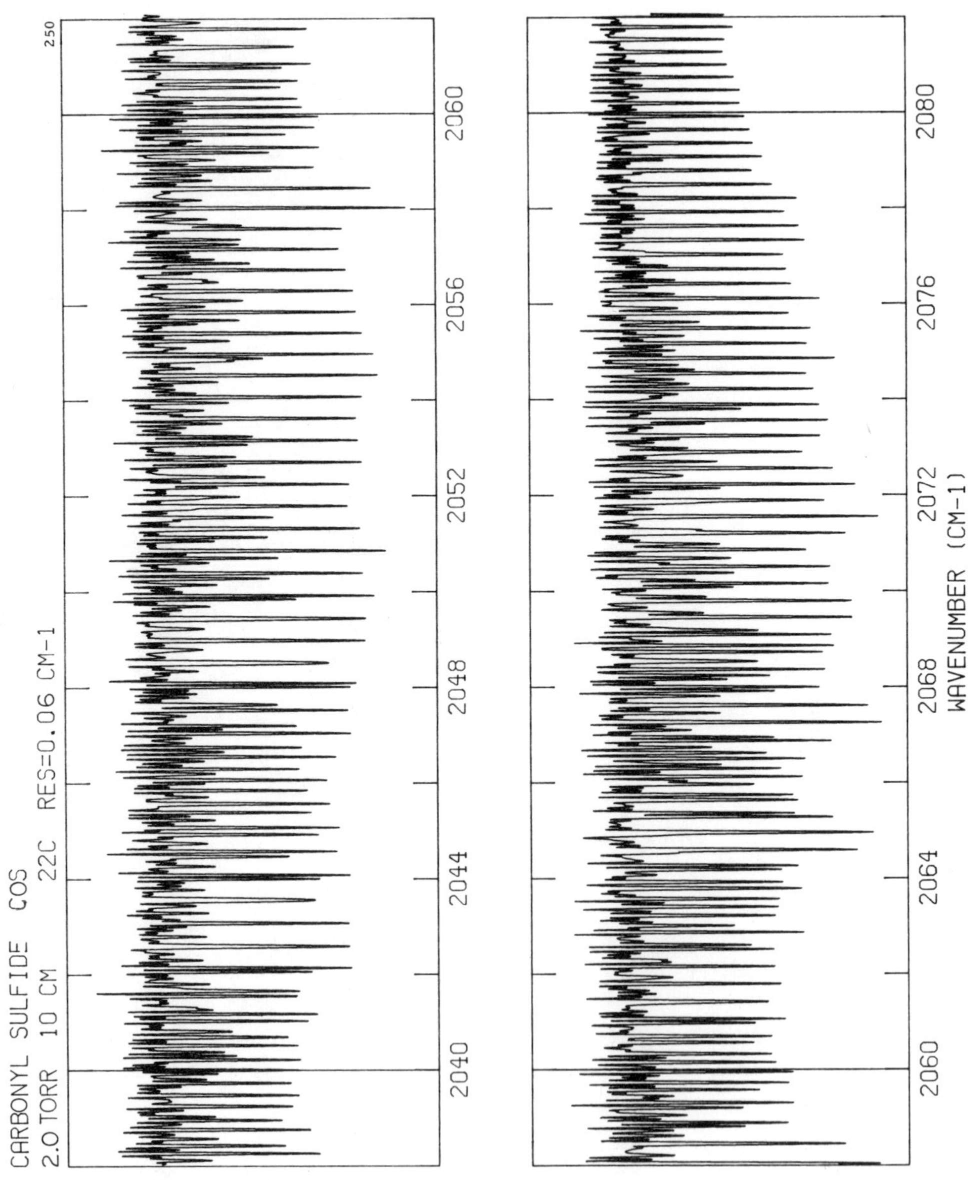
CARBONYL SULFIDE COS
2.0 TORR 10 CM 22C RES=0.06 CM-1
250
2040
2044
2048
2052
2056
2060
2060
2064
2068
2072
2076
2080
WAVENUMBER (CM-1)

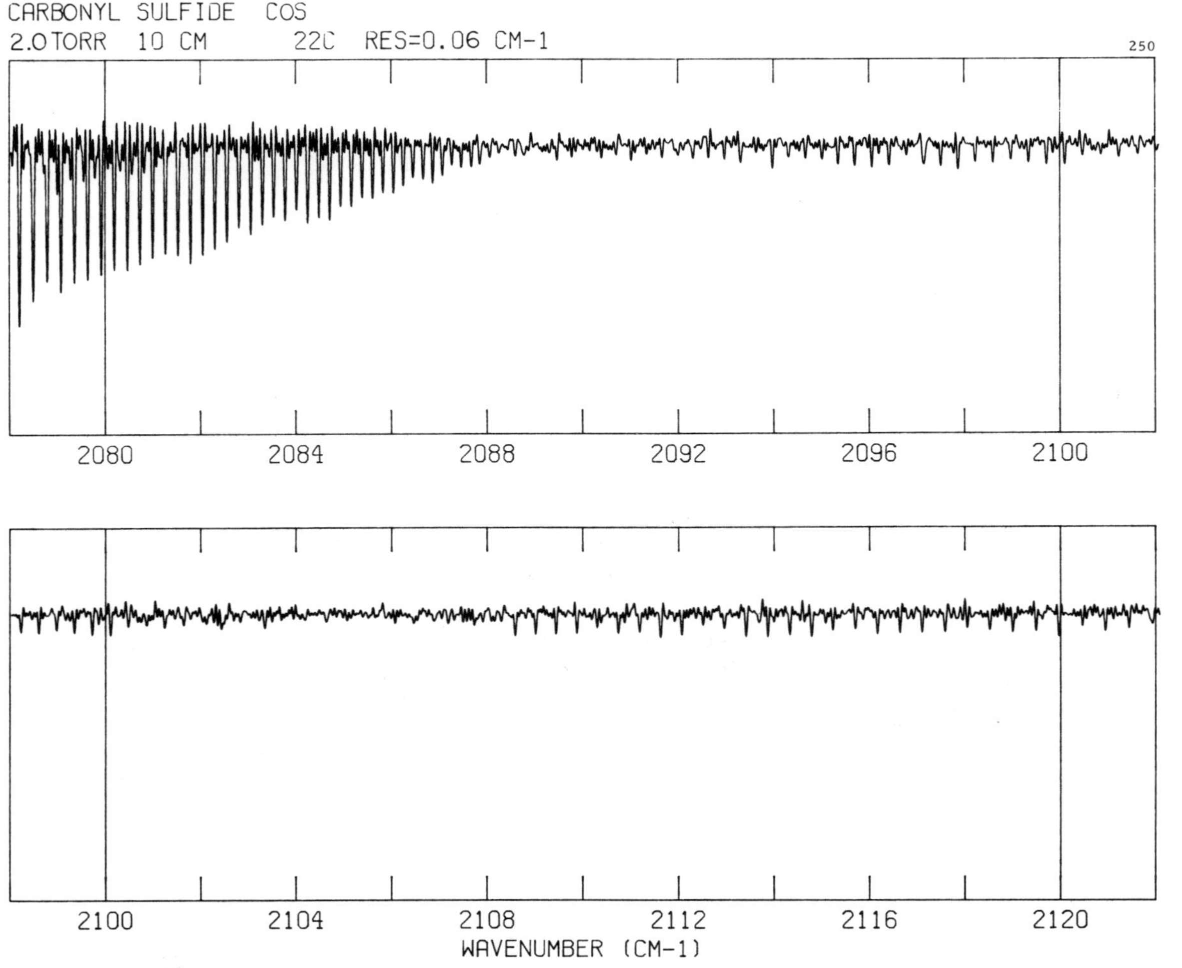
CARBONYL SULFIDE COS
2.0 TORR 10 CM 22C RES=0.06 CM-1
250
2080
2084
2088
2092
2096
2100
2100
2104
2108
2112
2116
2120
WAVENUMBER (CM-1)

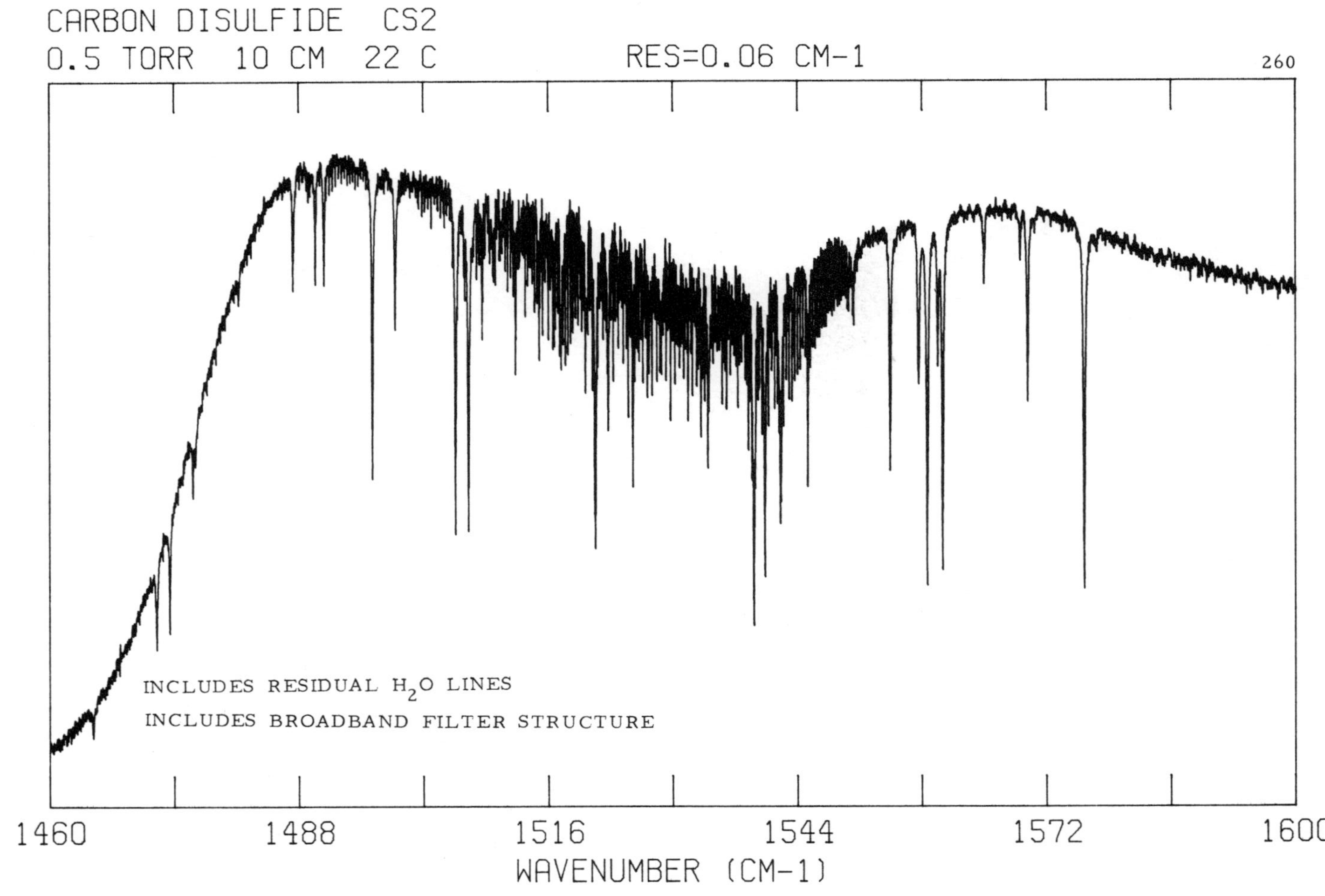
CARBON DISULFIDE CS2
0.5 TORR 10 CM 22 C
RES=0.06 CM-1
260
INCLUDES RESIDUAL H_2O LINES
INCLUDES BROADBAND FILTER STRUCTURE
1460
1488
1516
1544
1572
1600
WAVENUMBER (CM-1)

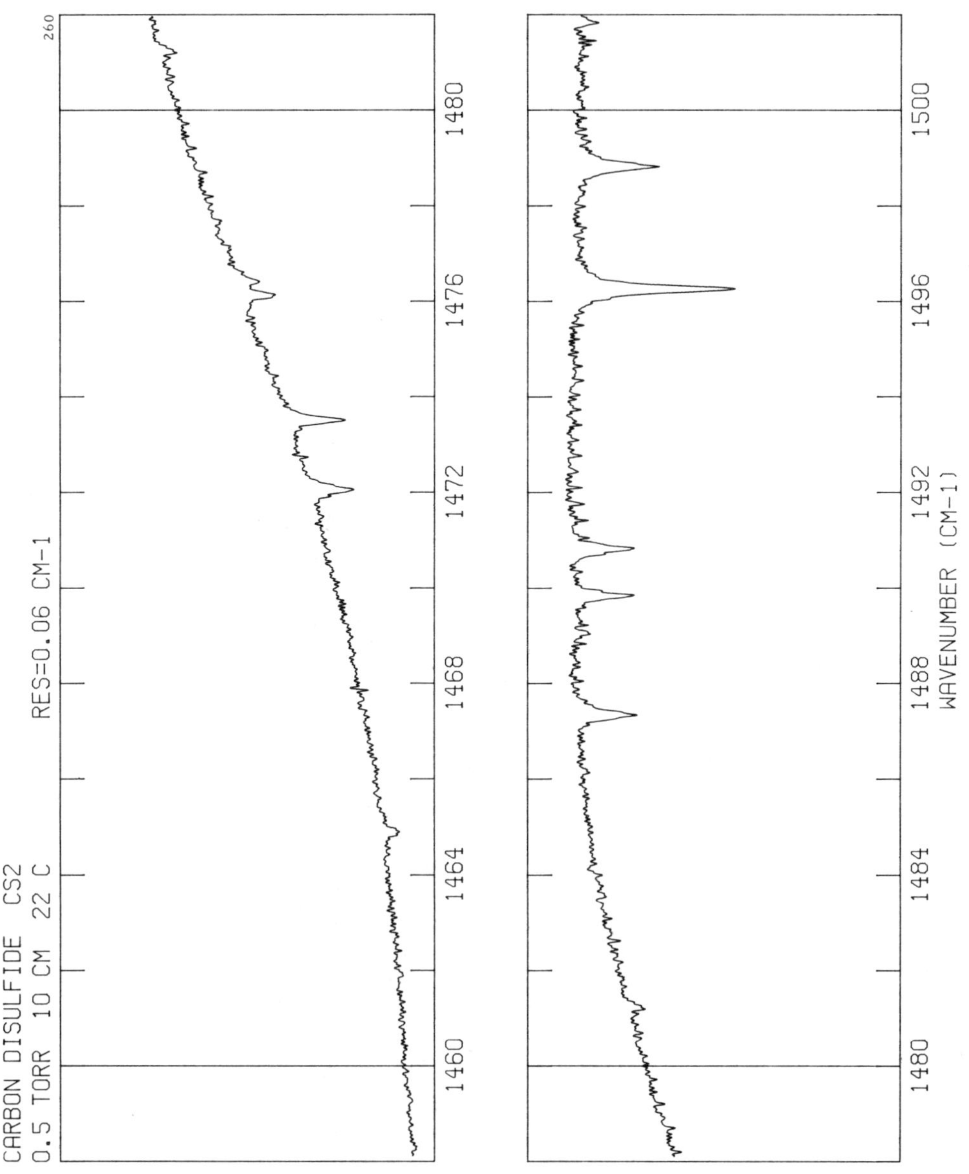
CARBON DISULFIDE CS2
0.5 TORR 10 CM 22 C
RES=0.06 CM-1
260
1460
1464
1468
1472
1476
1480
1480
1484
1488
1492
1496
1500
WAVENUMBER (CM-1)

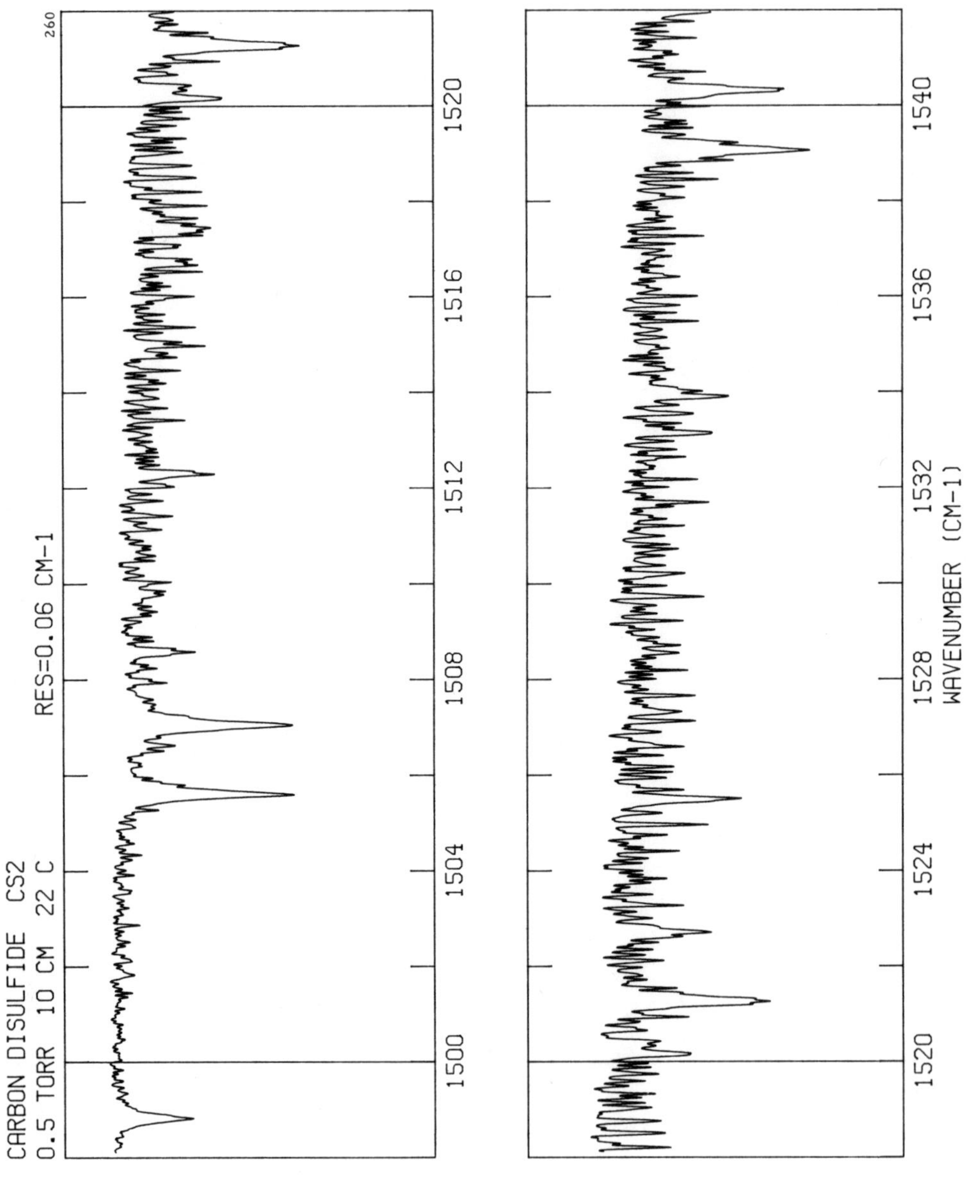
CARBON DISULFIDE CS2
0.5 TORR 10 CM 22 C
RES=0.06 CM-1
260
1500
1504
1508
1512
1516
1520
1520
1524
1528
1532
1536
1540
WAVENUMBER (CM-1)

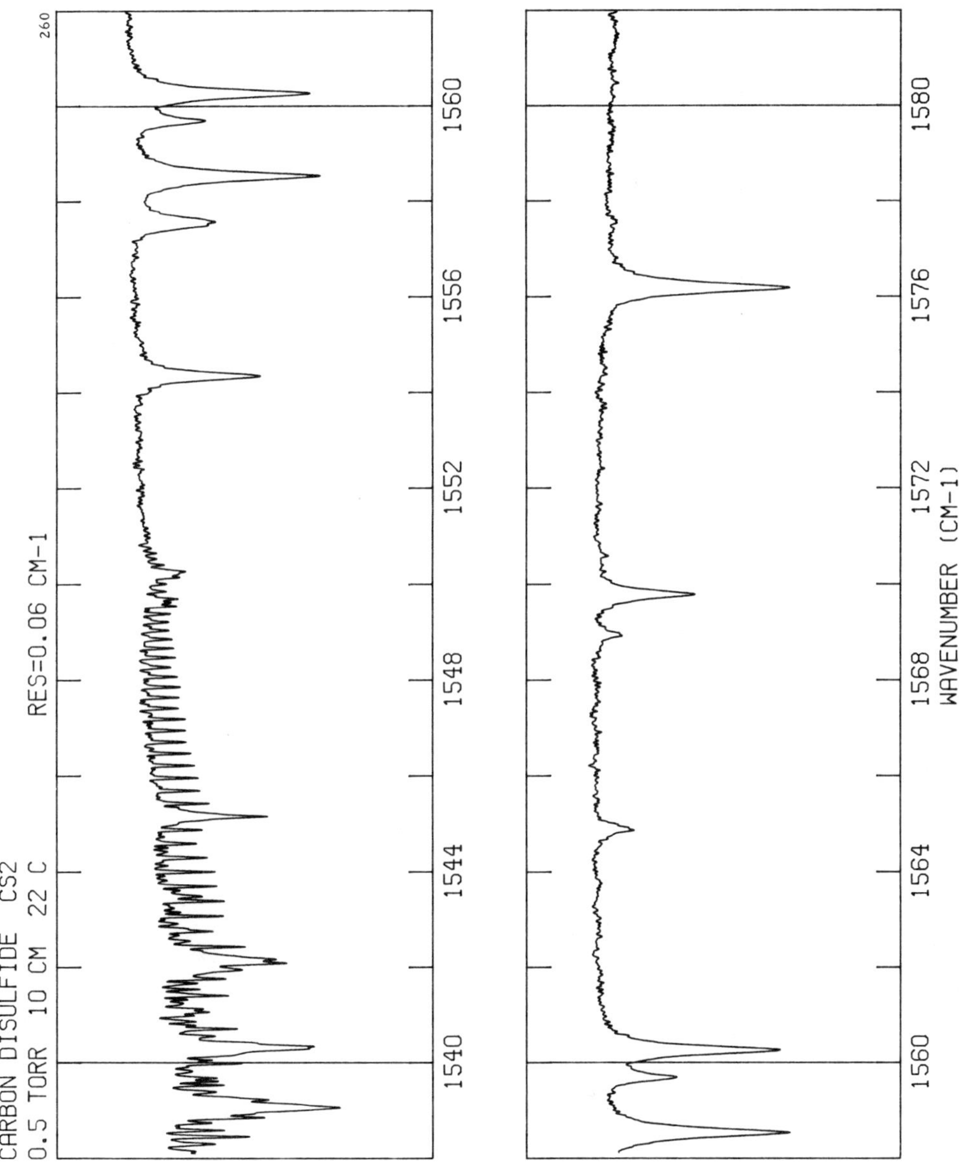
CARBON DISULFIDE CS2
0.5 TORR 10 CM 22 C
RES=0.06 CM-1
260
1540
1544
1548
1552
1556
1560
1560
1564
1568
1572
1576
1580
WAVENUMBER (CM-1)

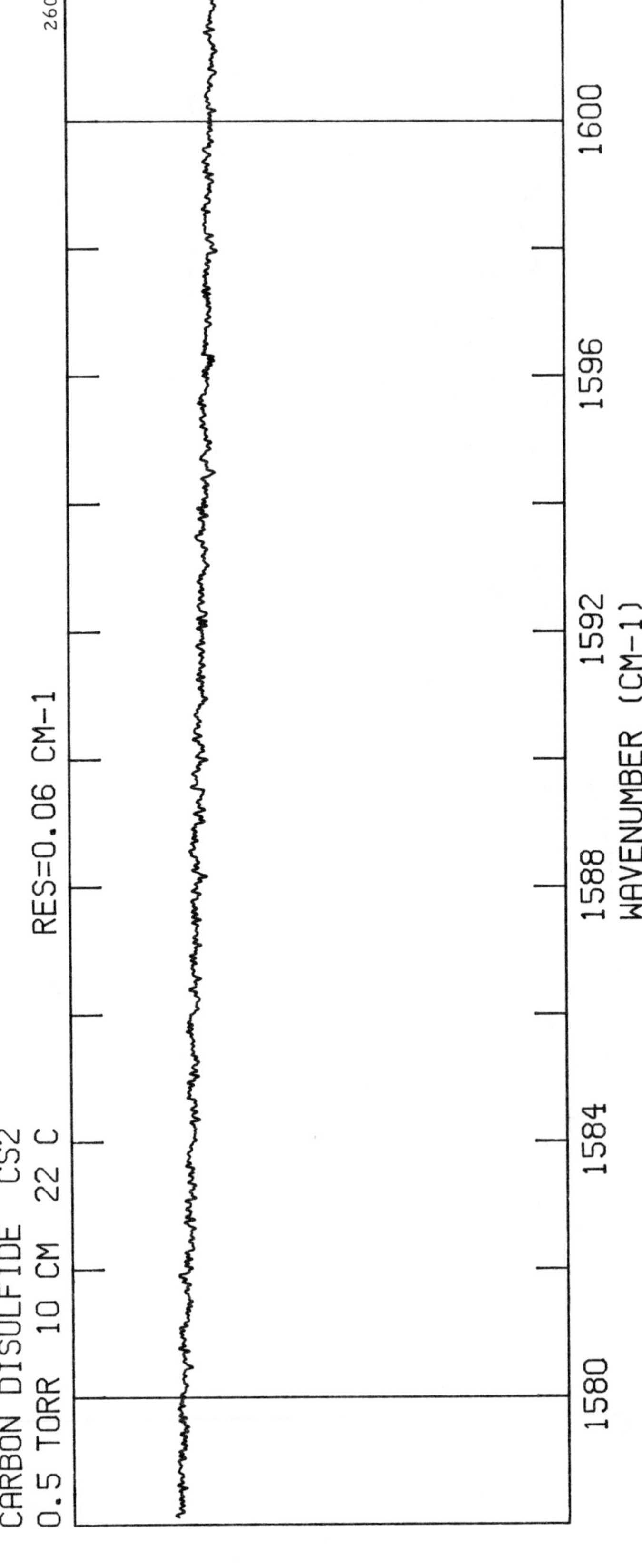
CARBON DISULFIDE CS2
0.5 TORR 10 CM 22 C
RES=0.06 CM-1
260
1580
1584
1588
1592
1596
1600
WAVENUMBER (CM-1)

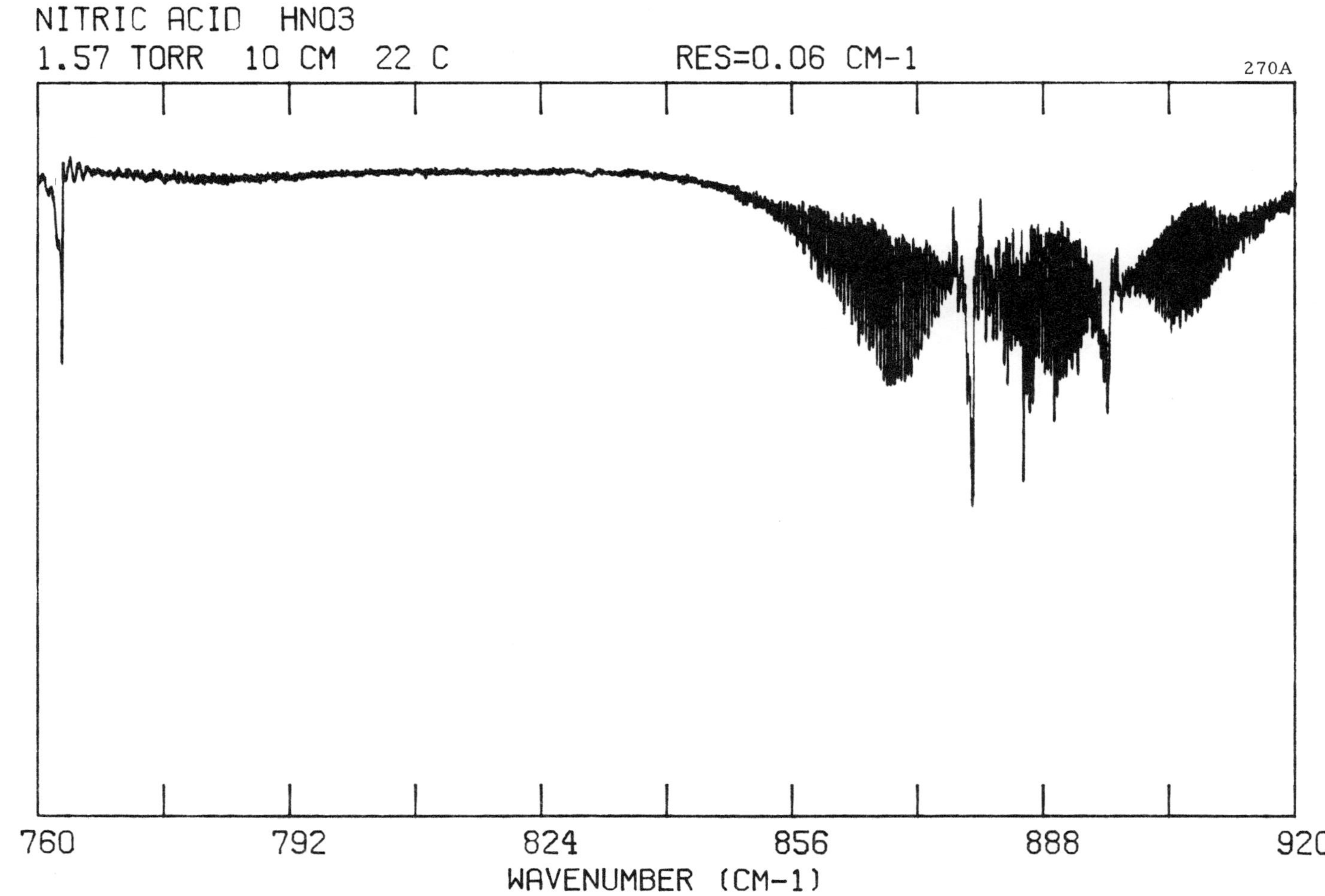

NITRIC ACID HNO3
1.57 TORR 10 CM 22 C
RES=0.06 CM-1
270A
760
792
824
856
888
920
WAVENUMBER (CM-1)

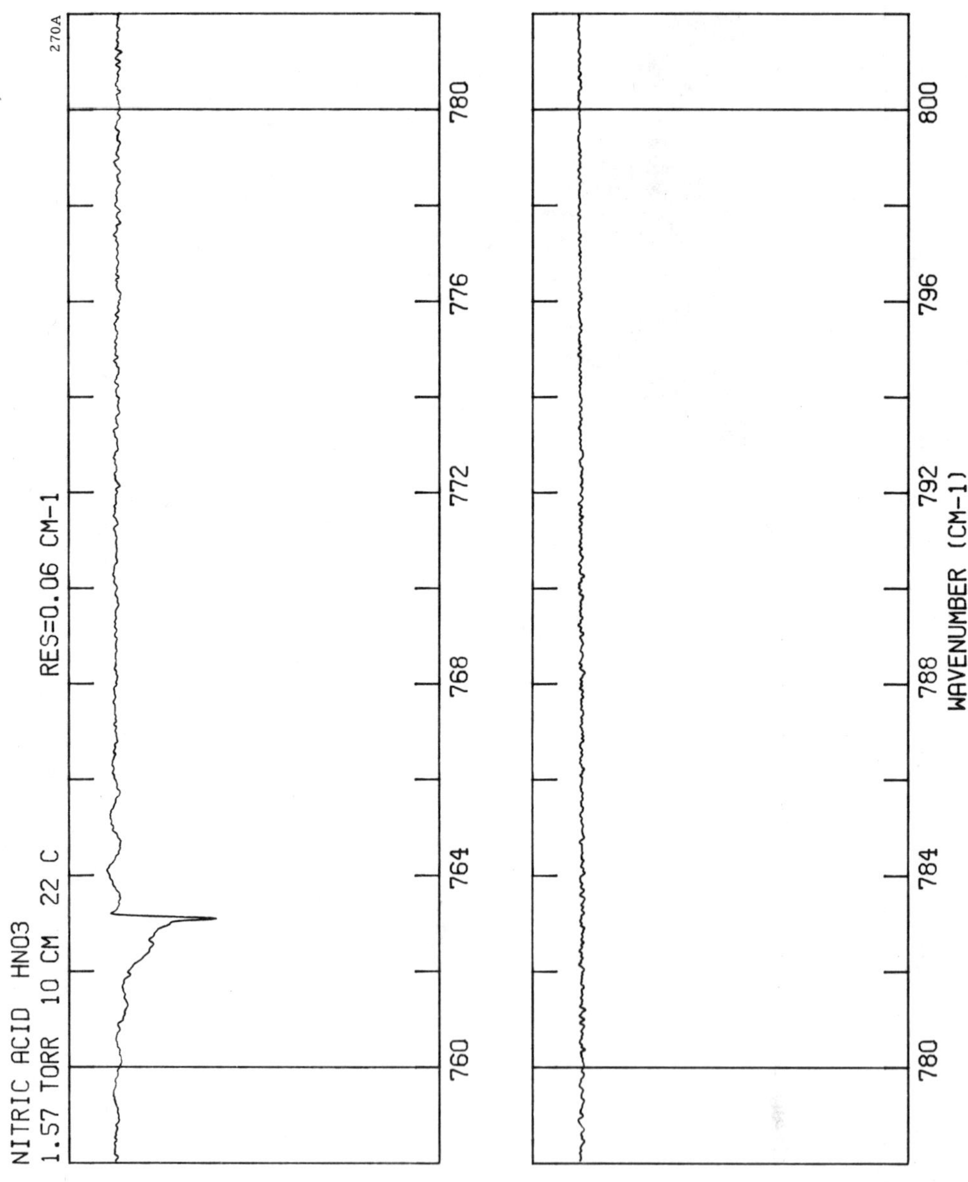
NITRIC ACID HNO3
1.57 TORR 10 CM 22 C
RES=0.06 CM-1
270A
760
764
768
772
776
780
780
784
788
792
796
800
WAVENUMBER (CM-1)

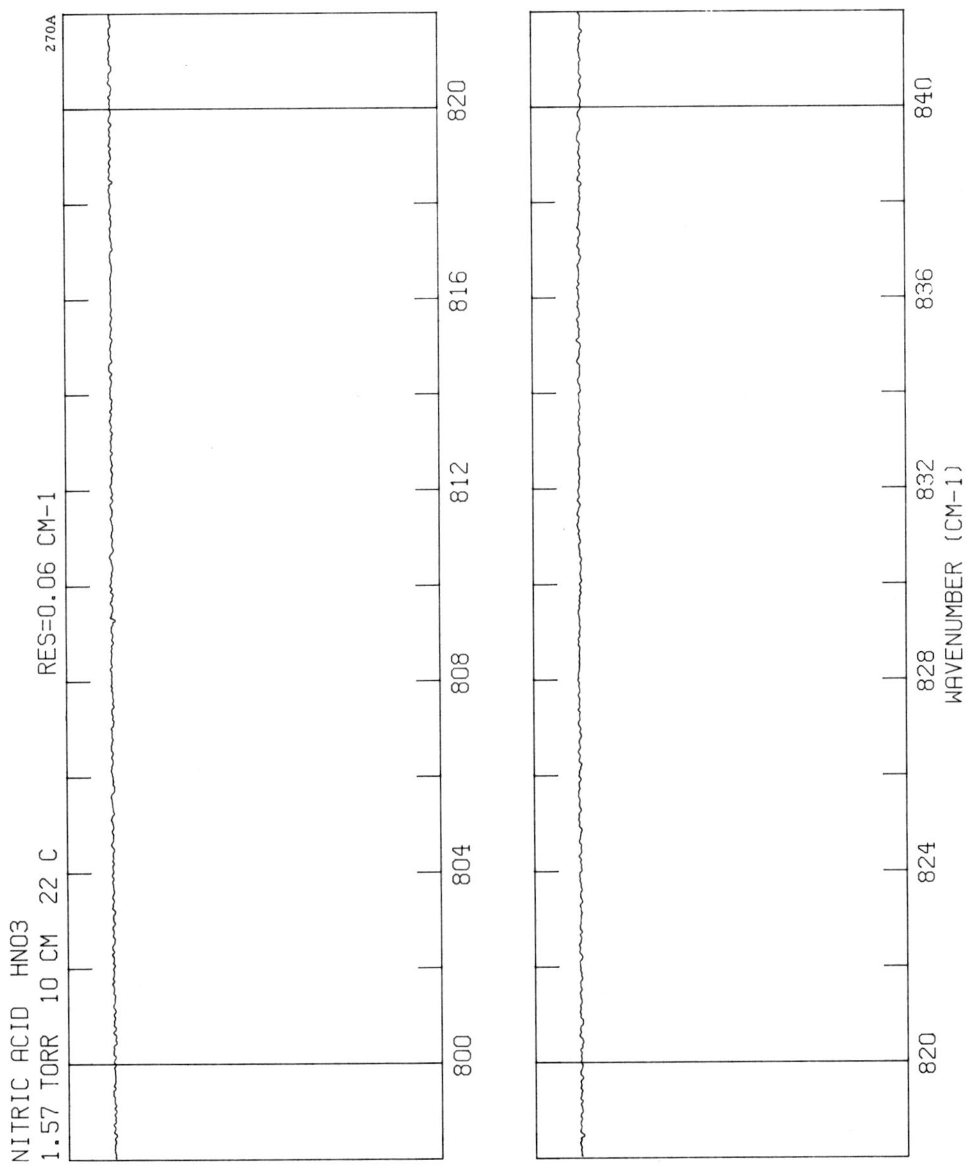
NITRIC ACID HNO3
1.57 TORR 10 CM 22 C
RES=0.06 CM-1
270A
800
804
808
812
816
820
820
824
828
832
836
840
WAVENUMBER (CM-1)

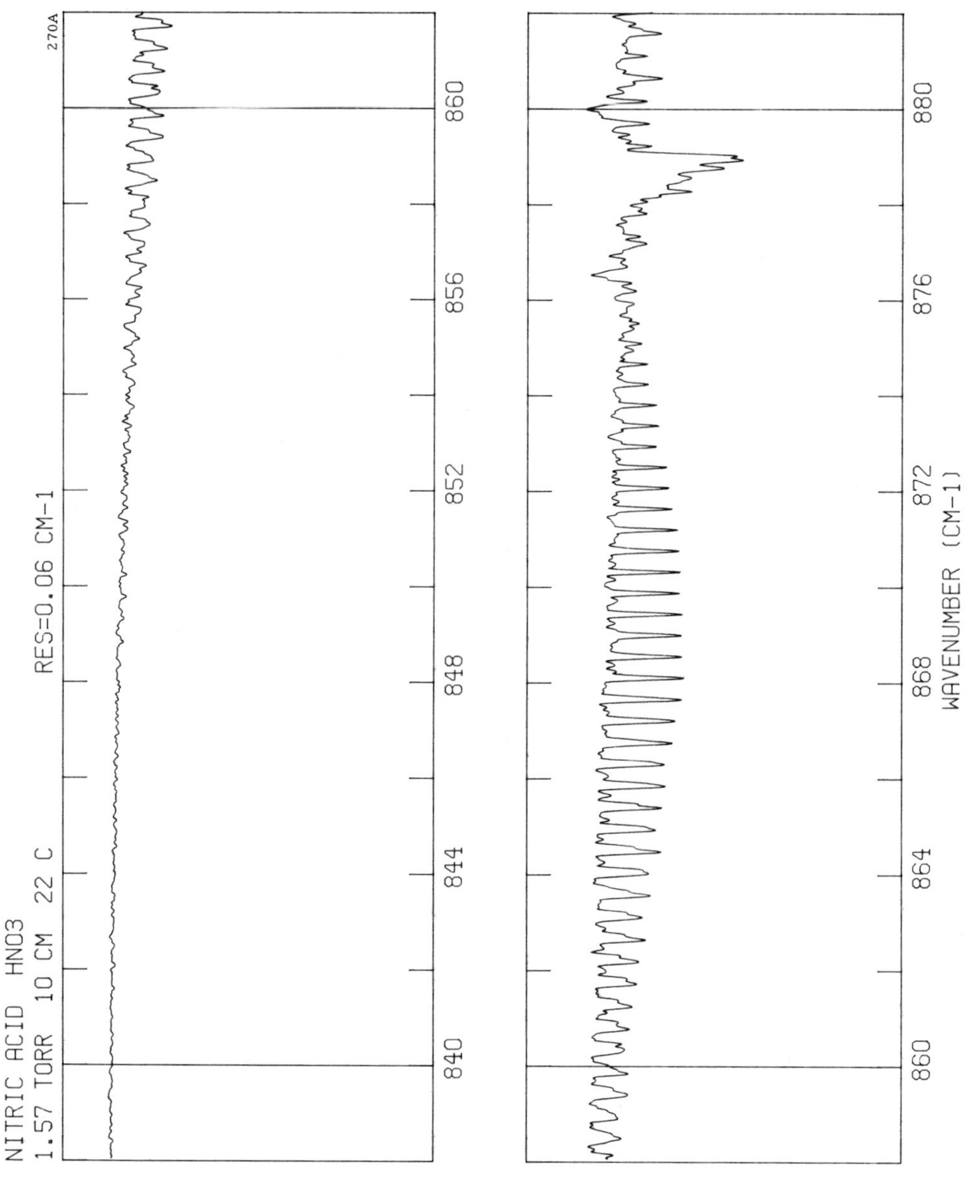
NITRIC ACID HNO3
1.57 TORR 10 CM 22 C
RES=0.06 CM-1
270A
840
844
848
852
856
860
860
864
868
872
876
880
WAVENUMBER (CM-1)

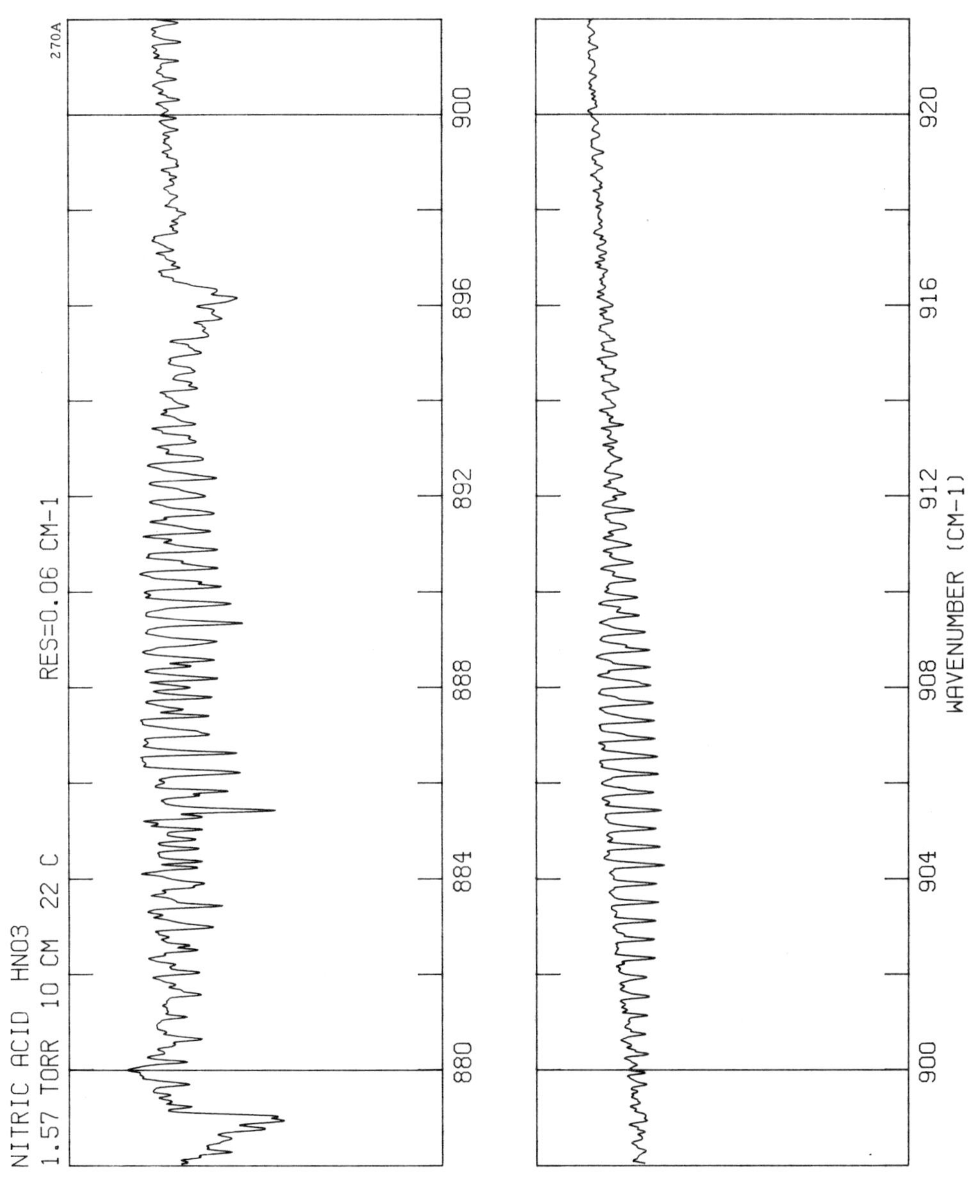
NITRIC ACID HNO3
1.57 TORR 10 CM 22 C
RES=0.06 CM-1
270A
880
884
888
892
896
900
904
908
912
916
920
WAVENUMBER (CM-1)

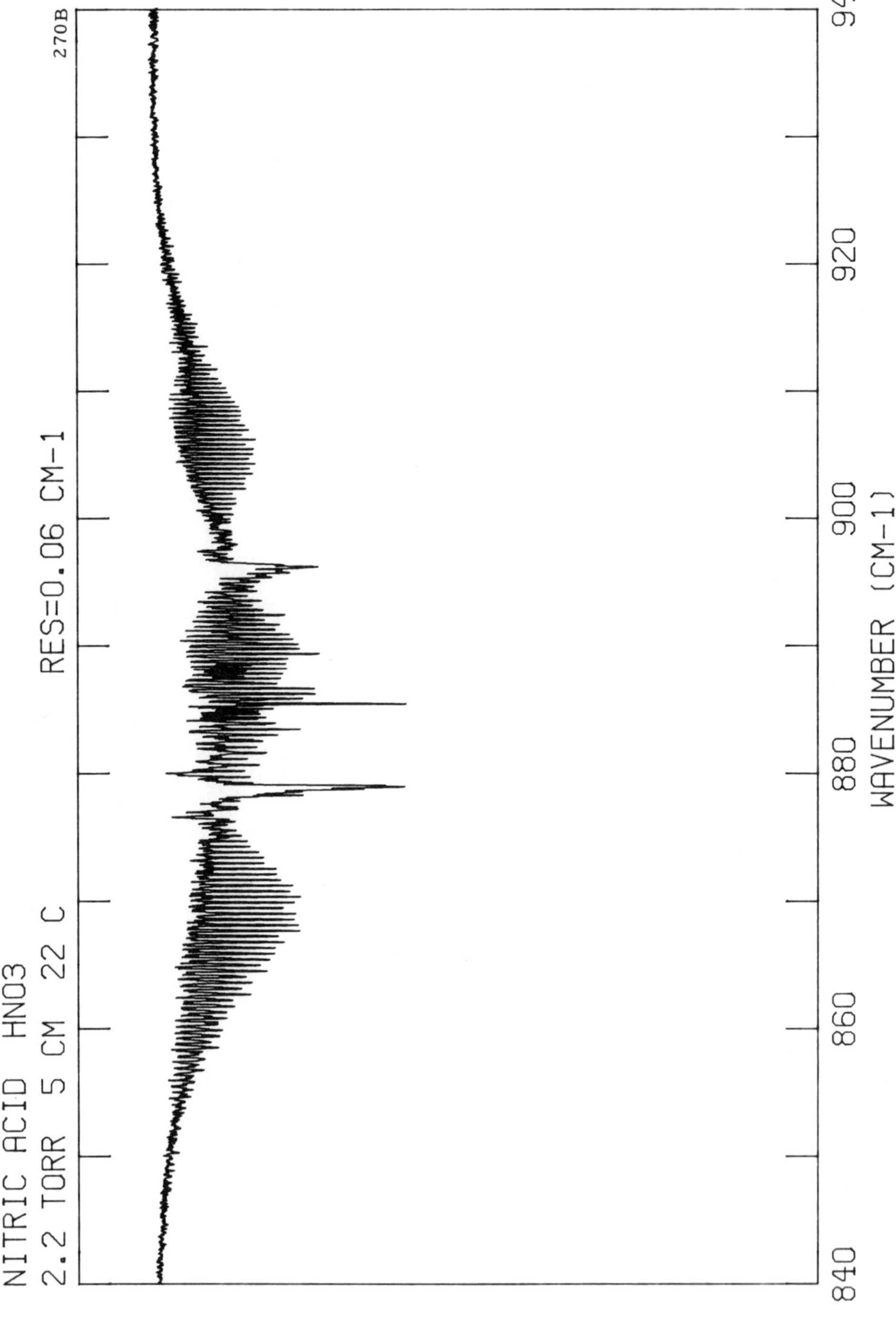
NITRIC ACID HNO3
2.2 TORR 5 CM 22 C
RES=0.06 CM-1
270B
840
860
880
900
920
940
WAVENUMBER (CM-1)

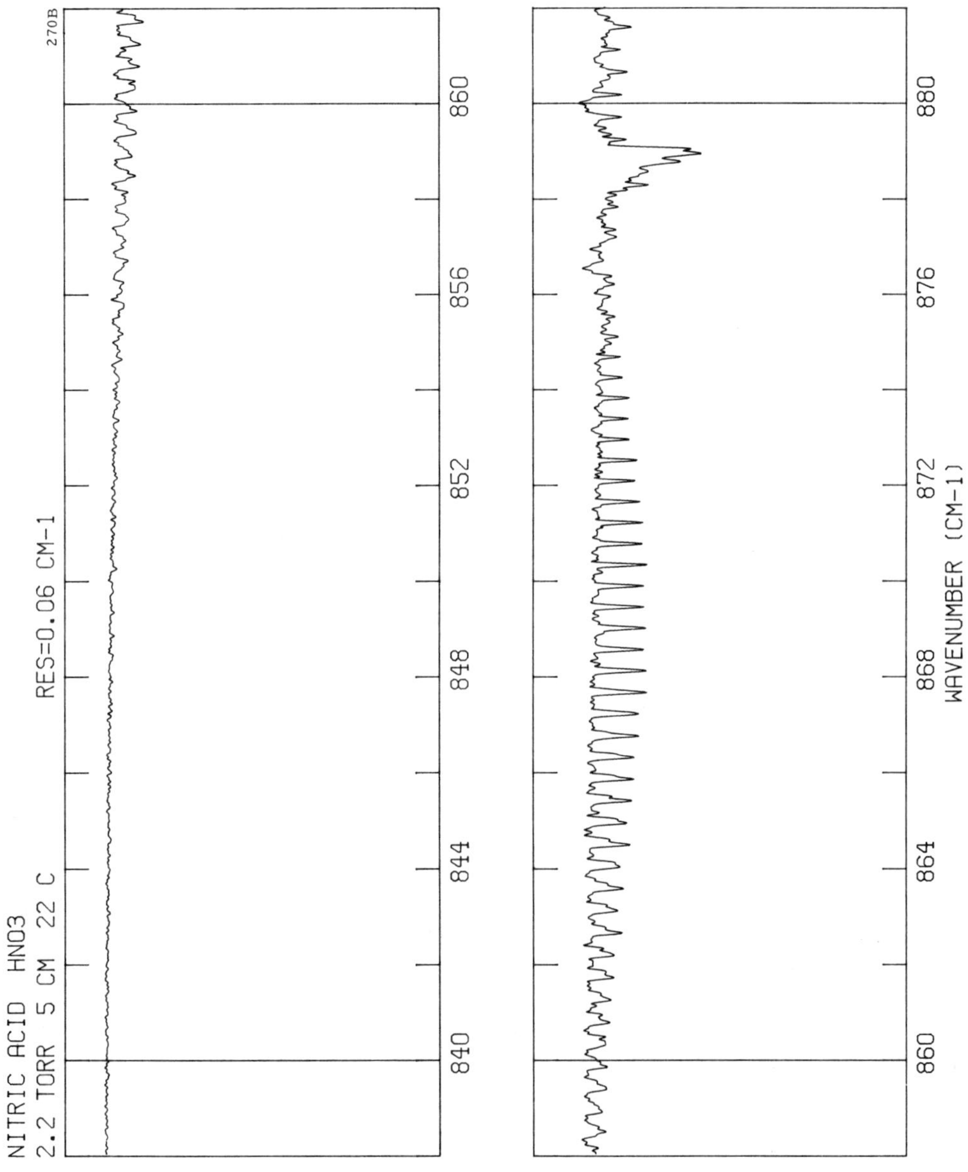

NITRIC ACID HNO3
2.2 TORR 5 CM 22 C
RES=0.06 CM-1
270B
840
844
848
852
856
860
860
864
868
872
876
880
WAVENUMBER (CM-1)

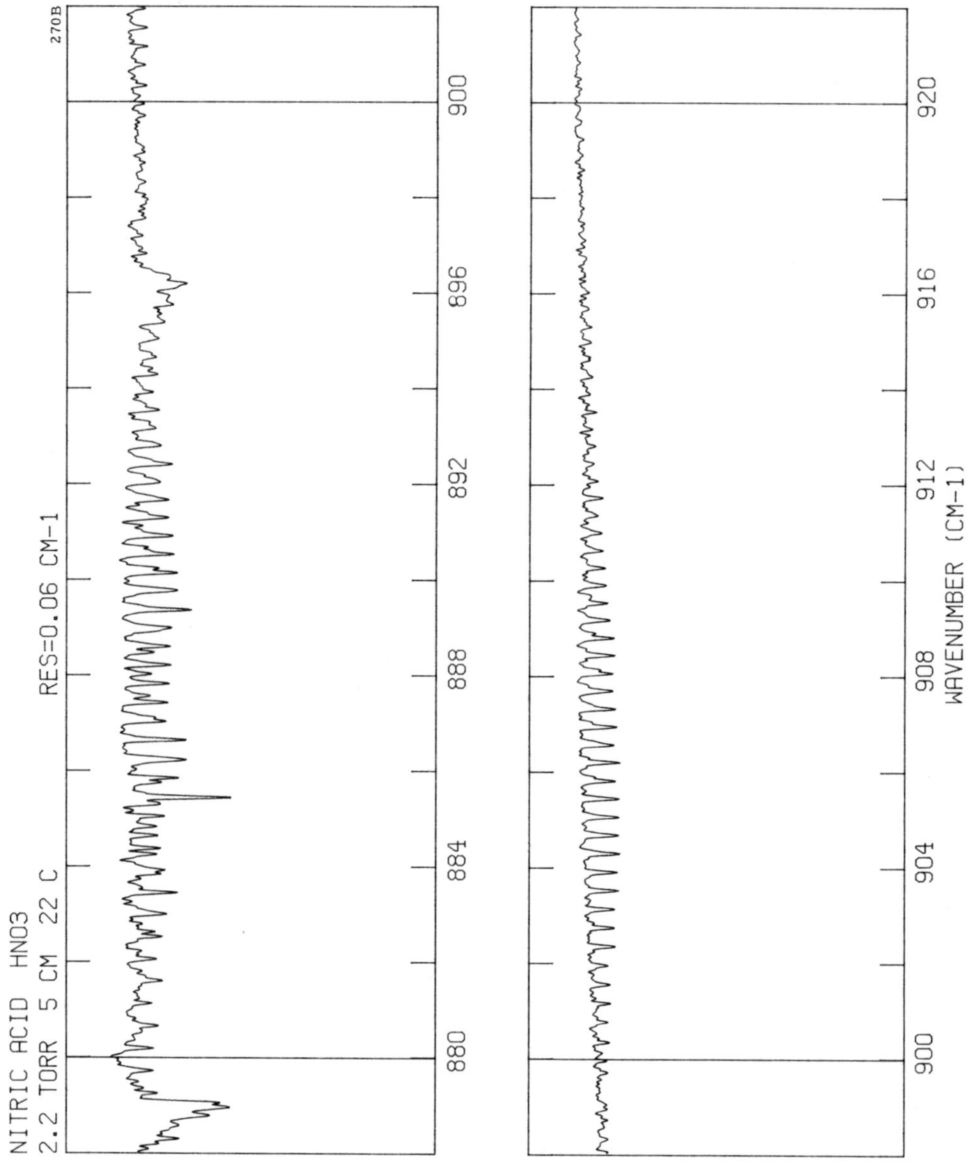
NITRIC ACID HNO3
2.2 TORR 5 CM 22 C
RES=0.06 CM-1
270B
880
884
888
892
896
900
900
904
908
912
916
920
WAVENUMBER (CM-1)

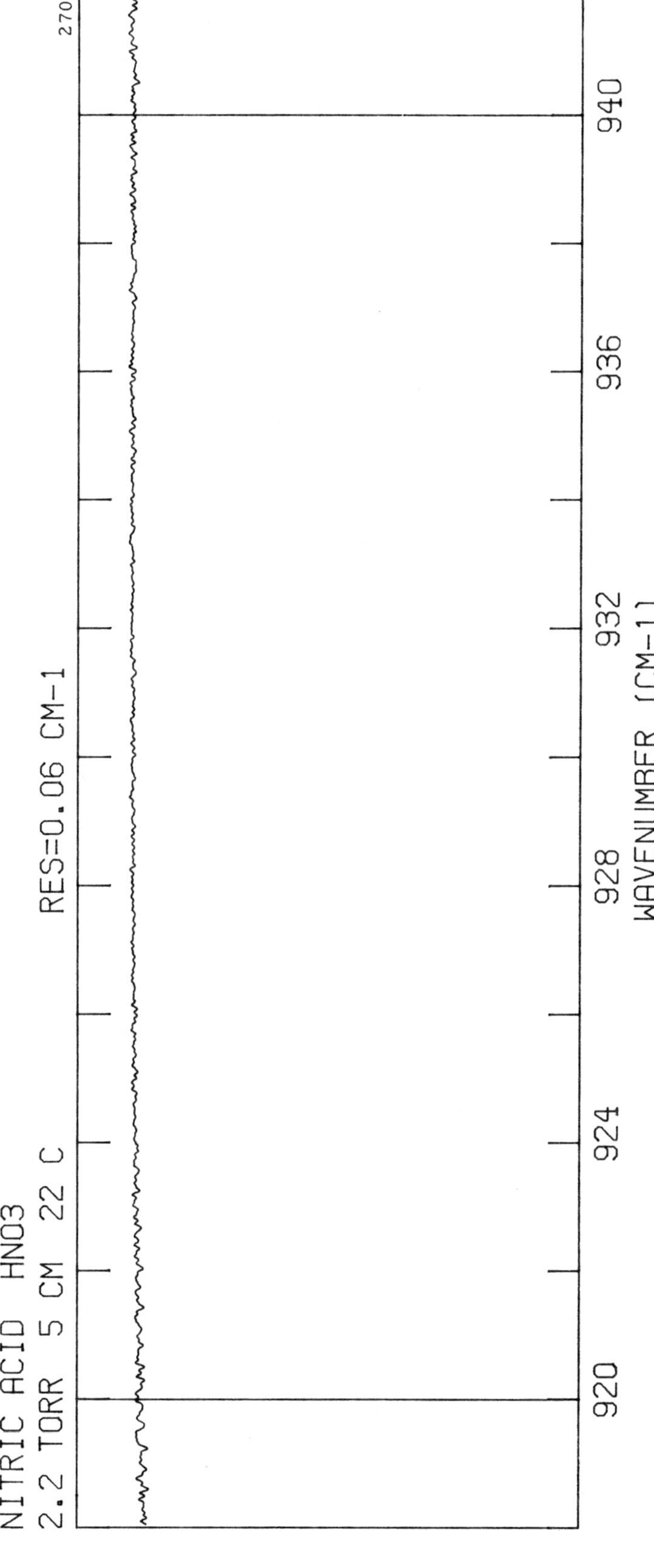

NITRIC ACID HNO3
2.2 TORR 5 CM 22 C
RES=0.06 CM-1
270B
920
924
928
932
936
940
WAVENUMBER (CM-1)

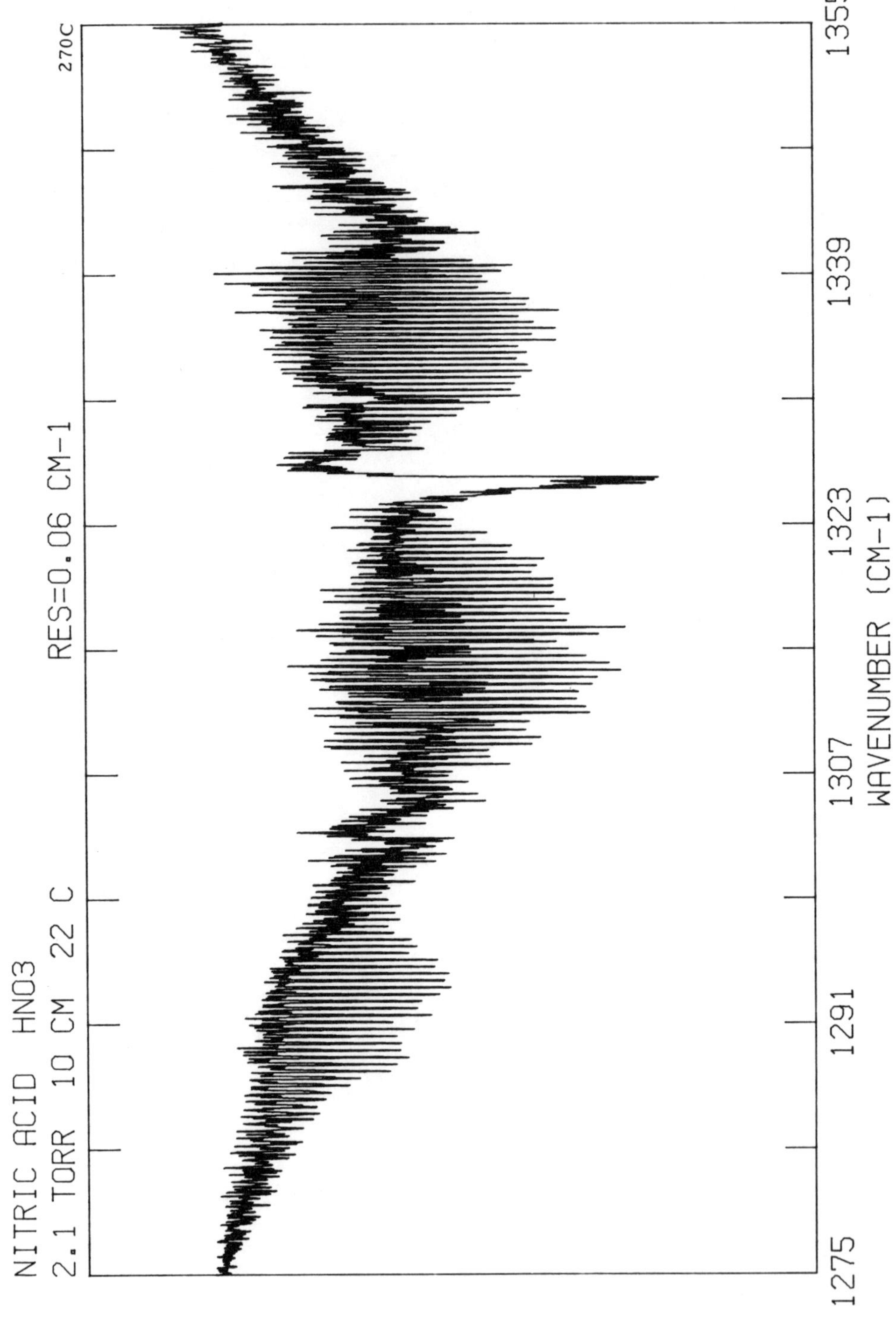
NITRIC ACID HNO3
2.1 TORR 10 CM 22 C
RES=0.06 CM-1
270C
1275
1291
1307
1323
1339
1355
WAVENUMBER (CM-1)

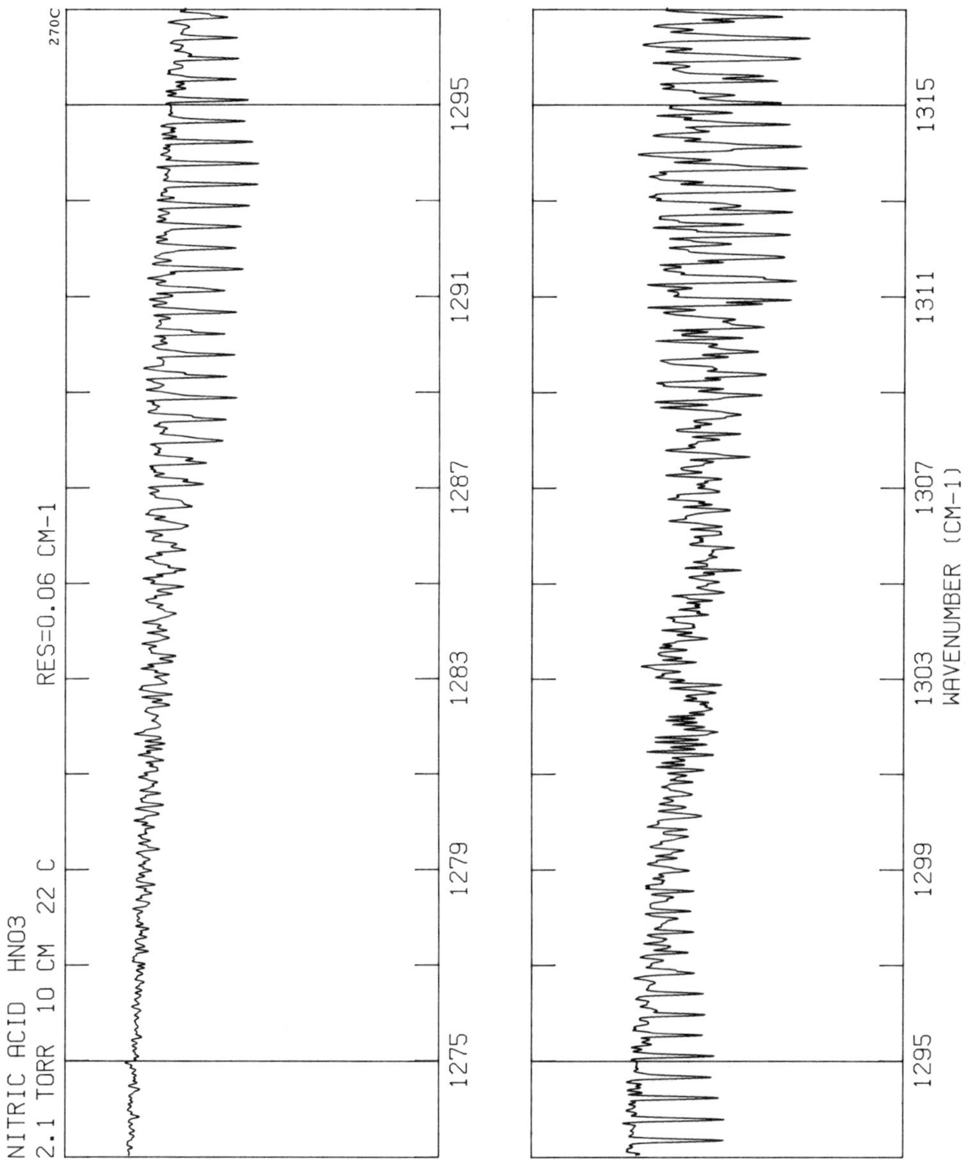
NITRIC ACID HNO3
2.1 TORR 10 CM 22 C
RES=0.06 CM-1
270C
1275
1279
1283
1287
1291
1295
1295
1299
1303
1307
1311
1315
WAVENUMBER (CM-1)

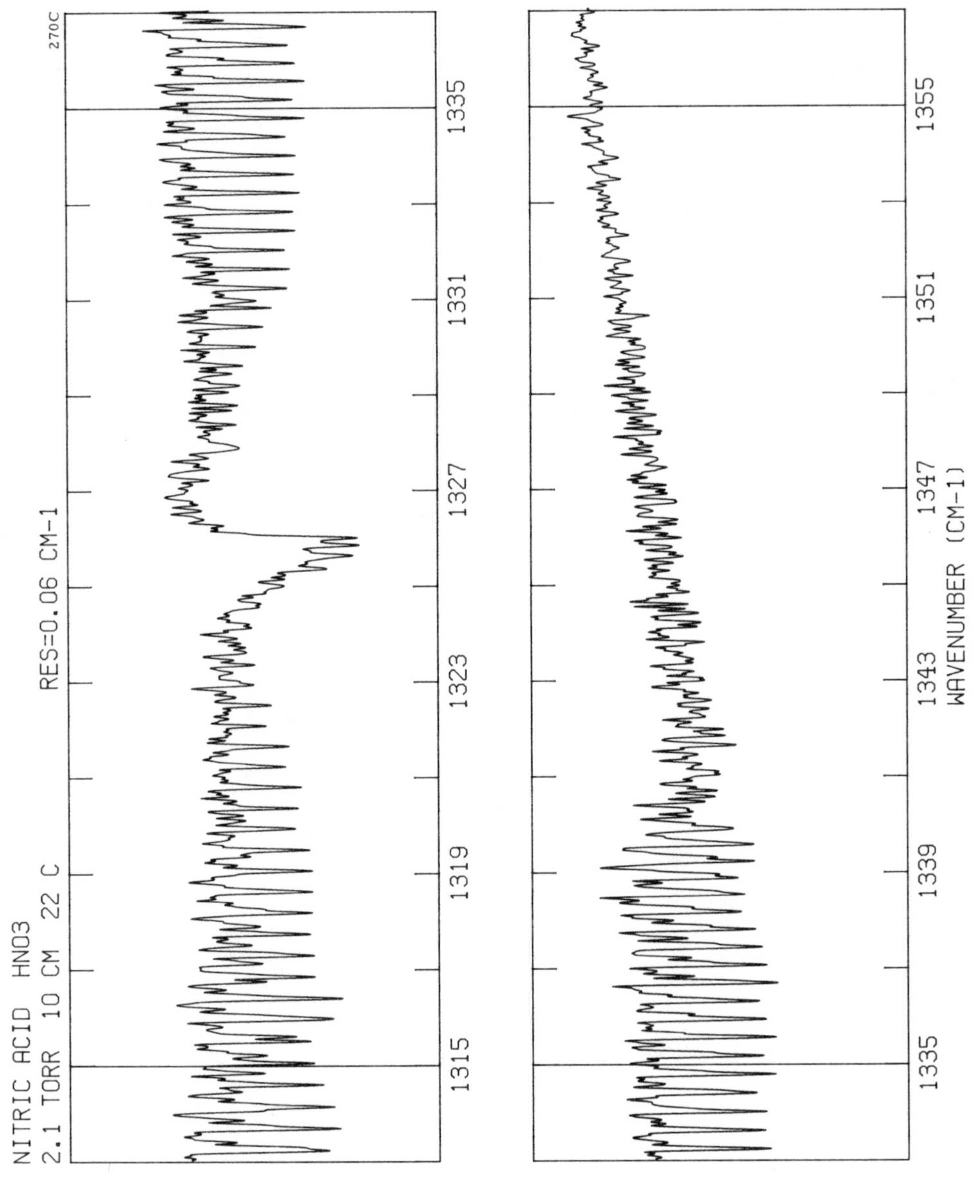
NITRIC ACID HNO3
2.1 TORR 10 CM 22 C
RES=0.06 CM-1
270C
1315
1319
1323
1327
1331
1335
1335
1339
1343
1347
1351
1355
WAVENUMBER (CM-1)

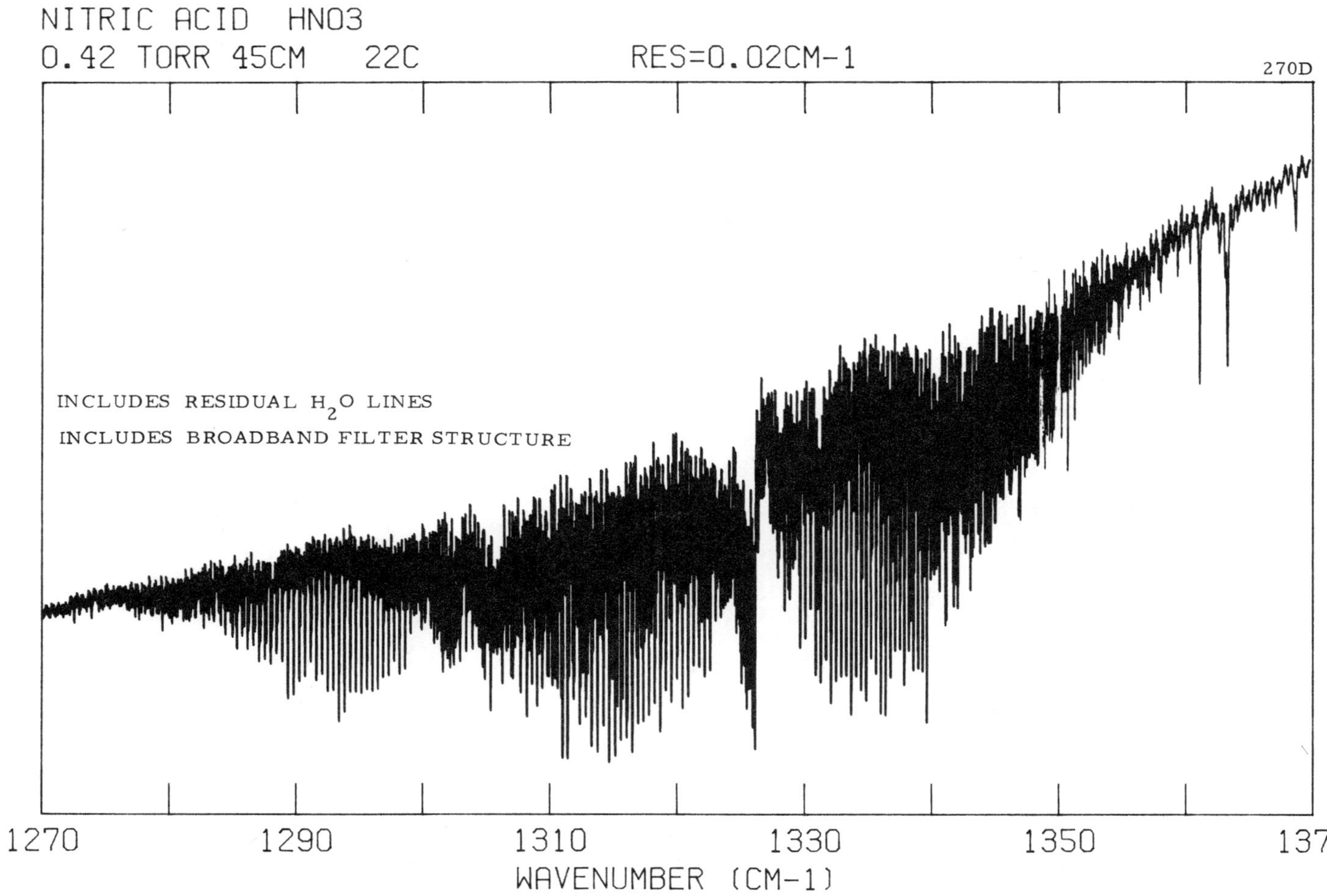
NITRIC ACID HNO3
0.42 TORR 45CM 22C
RES=0.02CM-1
270D
INCLUDES RESIDUAL H2O LINES
INCLUDES BROADBAND FILTER STRUCTURE
1270
1290
1310
1330
1350
1370
WAVENUMBER (CM-1)

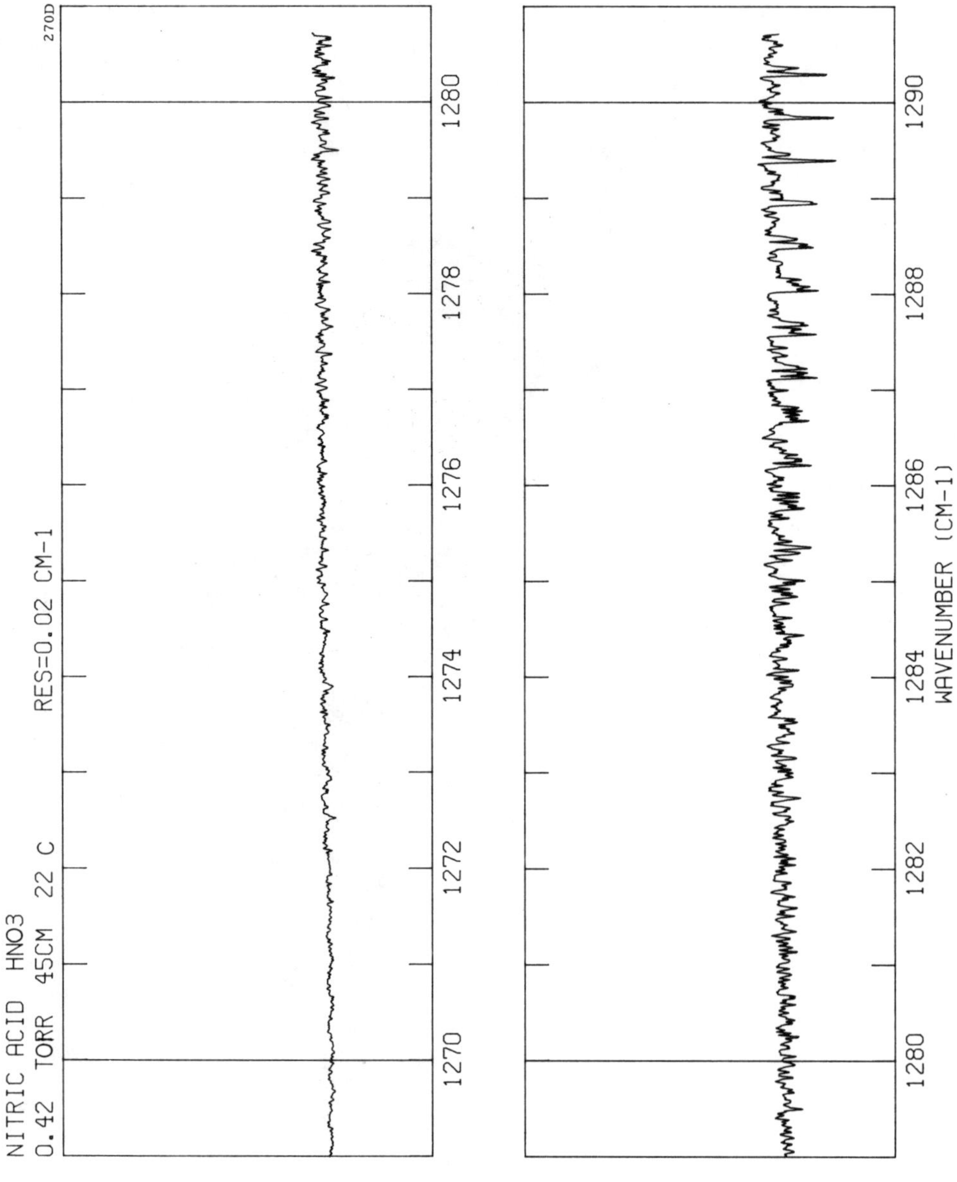
NITRIC ACID HNO3
0.42 TORR 45CM 22 C
RES=0.02 CM-1
270D
1270
1272
1274
1276
1278
1280
1280
1282
1284
1286
1288
1290
WAVENUMBER (CM-1)

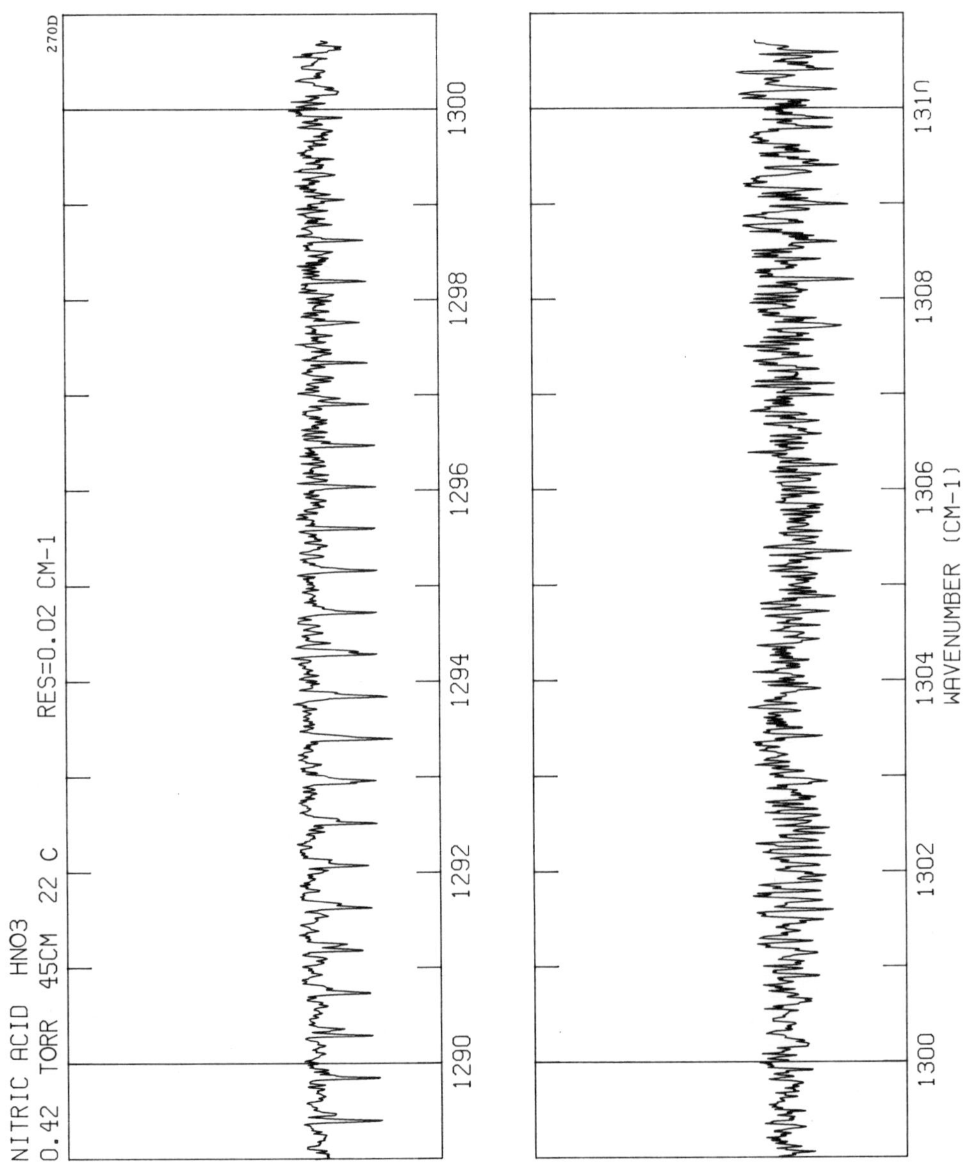

NITRIC ACID HNO3
0.42 TORR 45CM 22 C
RES=0.02 CM-1
270D
1290
1292
1294
1296
1298
1300
1300
1302
1304
1306
1308
1310
WAVENUMBER (CM-1)

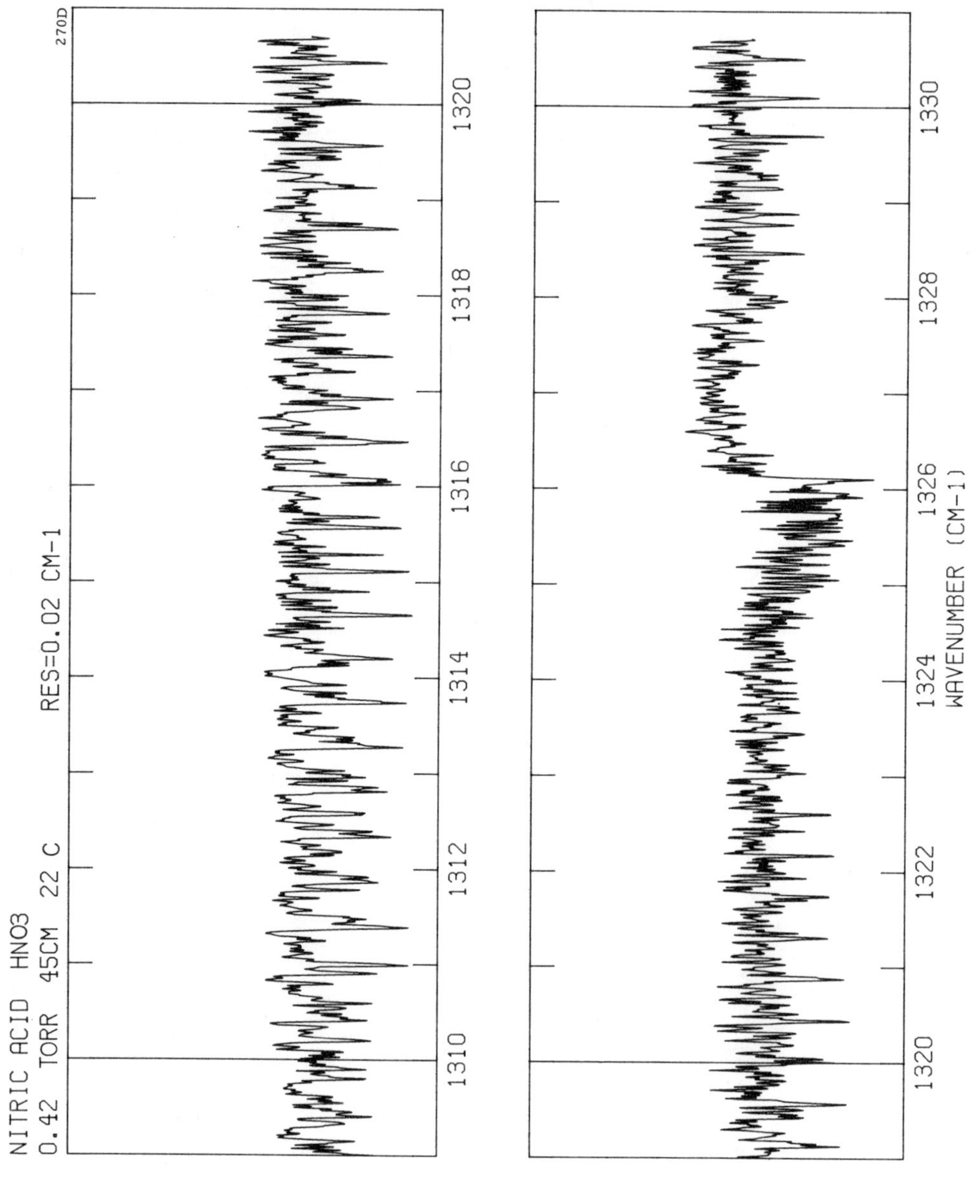
270D
NITRIC ACID HNO3
0.42 TORR 45CM 22 C
RES=0.02 CM-1
1310
1312
1314
1316
1318
1320
1322
1324
1326
1328
1330
WAVENUMBER (CM-1)

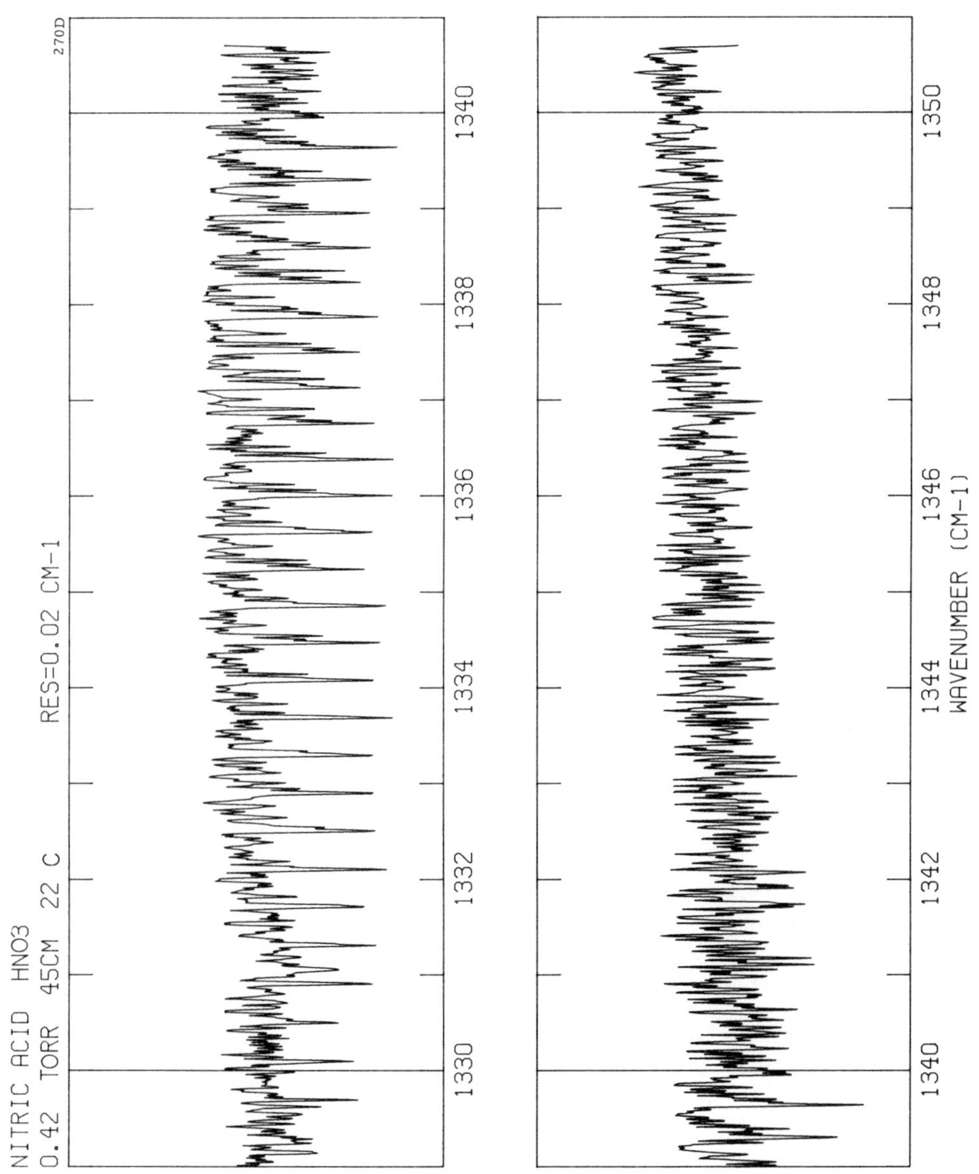
NITRIC ACID HNO3
0.42 TORR 45CM 22 C
RES=0.02 CM-1
270D
1330
1332
1334
1336
1338
1340
1342
1344
1346
1348
1350
WAVENUMBER (CM-1)

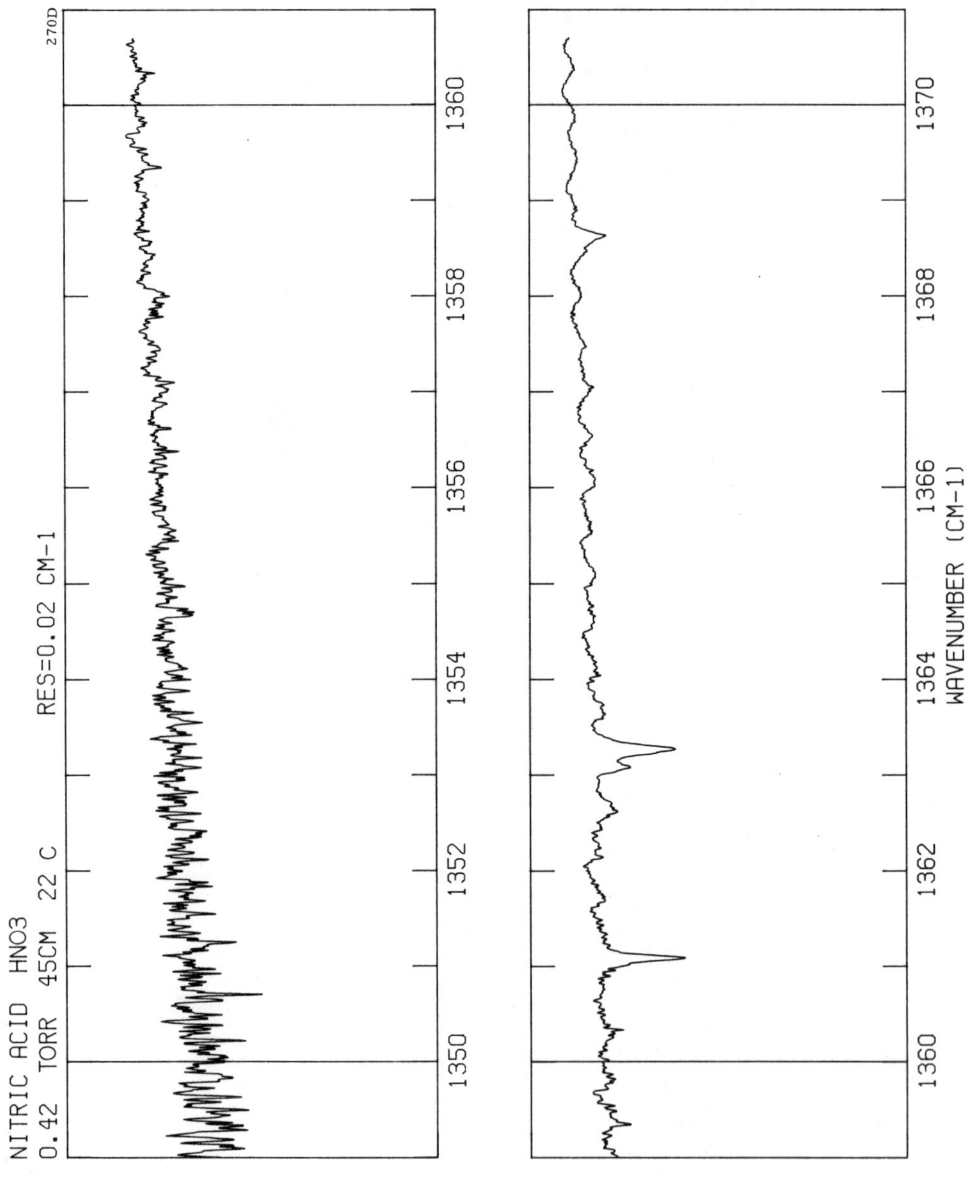
NITRIC ACID HNO3
0.42 TORR 45CM 22 C
RES=0.02 CM-1
270D
1350
1352
1354
1356
1358
1360
1360
1362
1364
1366
1368
1370
WAVENUMBER (CM-1)

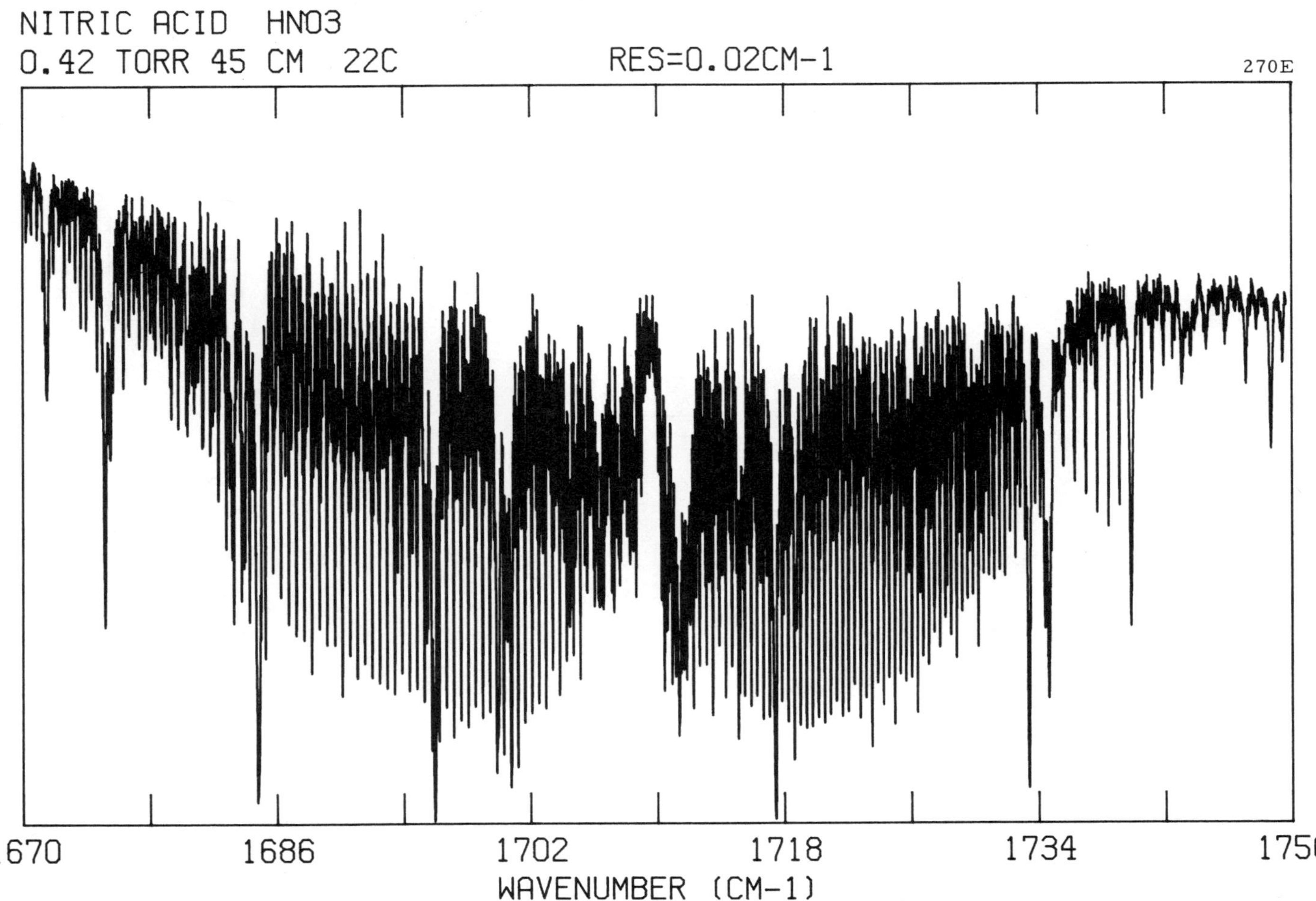
NITRIC ACID HNO3
0.42 TORR 45 CM 22C
RES=0.02CM-1
270E
1670
1686
1702
1718
1734
1750
WAVENUMBER (CM-1)

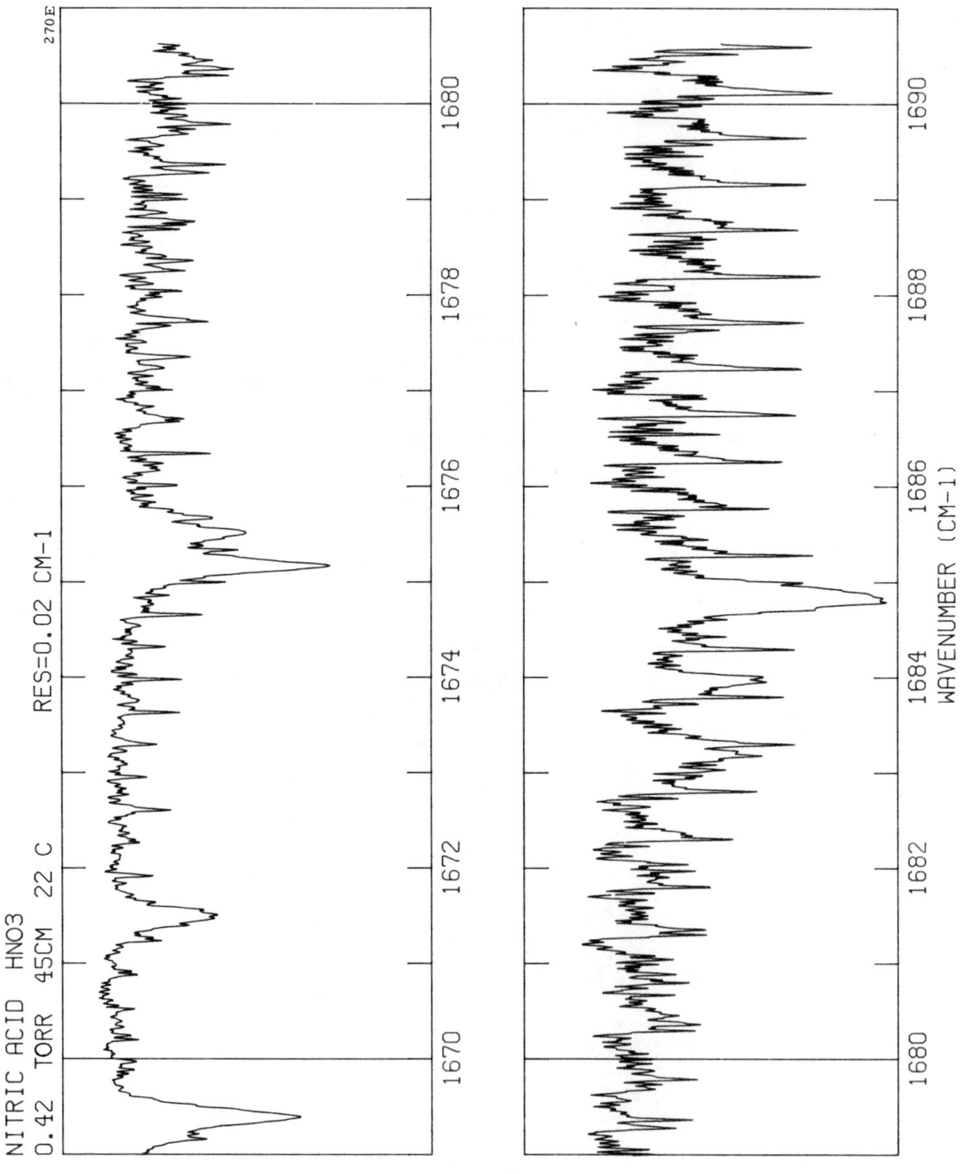
NITRIC ACID HNO3
0.42 TORR 45CM 22 C
RES=0.02 CM-1
270E
1670
1672
1674
1676
1678
1680
1680
1682
1684
1686
1688
1690
WAVENUMBER (CM-1)

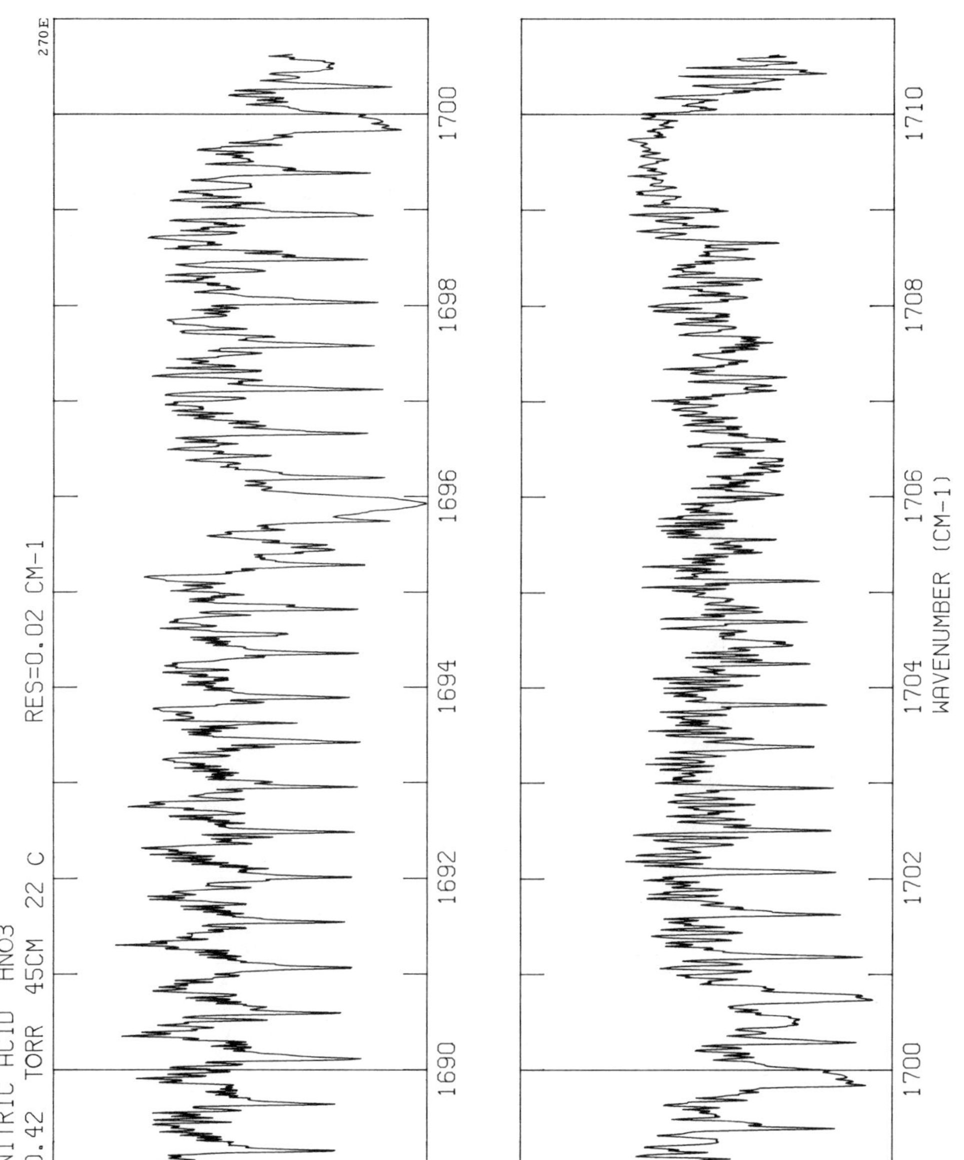

NITRIC ACID HNO3
0.42 TORR 45CM 22 C
RES=0.02 CM-1
270E
1690
1692
1694
1696
1698
1700
1700
1702
1704
1706
1708
1710
WAVENUMBER (CM-1)

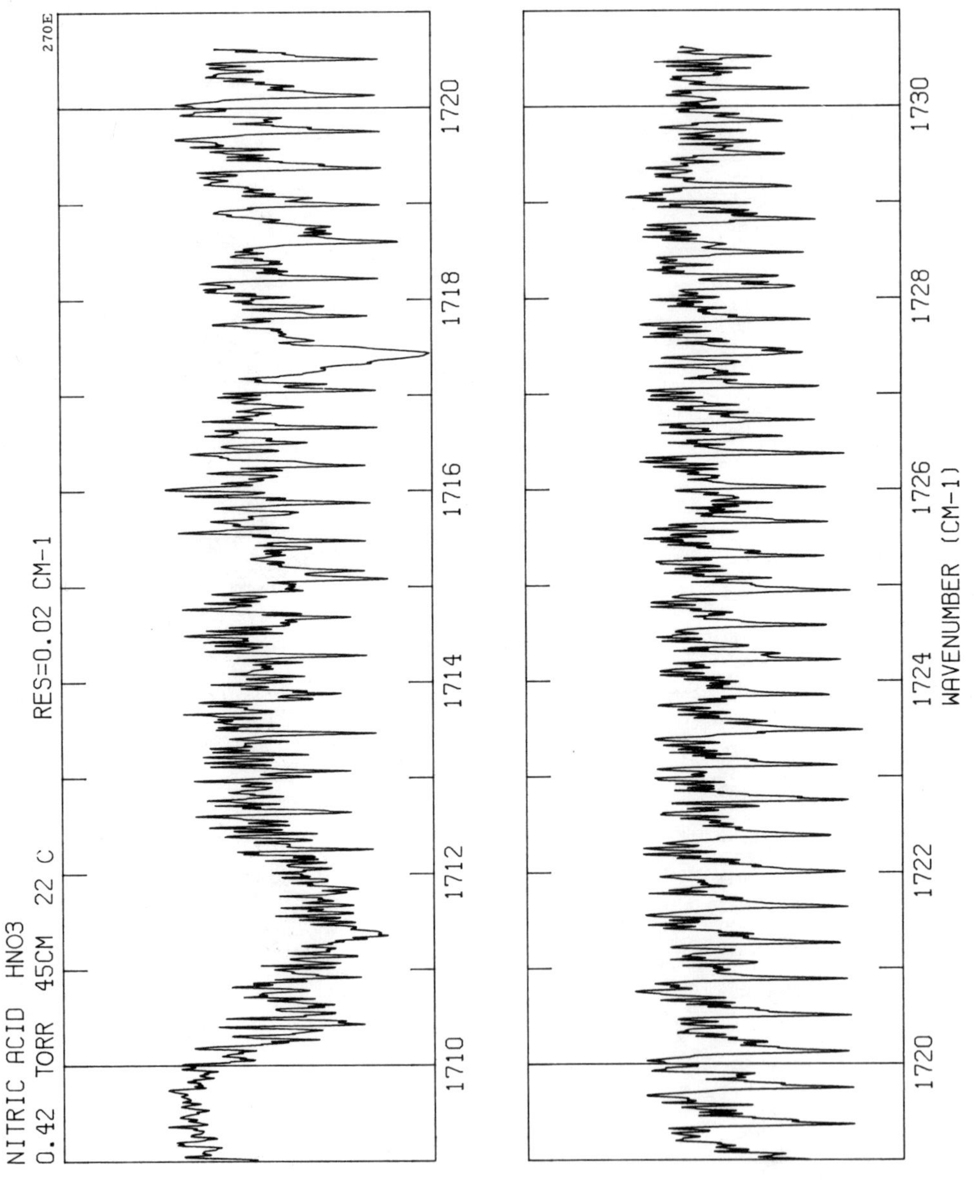
NITRIC ACID HNO3
0.42 TORR 45CM 22 C
RES=0.02 CM-1
270E
1710
1712
1714
1716
1718
1720
1722
1724
1726
1728
1730
WAVENUMBER (CM-1)

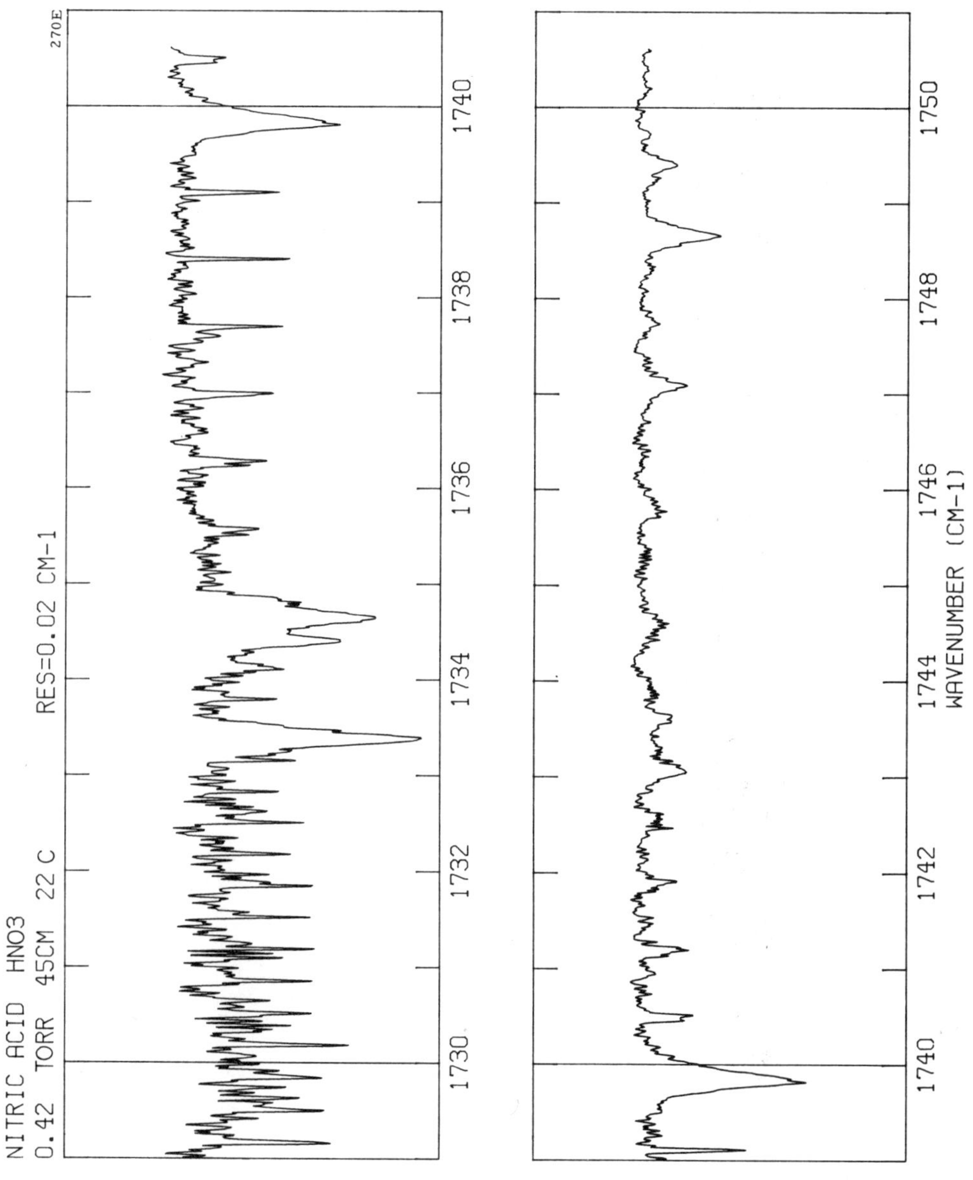

NITRIC ACID HNO3
0.42 TORR 45CM 22 C
RES=0.02 CM-1
270E
1730
1732
1734
1736
1738
1740
1742
1744
1746
1748
1750
WAVENUMBER (CM-1)

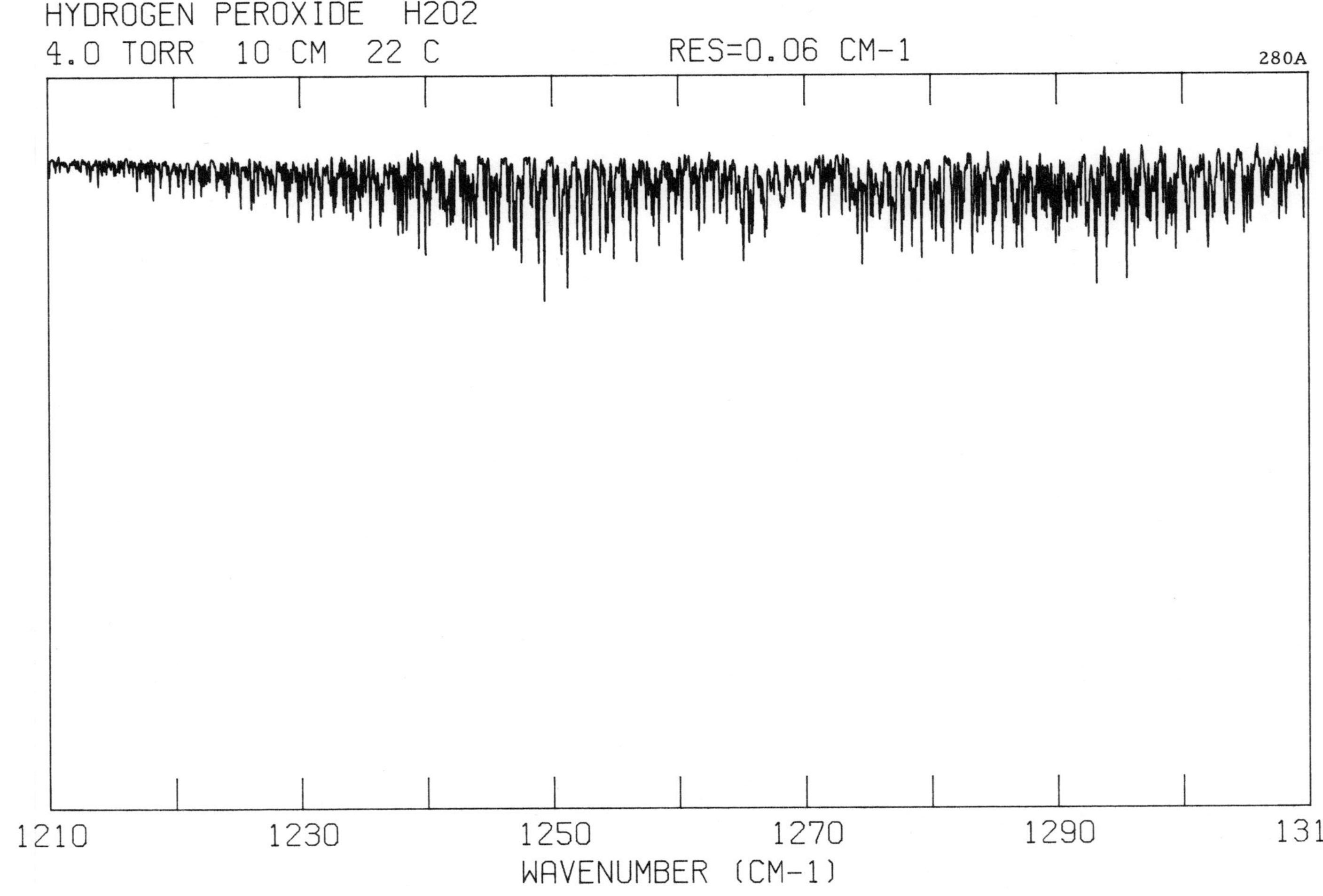
HYDROGEN PEROXIDE H2O2
4.0 TORR 10 CM 22 C
RES=0.06 CM-1
280A
1210
1230
1250
1270
1290
1310
WAVENUMBER (CM-1)

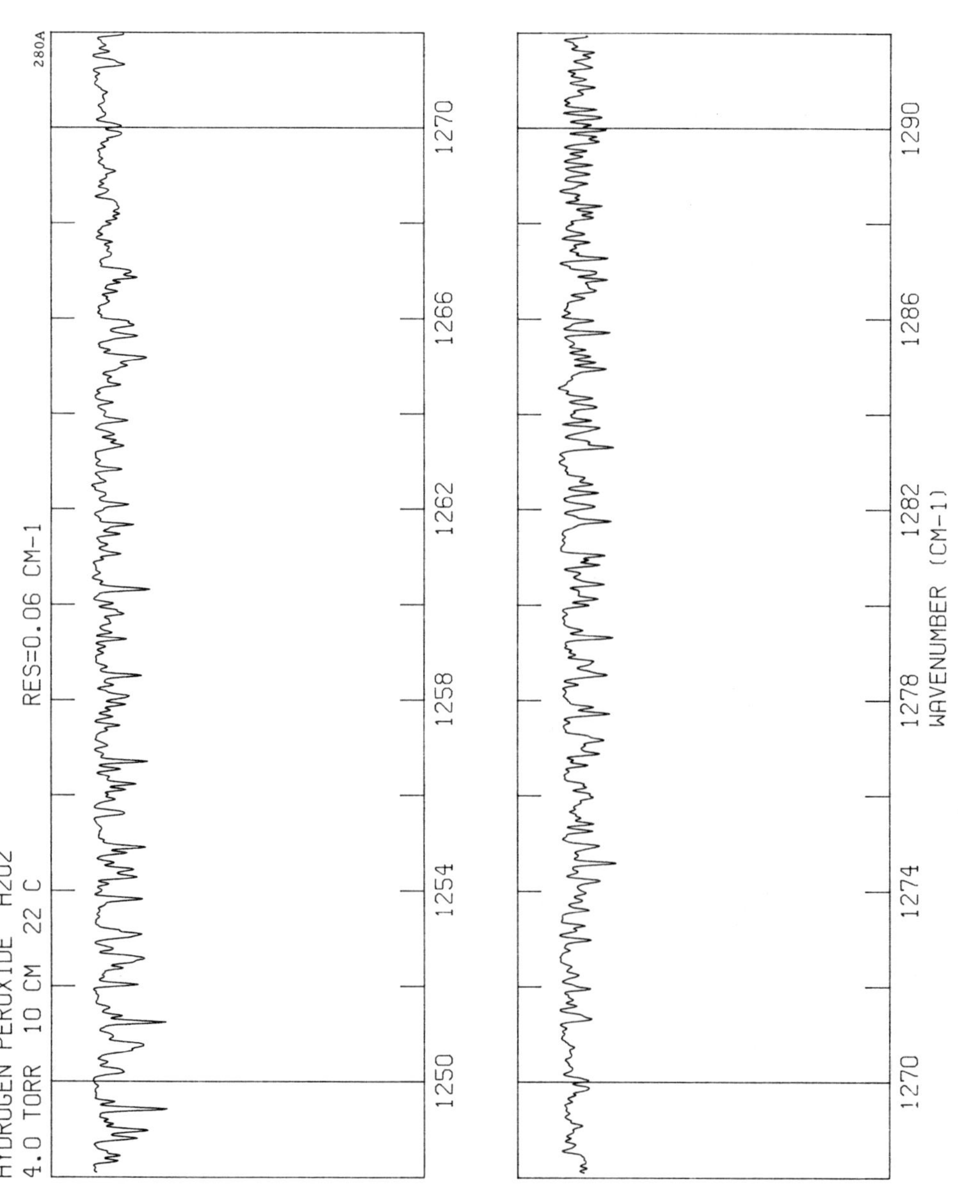
HYDROGEN PEROXIDE H2O2
4.0 TORR 10 CM 22 C
RES=0.06 CM-1
280A
1250
1254
1258
1262
1266
1270
1274
1278
1282
1286
1290
WAVENUMBER (CM-1)

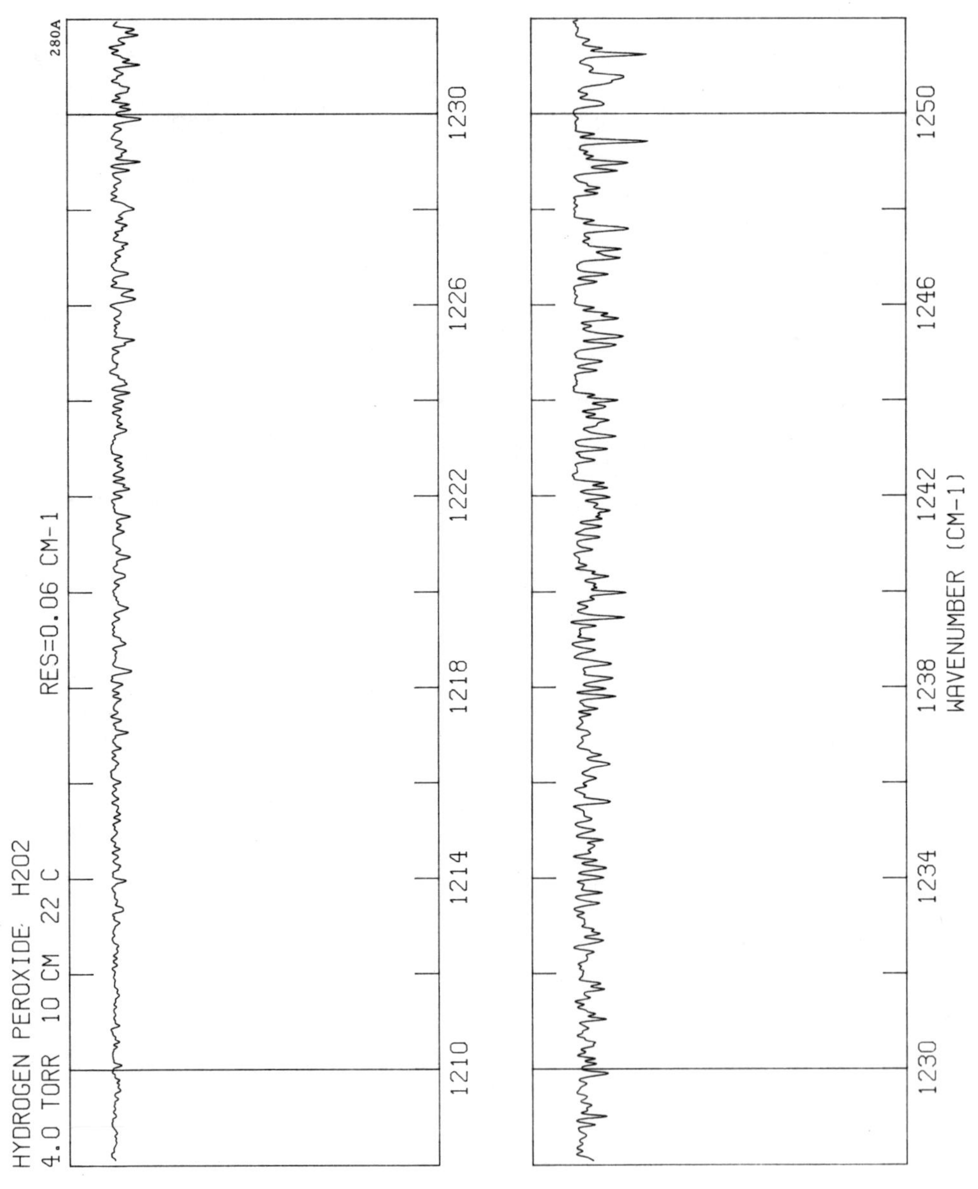
HYDROGEN PEROXIDE H2O2
4.0 TORR 10 CM 22 C
RES=0.06 CM-1
280A
1210
1214
1218
1222
1226
1230
1234
1238
1242
1246
1250
WAVENUMBER (CM-1)

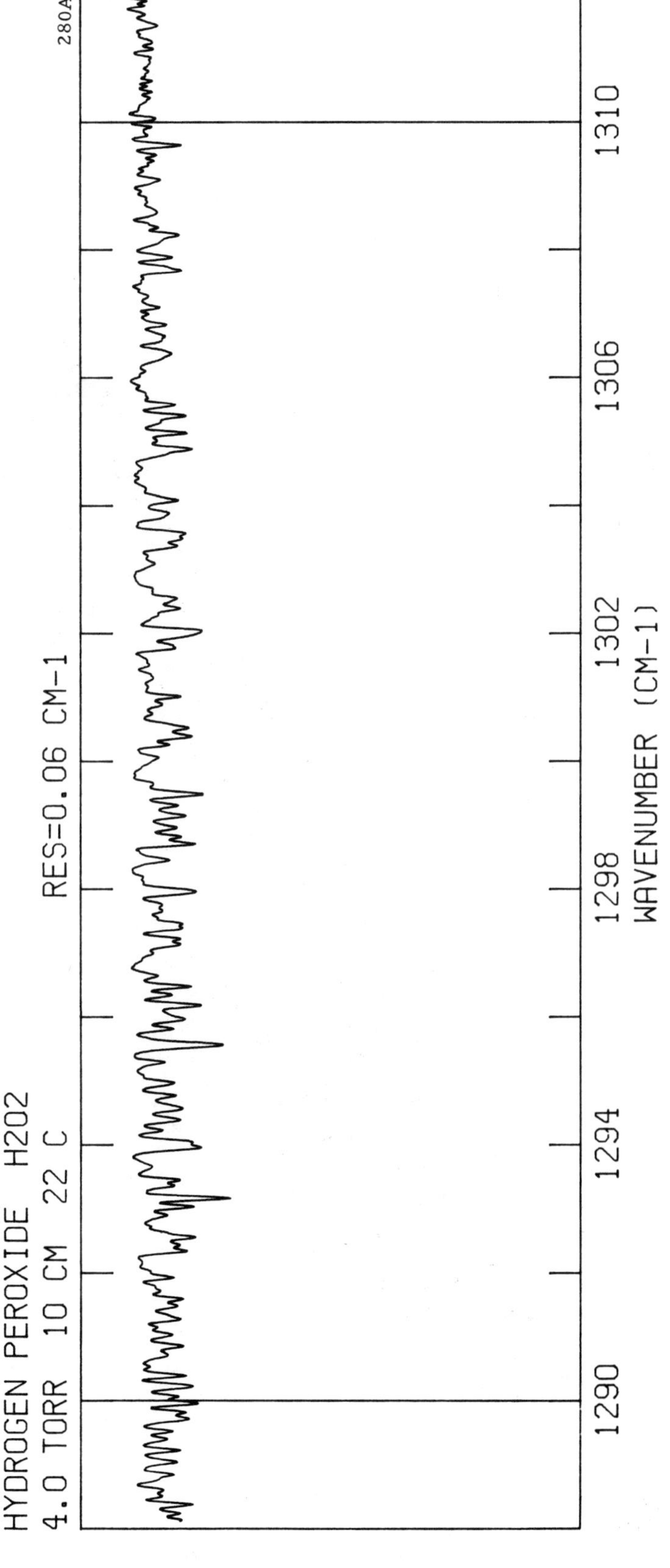
HYDROGEN PEROXIDE H2O2
4.0 TORR 10 CM 22 C
RES=0.06 CM-1
280A
1290
1294
1298
1302
1306
1310
WAVENUMBER (CM-1)

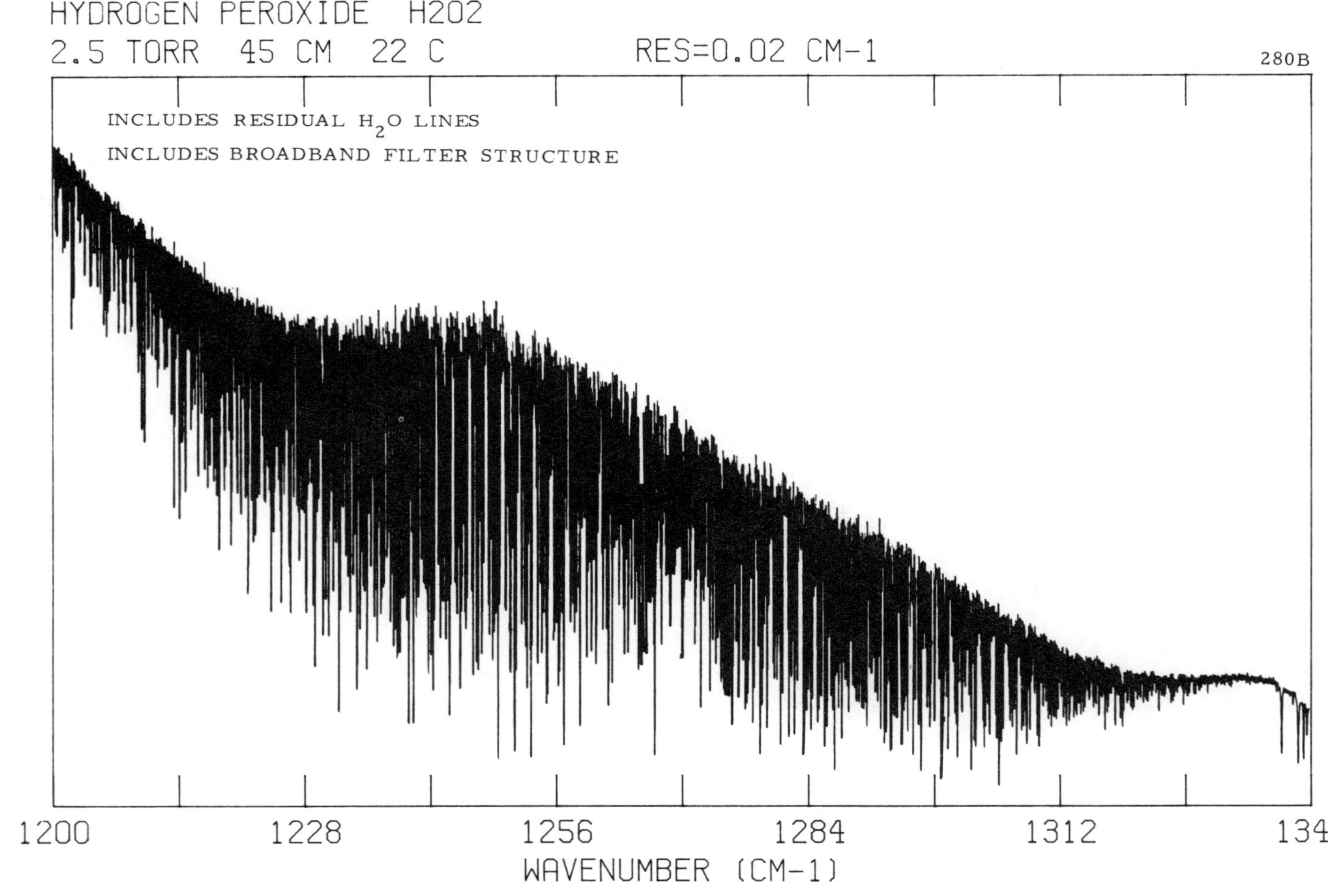
HYDROGEN PEROXIDE H2O2
2.5 TORR 45 CM 22 C
RES=0.02 CM-1
280B
INCLUDES RESIDUAL H_2O LINES
INCLUDES BROADBAND FILTER STRUCTURE
1200
1228
1256
1284
1312
1340
WAVENUMBER (CM-1)

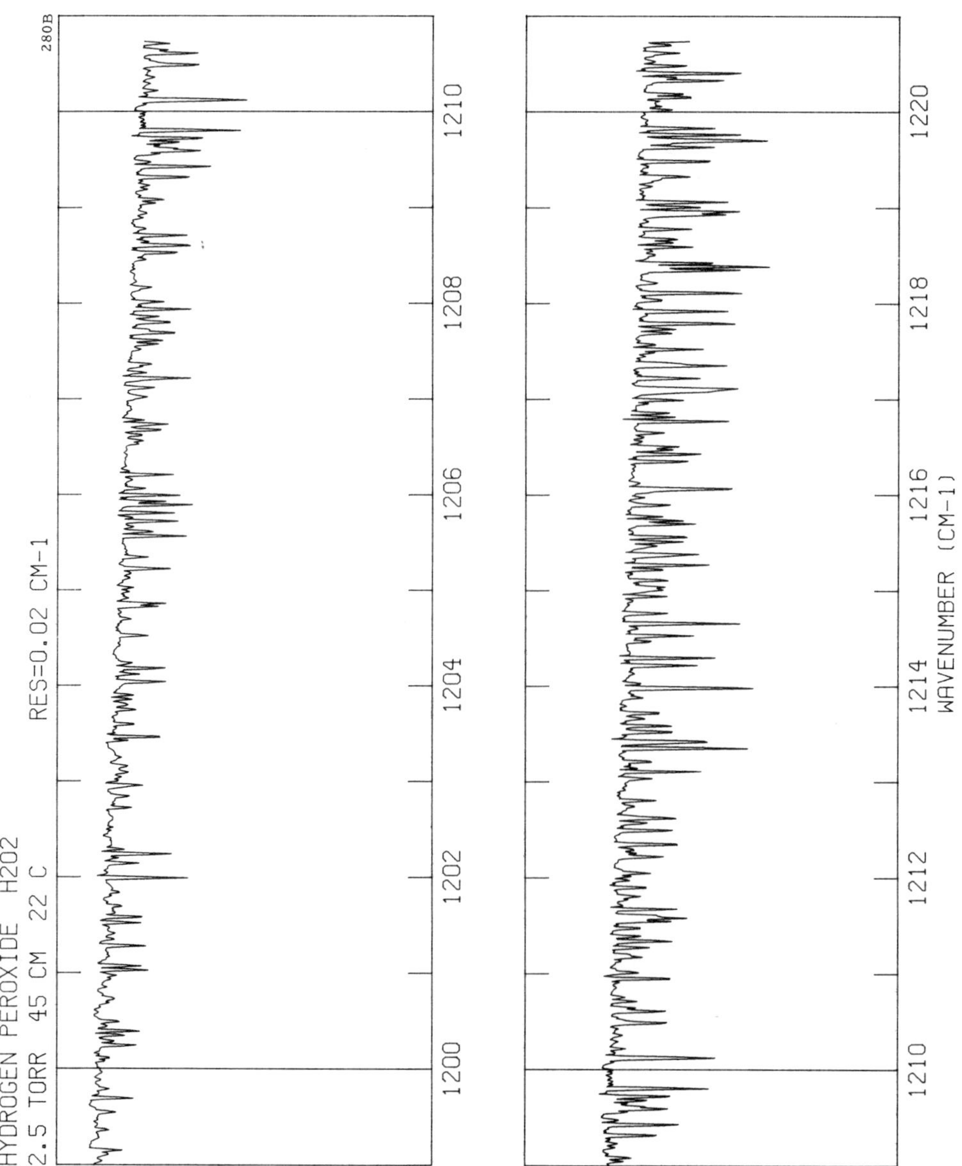

280B
HYDROGEN PEROXIDE H2O2
2.5 TORR 45 CM 22 C
RES=0.02 CM-1
1200
1202
1204
1206
1208
1210
1212
1214
1216
1218
1220
WAVENUMBER (CM-1)

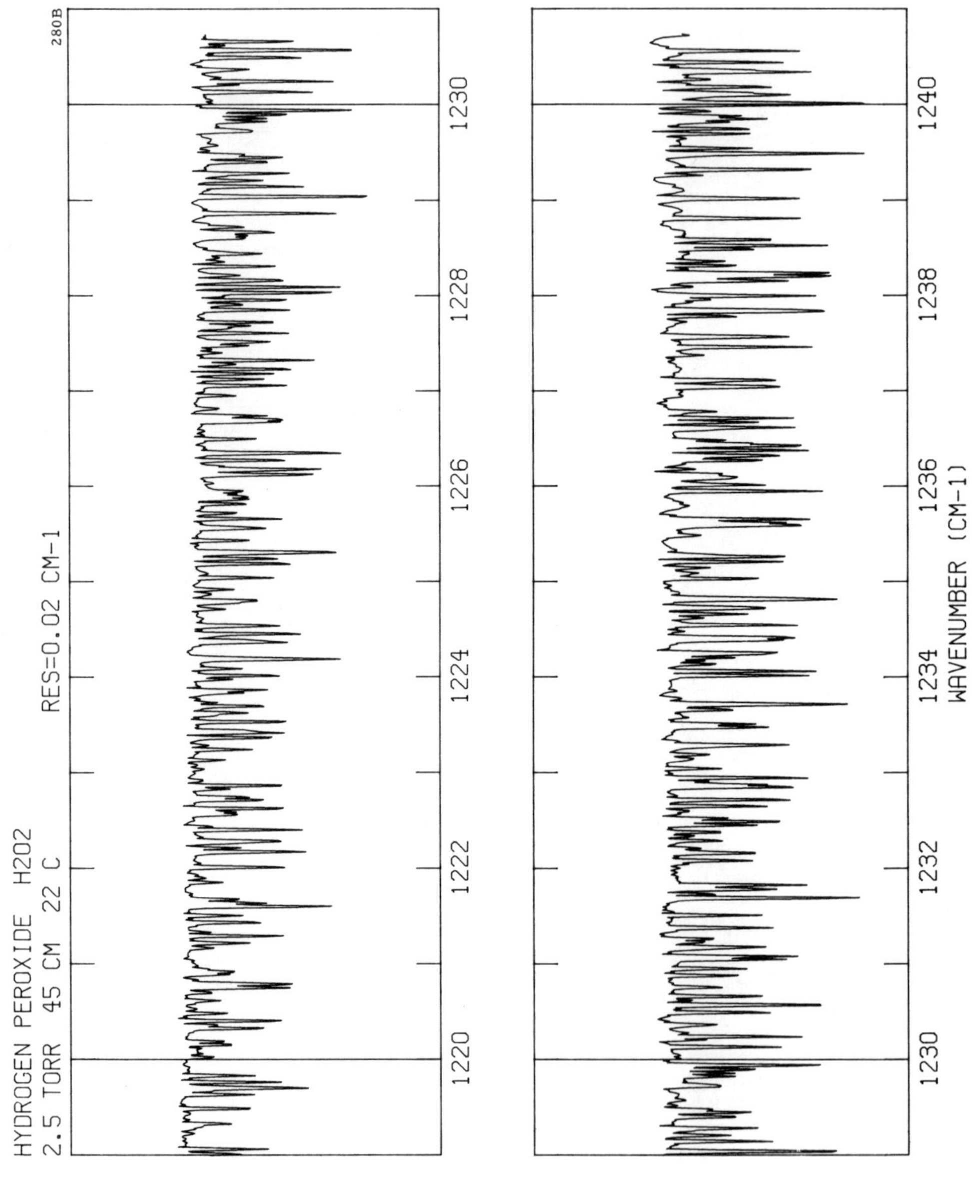
HYDROGEN PEROXIDE H2O2
2.5 TORR 45 CM 22 C
RES=0.02 CM-1
280B
1220
1222
1224
1226
1228
1230
1232
1234
1236
1238
1240
WAVENUMBER (CM-1)

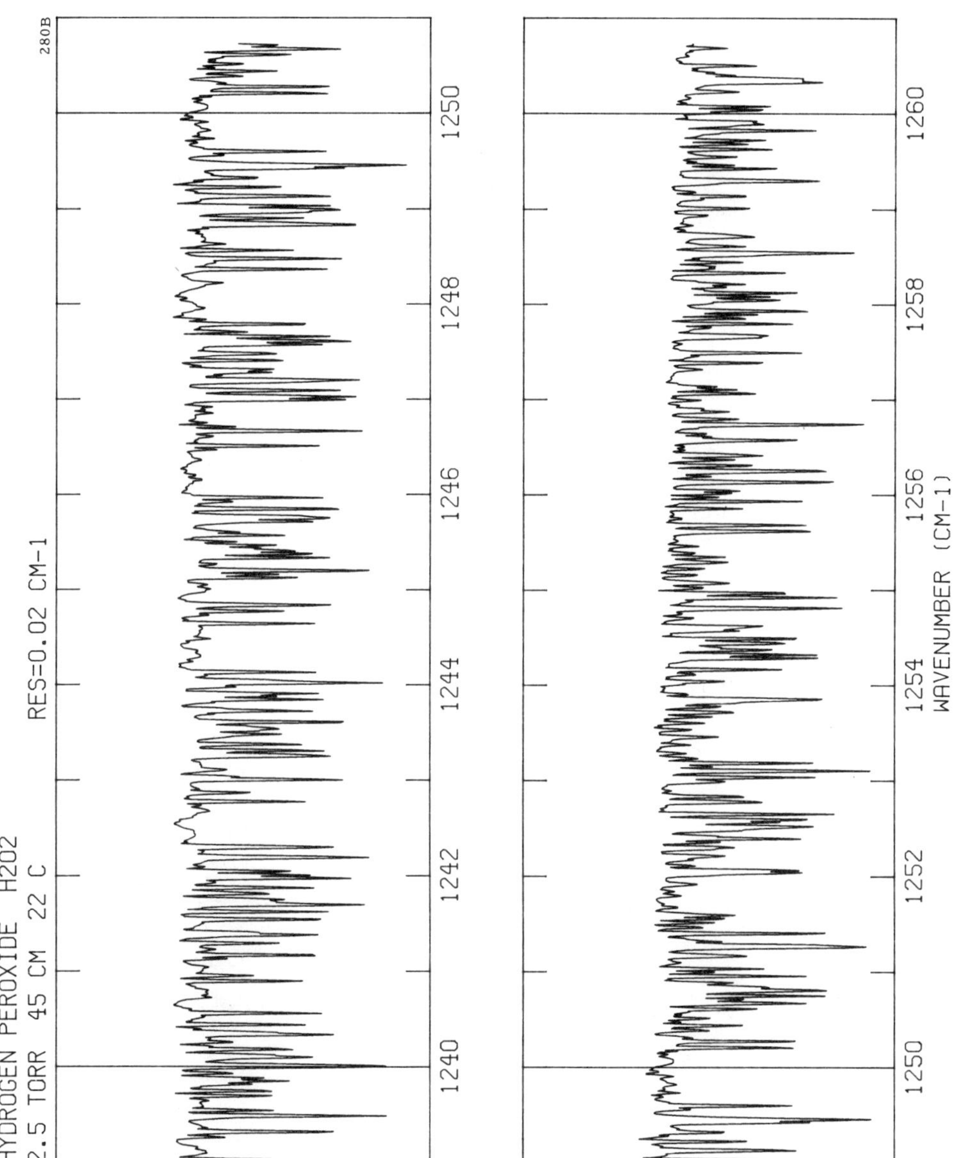
HYDROGEN PEROXIDE H2O2
2.5 TORR 45 CM 22 C
RES=0.02 CM-1
280B
1240
1242
1244
1246
1248
1250
1252
1254
1256
1258
1260
WAVENUMBER (CM-1)

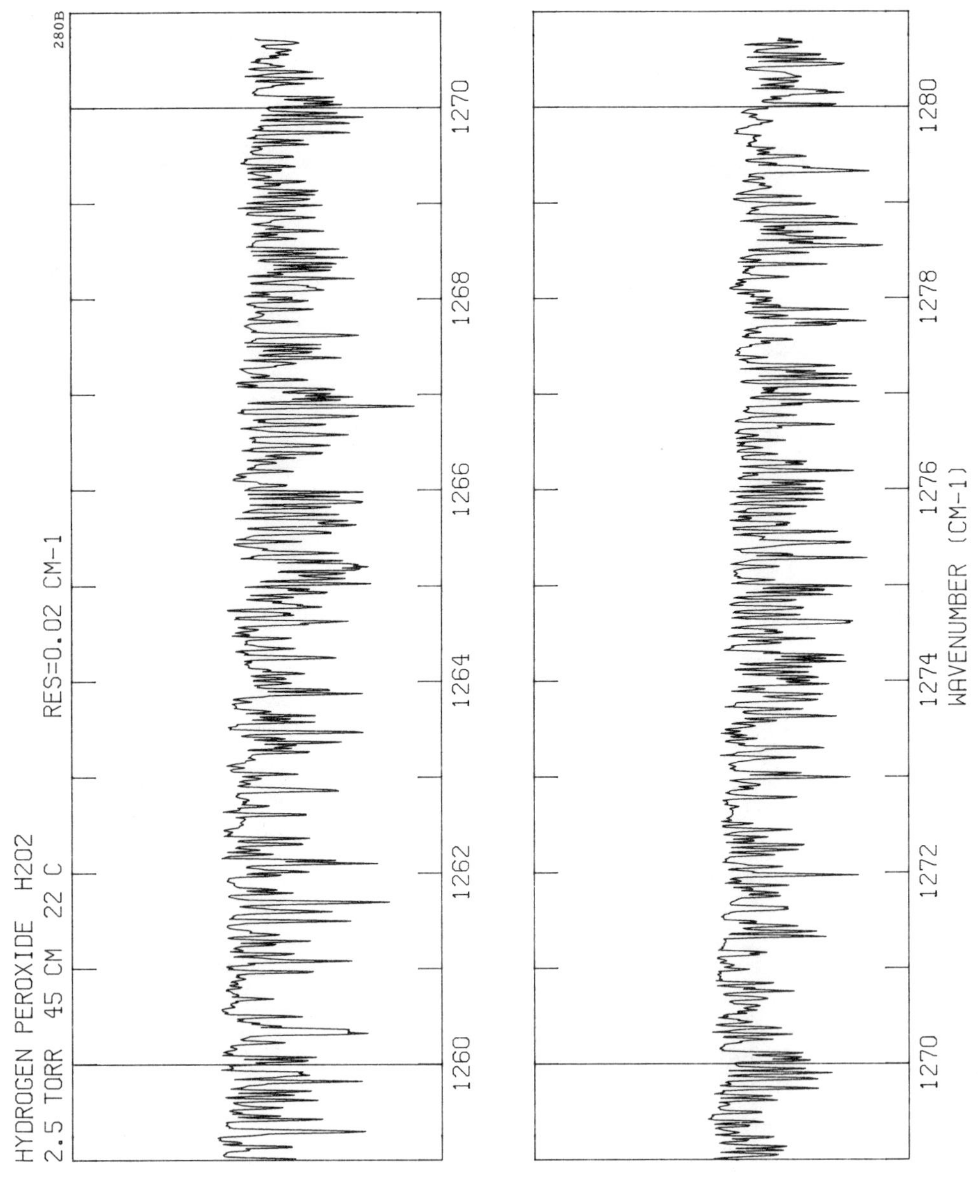
HYDROGEN PEROXIDE H2O2
2.5 TORR 45 CM 22 C
RES=0.02 CM-1
280B
1260
1262
1264
1266
1268
1270
1272
1274
1276
1278
1280
WAVENUMBER (CM-1)

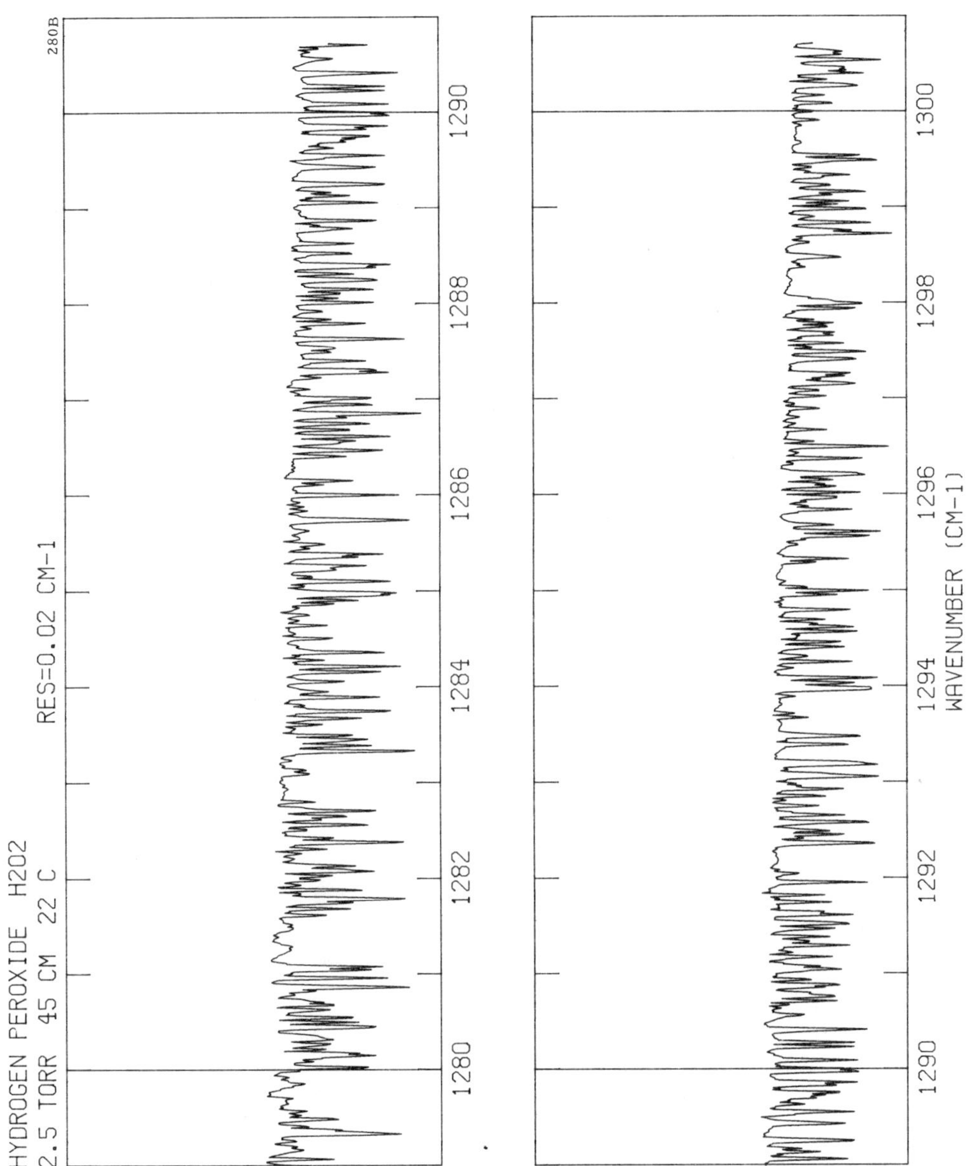
HYDROGEN PEROXIDE H202
2.5 TORR 45 CM 22 C
RES=0.02 CM-1
280B
1280
1282
1284
1286
1288
1290
1290
1292
1294
1296
1298
1300
WAVENUMBER (CM-1)

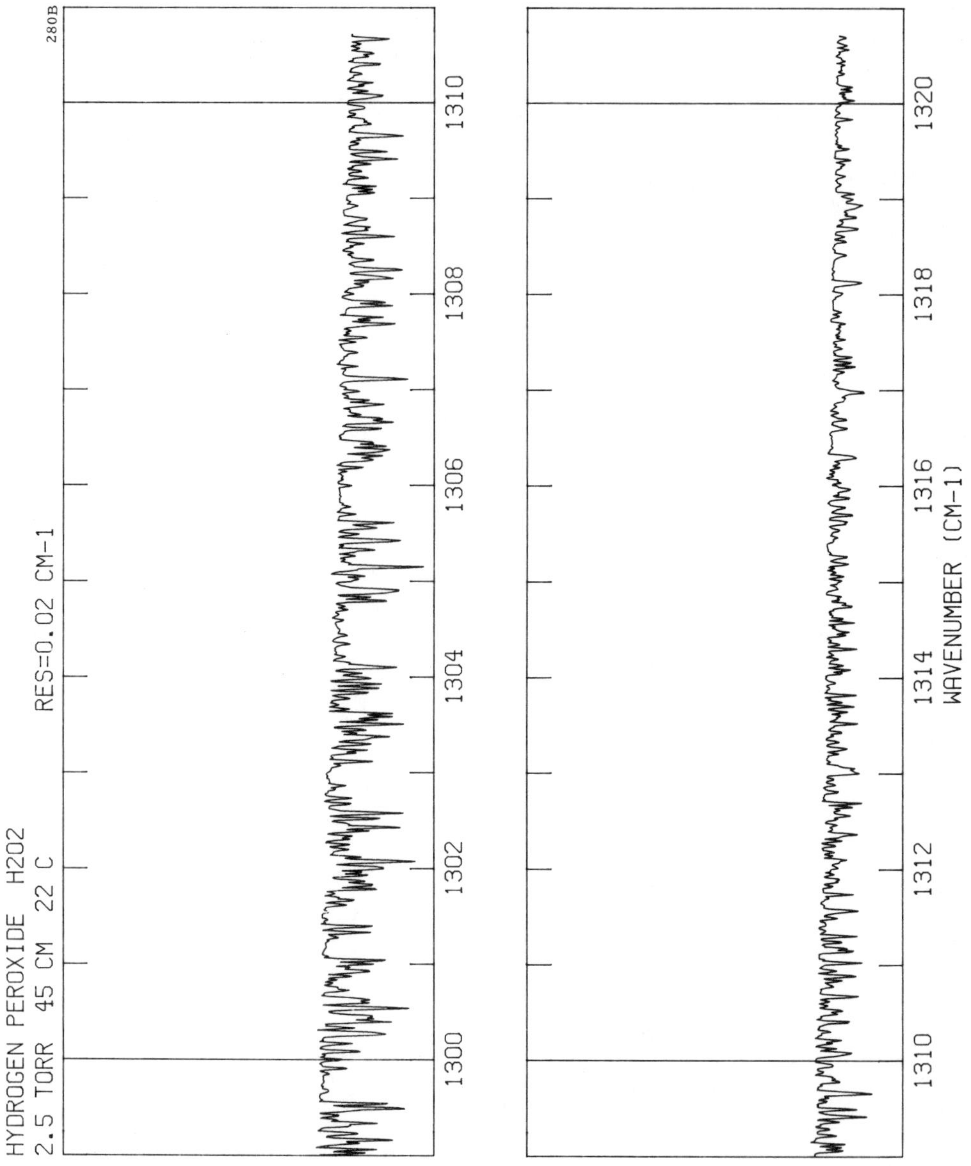
280B
HYDROGEN PEROXIDE H2O2
2.5 TORR 45 CM 22 C
RES=0.02 CM-1
1300
1302
1304
1306
1308
1310
1312
1314
1316
1318
1320
WAVENUMBER (CM-1)

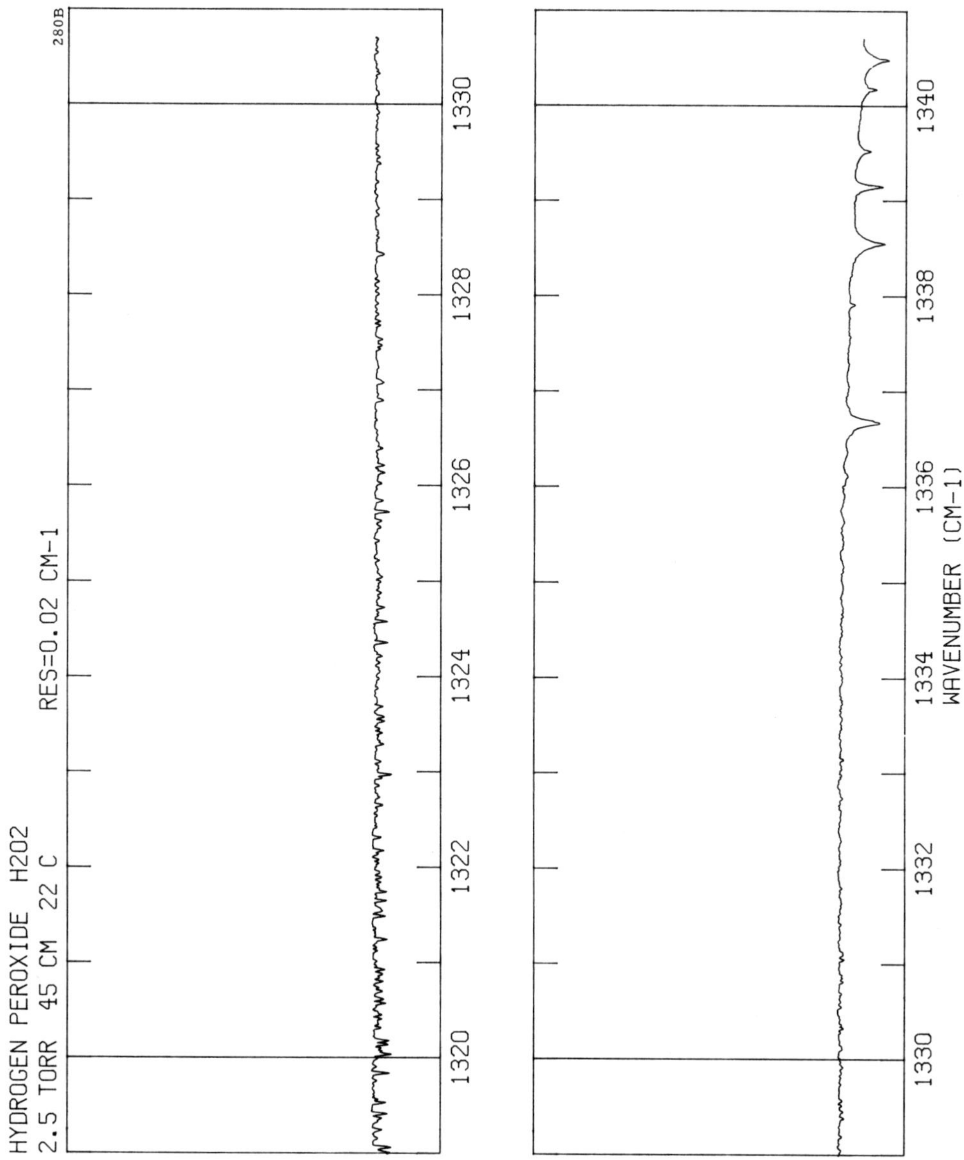
HYDROGEN PEROXIDE H2O2
2.5 TORR 45 CM 22 C
RES=0.02 CM-1
280B
1320
1322
1324
1326
1328
1330
1330
1332
1334
1336
1338
1340
WAVENUMBER (CM-1)

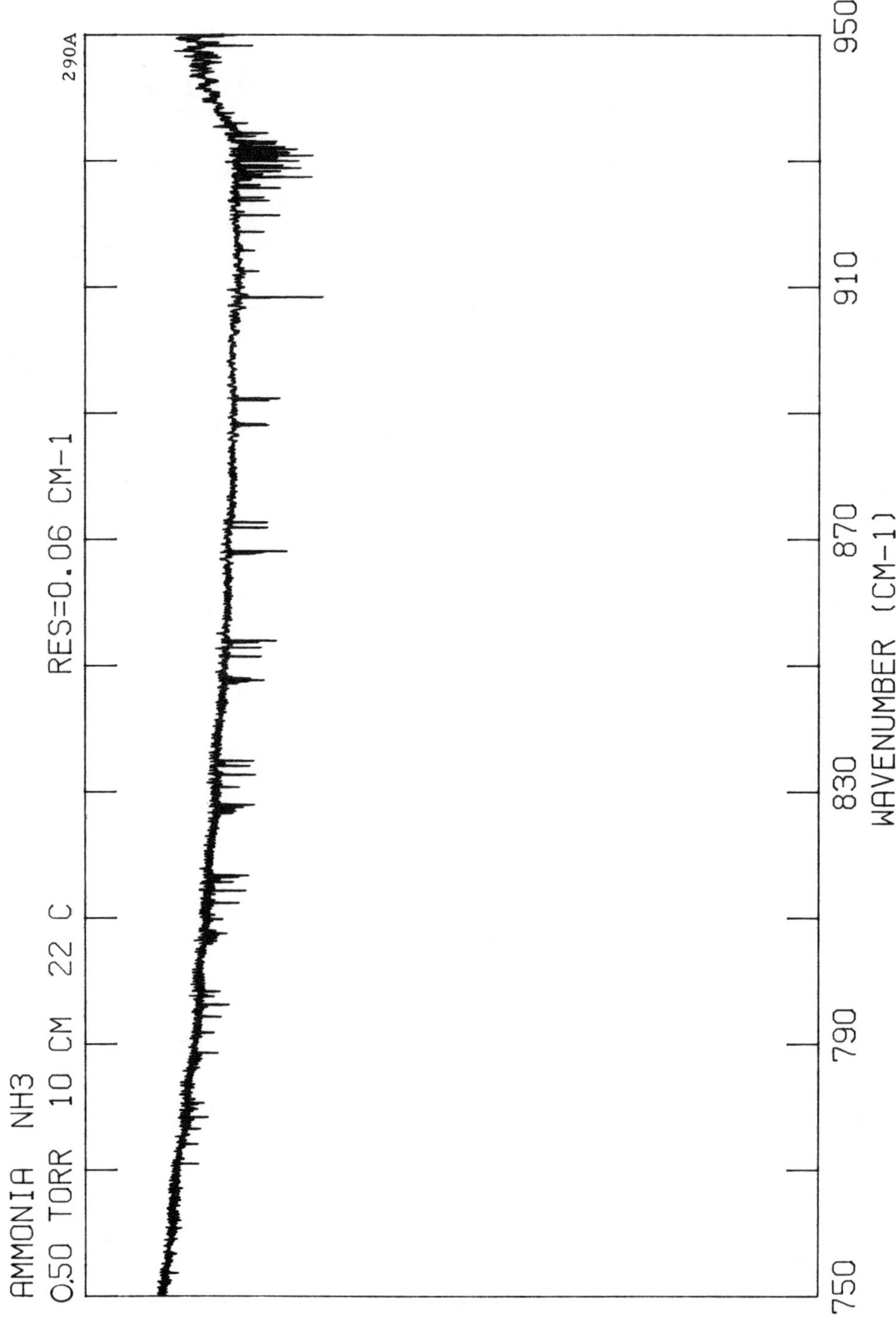
AMMONIA NH3
0.50 TORR 10 CM 22 C
RES=0.06 CM-1
290A
750
790
830
870
910
950
WAVENUMBER (CM-1)

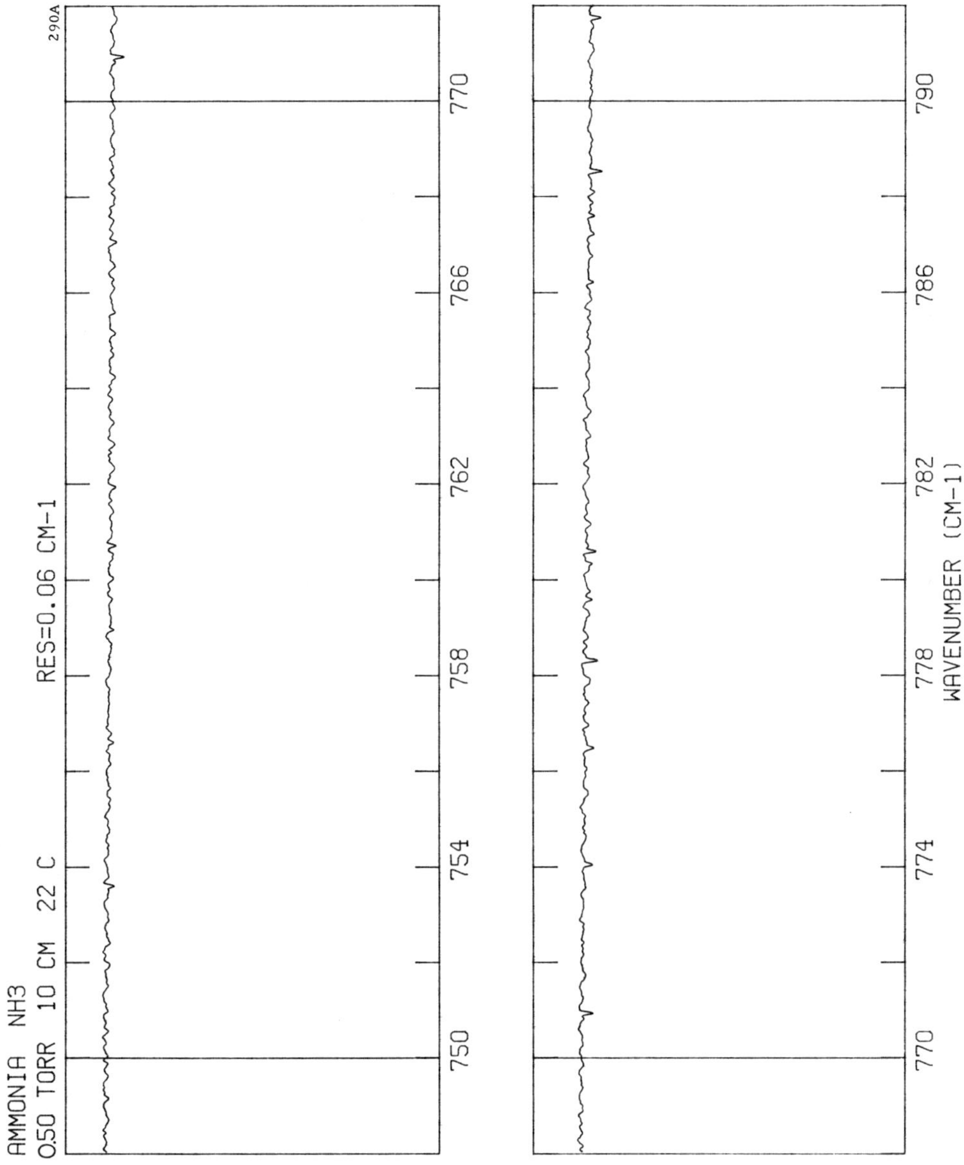
AMMONIA NH3
0.50 TORR 10 CM 22 C
RES=0.06 CM-1
290A
750
754
758
762
766
770
770
774
778
782
786
790
WAVENUMBER (CM-1)

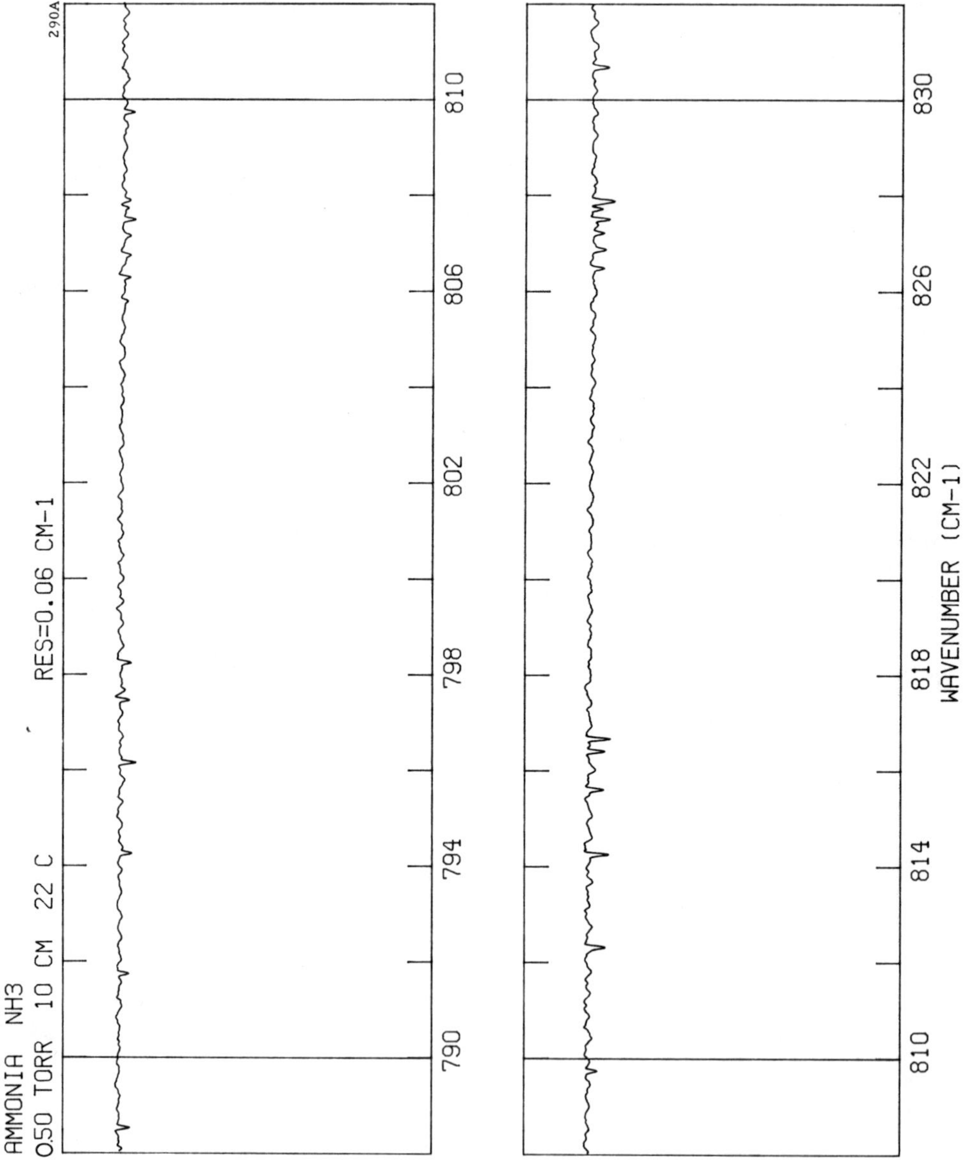
AMMONIA NH3
0.50 TORR 10 CM 22 C
RES=0.06 CM-1
290A
790
794
798
802
806
810
810
814
818
822
826
830
WAVENUMBER (CM-1)

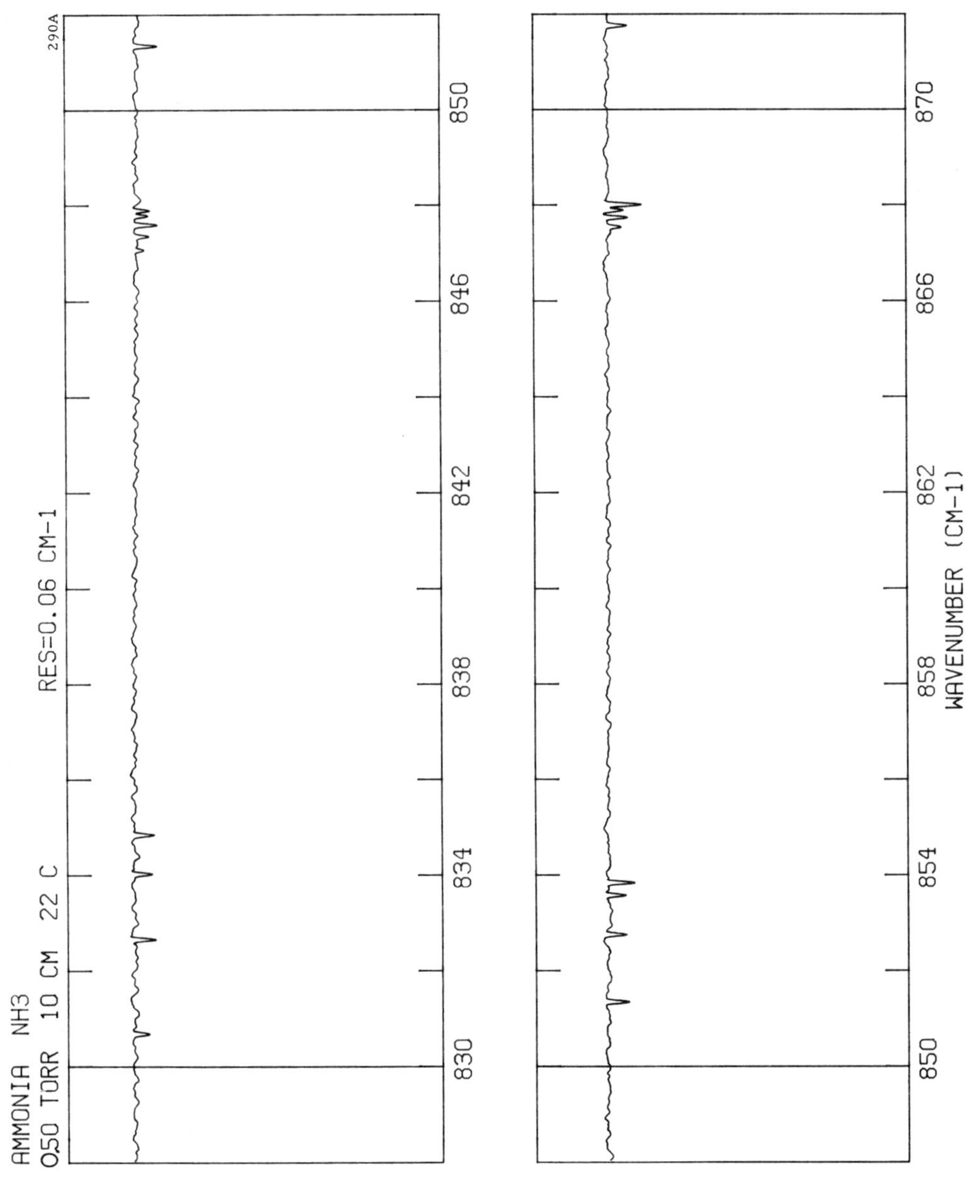
AMMONIA NH3
0.50 TORR 10 CM 22 C
RES=0.06 CM-1
290A
830
834
838
842
846
850
850
854
858
862
866
870
WAVENUMBER (CM-1)

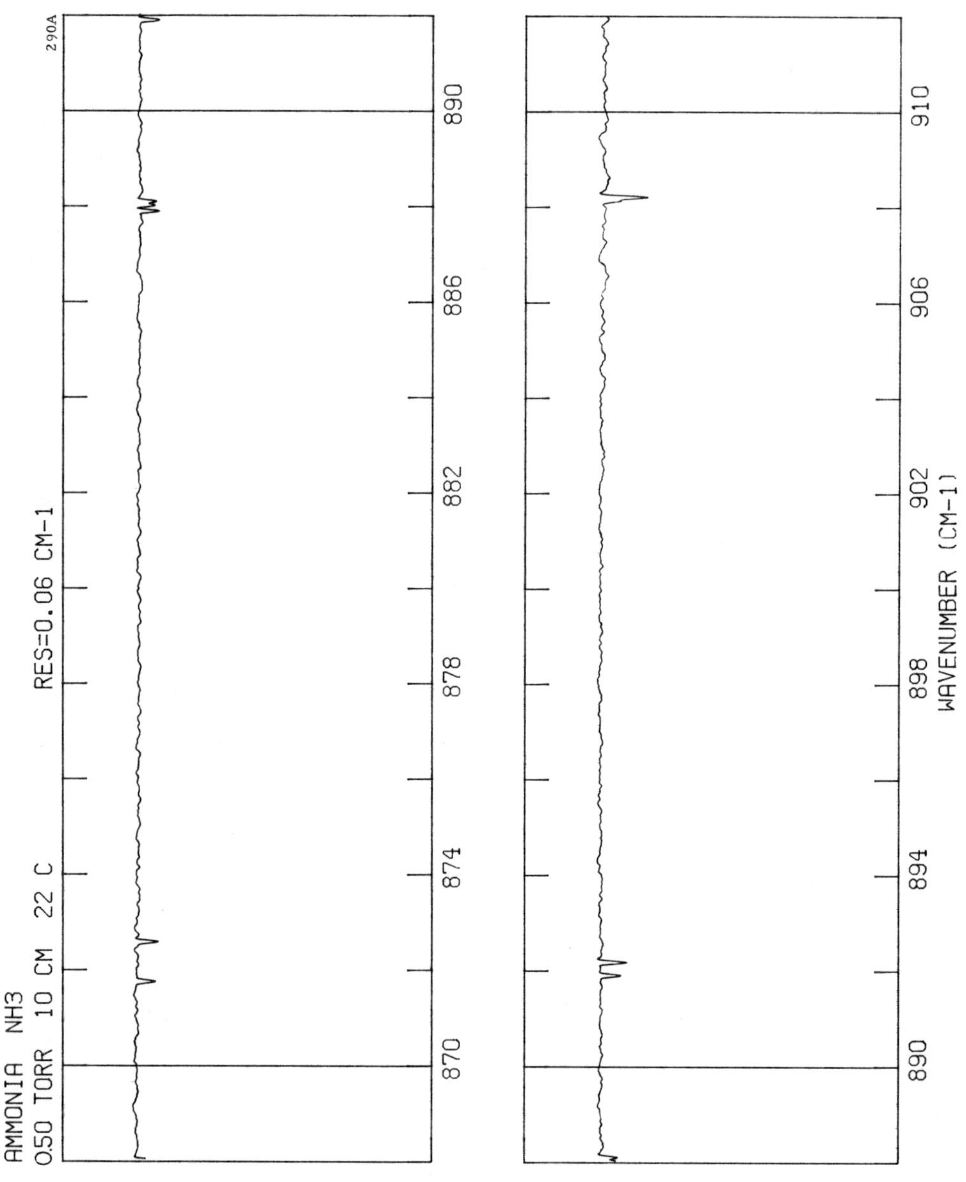
290A
AMMONIA NH3
0.50 TORR 10 CM 22 C
RES=0.06 CM-1
870
874
878
882
886
890
890
894
898
902
906
910
WAVENUMBER (CM-1)

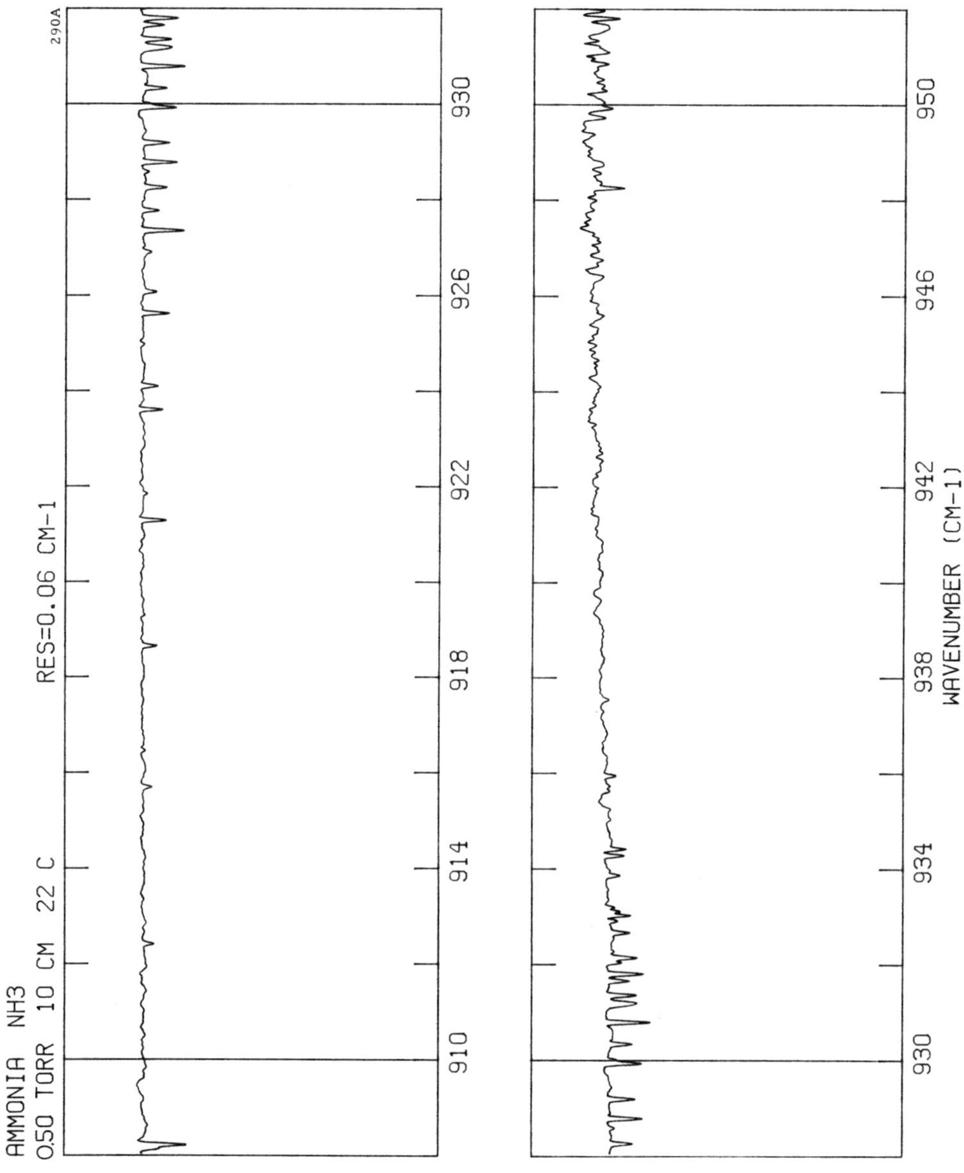
AMMONIA NH3
0.50 TORR 10 CM 22 C
RES=0.06 CM-1
290A
910
914
918
922
926
930
934
938
942
946
950
WAVENUMBER (CM-1)

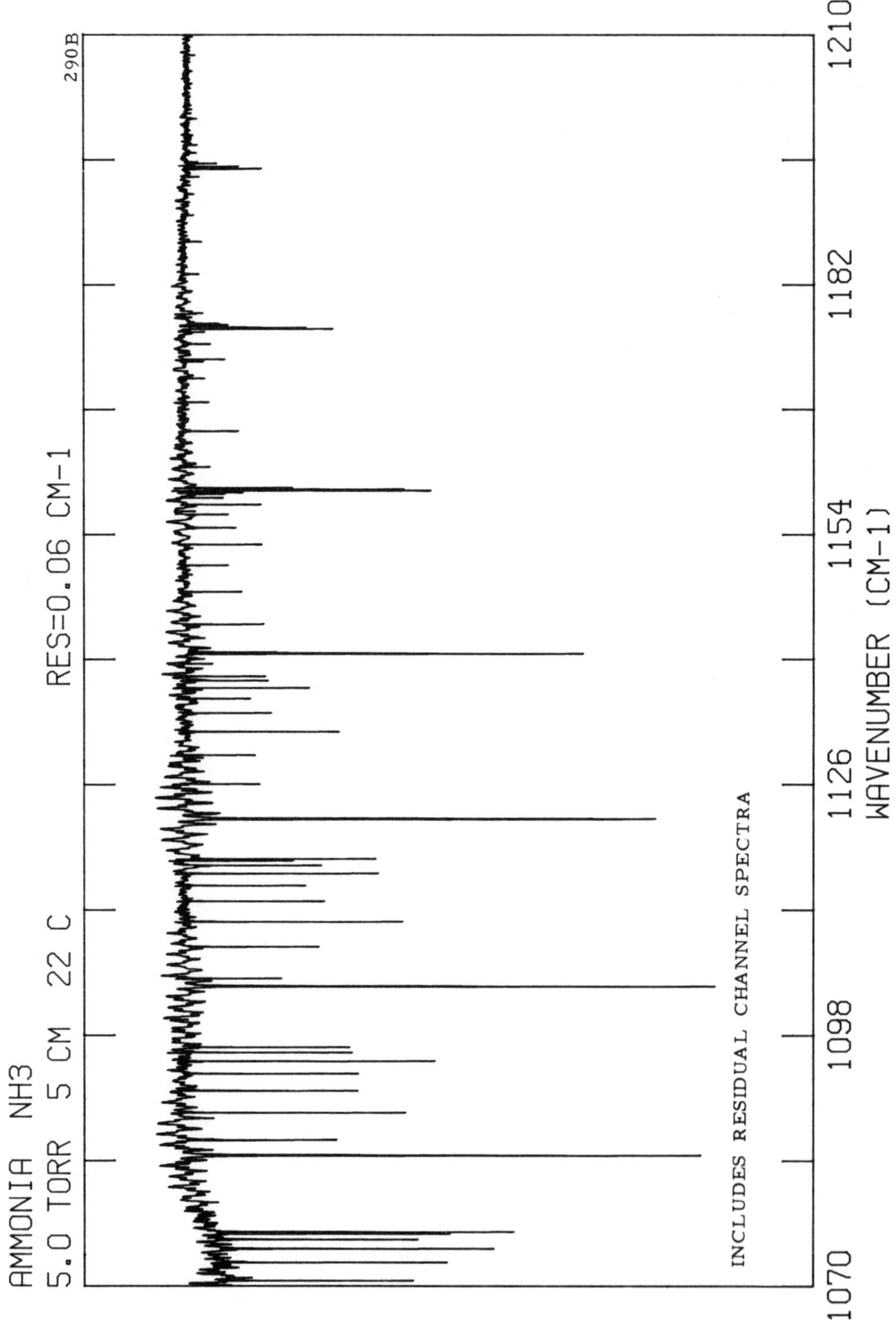

AMMONIA NH3
5.0 TORR 5 CM 22 C
RES=0.06 CM-1
290B
INCLUDES RESIDUAL CHANNEL SPECTRA
1070
1098
1126
1154
1182
1210
WAVENUMBER (CM-1)

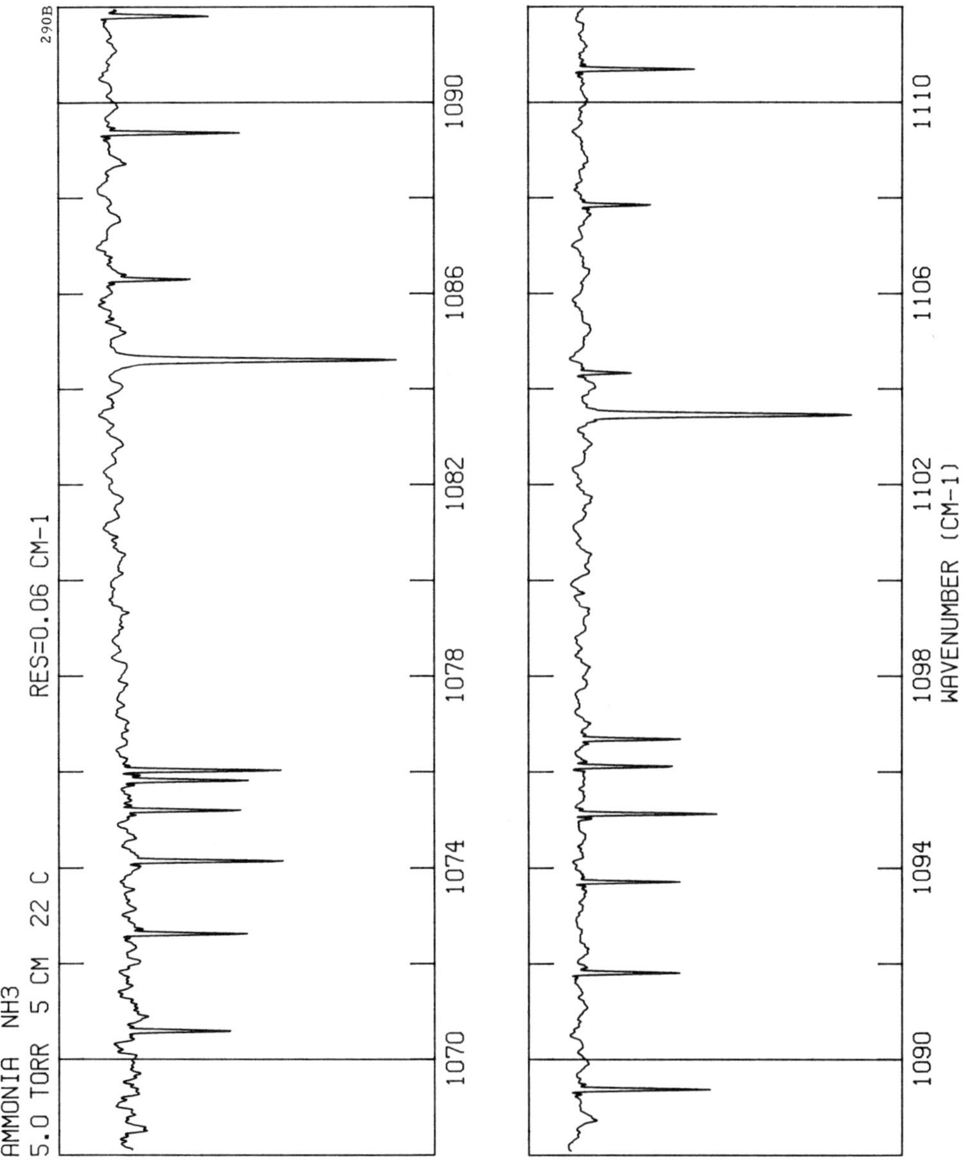
AMMONIA NH3
5.0 TORR 5 CM 22 C
RES=0.06 CM-1
290B
1070
1074
1078
1082
1086
1090
1090
1094
1098
1102
1106
1110
WAVENUMBER (CM-1)

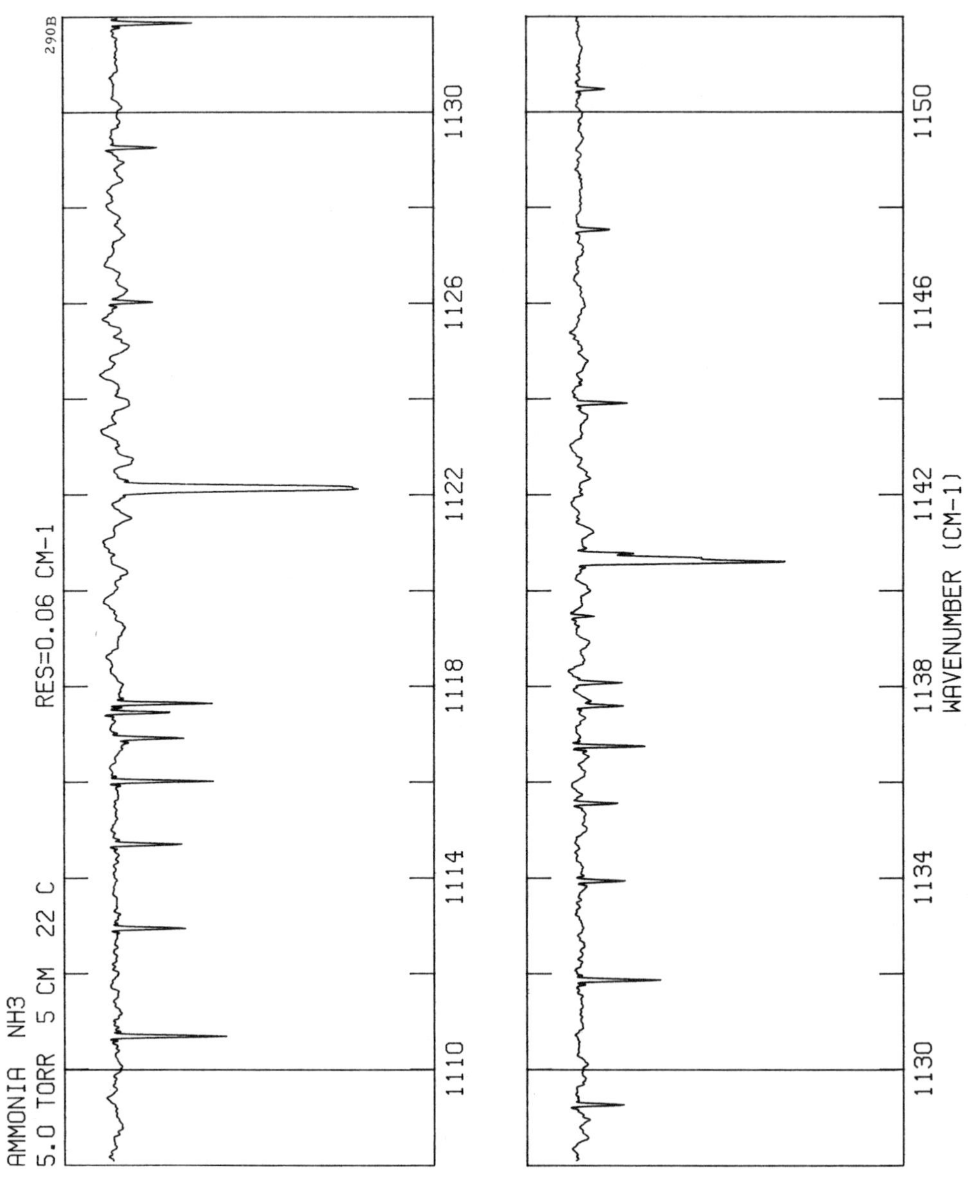
AMMONIA NH3
5.0 TORR 5 CM 22 C
RES=0.06 CM-1
290B
1110
1114
1118
1122
1126
1130
1134
1138
1142
1146
1150
WAVENUMBER (CM-1)

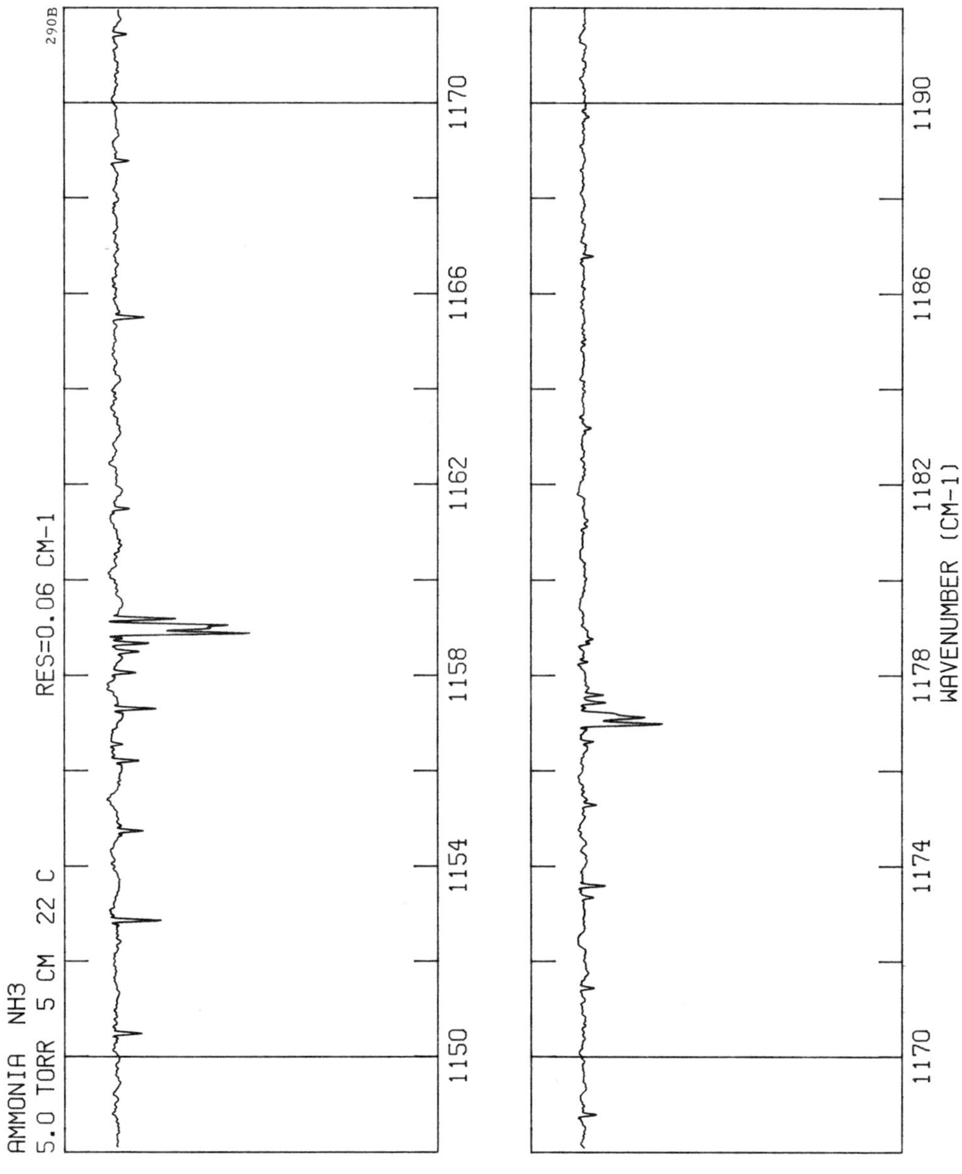
AMMONIA NH3
5.0 TORR 5 CM 22 C
RES=0.06 CM-1
290B
1150
1154
1158
1162
1166
1170
1170
1174
1178
1182
1186
1190
WAVENUMBER (CM-1)

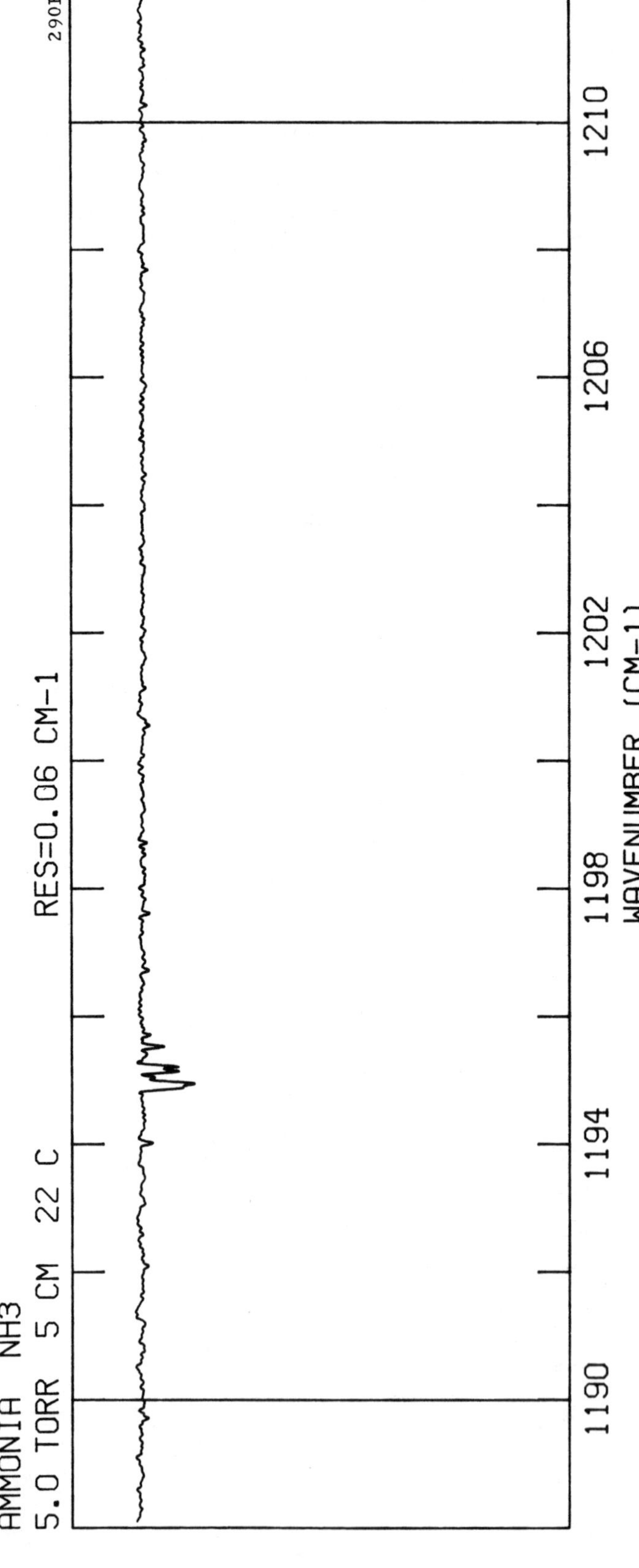

290B
AMMONIA NH3
5.0 TORR 5 CM 22 C
RES=0.06 CM-1
1190
1194
1198
1202
1206
1210
WAVENUMBER (CM-1)

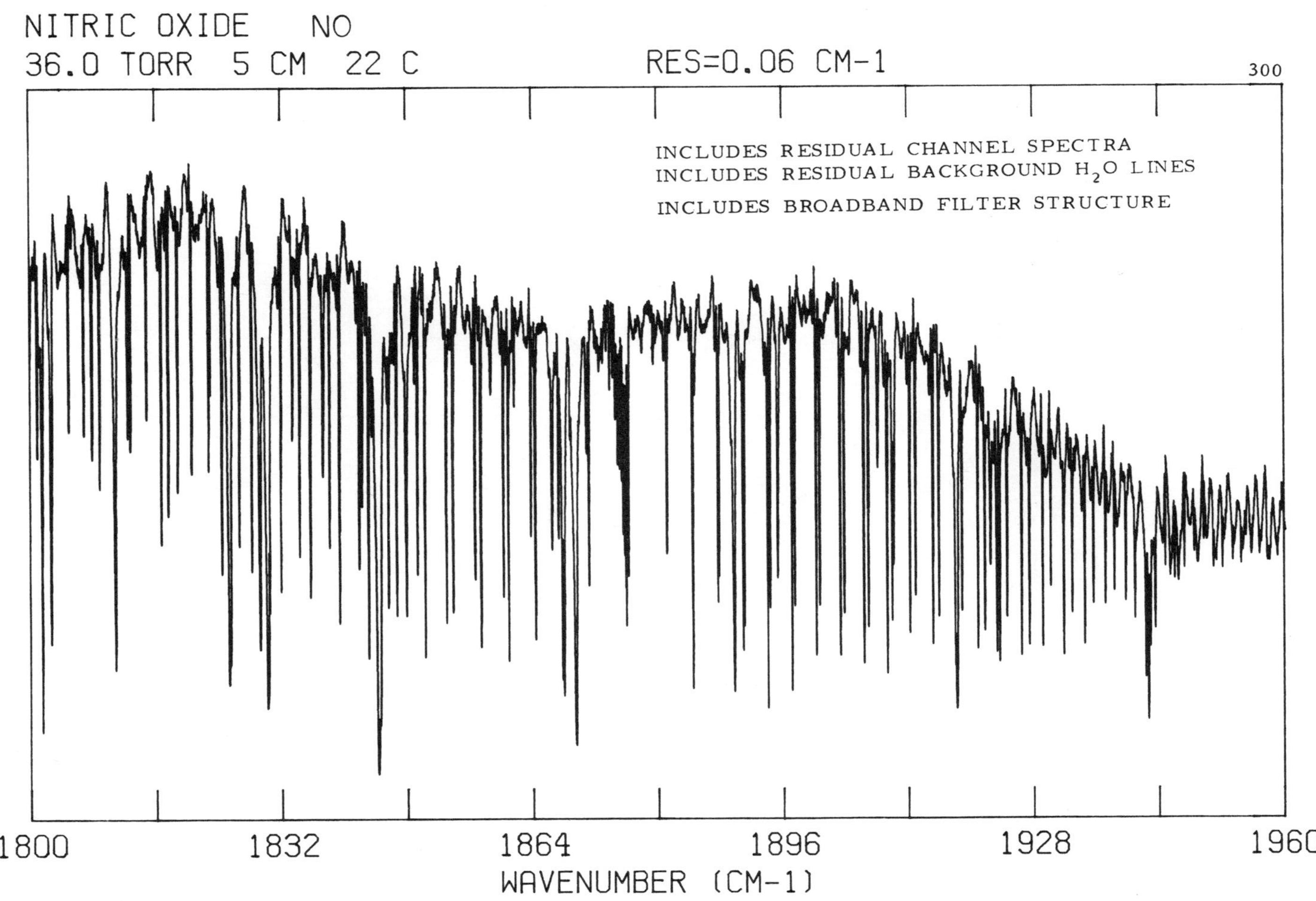
NITRIC OXIDE NO
36.0 TORR 5 CM 22 C
RES=0.06 CM-1
300
INCLUDES RESIDUAL CHANNEL SPECTRA
INCLUDES RESIDUAL BACKGROUND H_2O LINES
INCLUDES BROADBAND FILTER STRUCTURE
1800
1832
1864
1896
1928
1960
WAVENUMBER (CM-1)

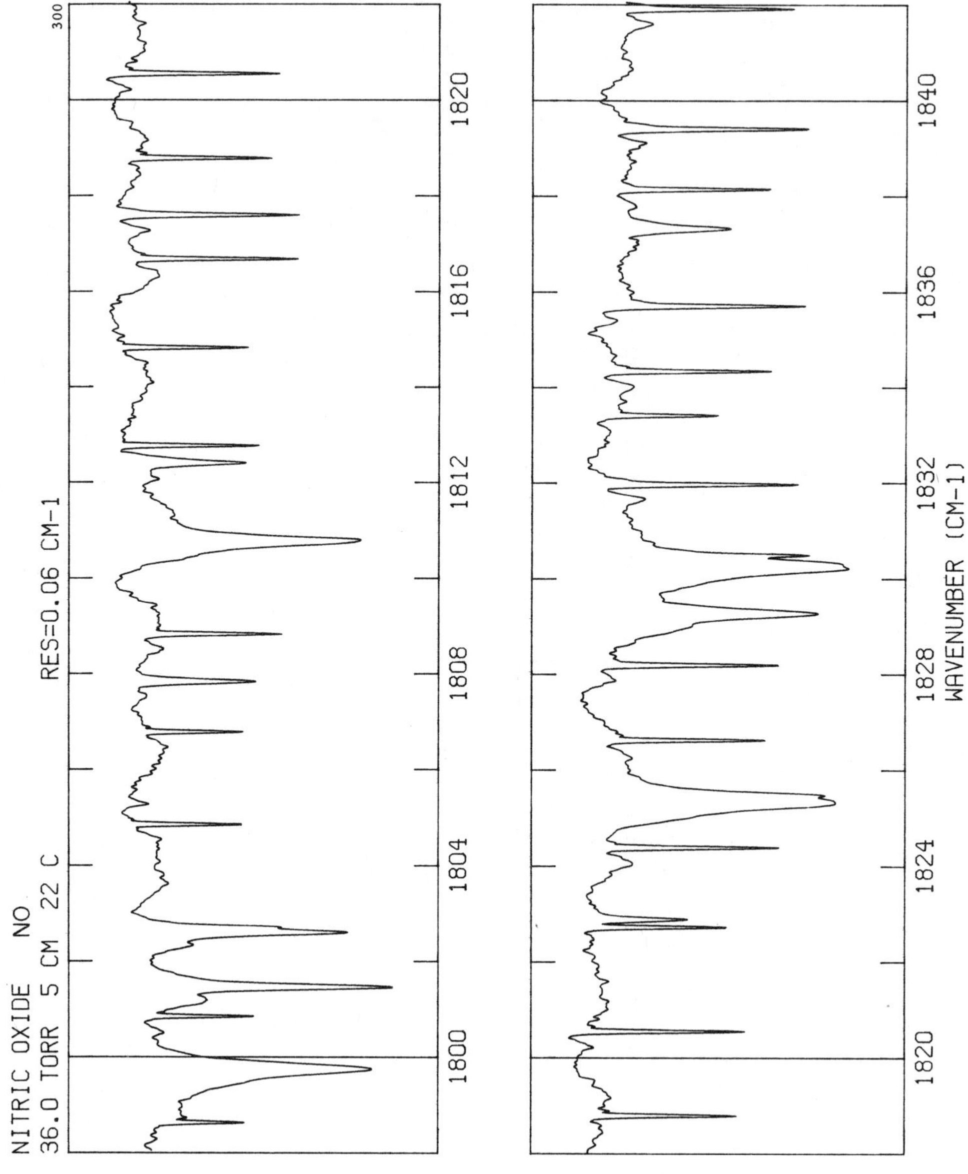
NITRIC OXIDE NO
36.0 TORR 5 CM 22 C
RES=0.06 CM-1
300
1800
1804
1808
1812
1816
1820
1820
1824
1828
1832
1836
1840
WAVENUMBER (CM-1)

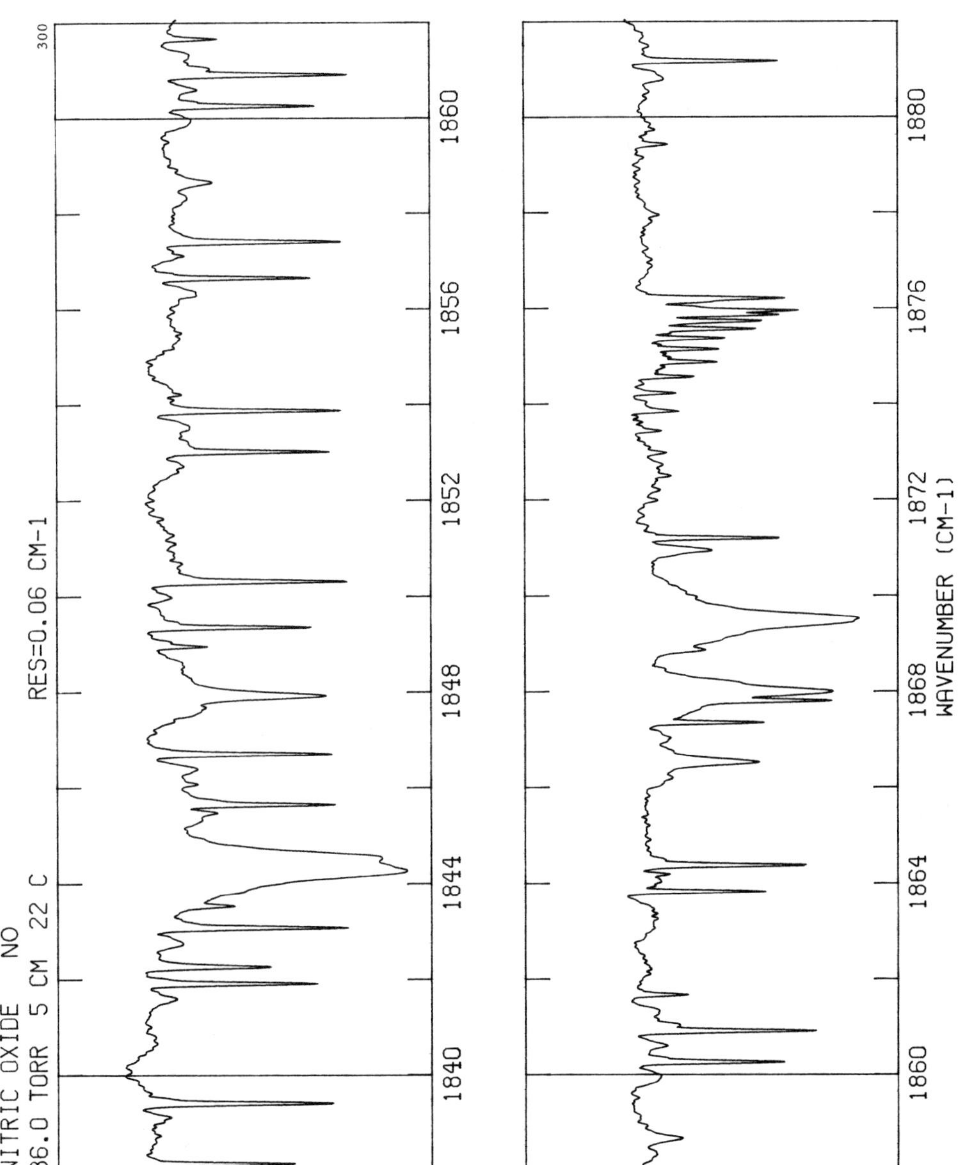

NITRIC OXIDE NO
36.0 TORR 5 CM 22 C
RES=0.06 CM-1
300
1840
1844
1848
1852
1856
1860
1860
1864
1868
1872
1876
1880
WAVENUMBER (CM-1)

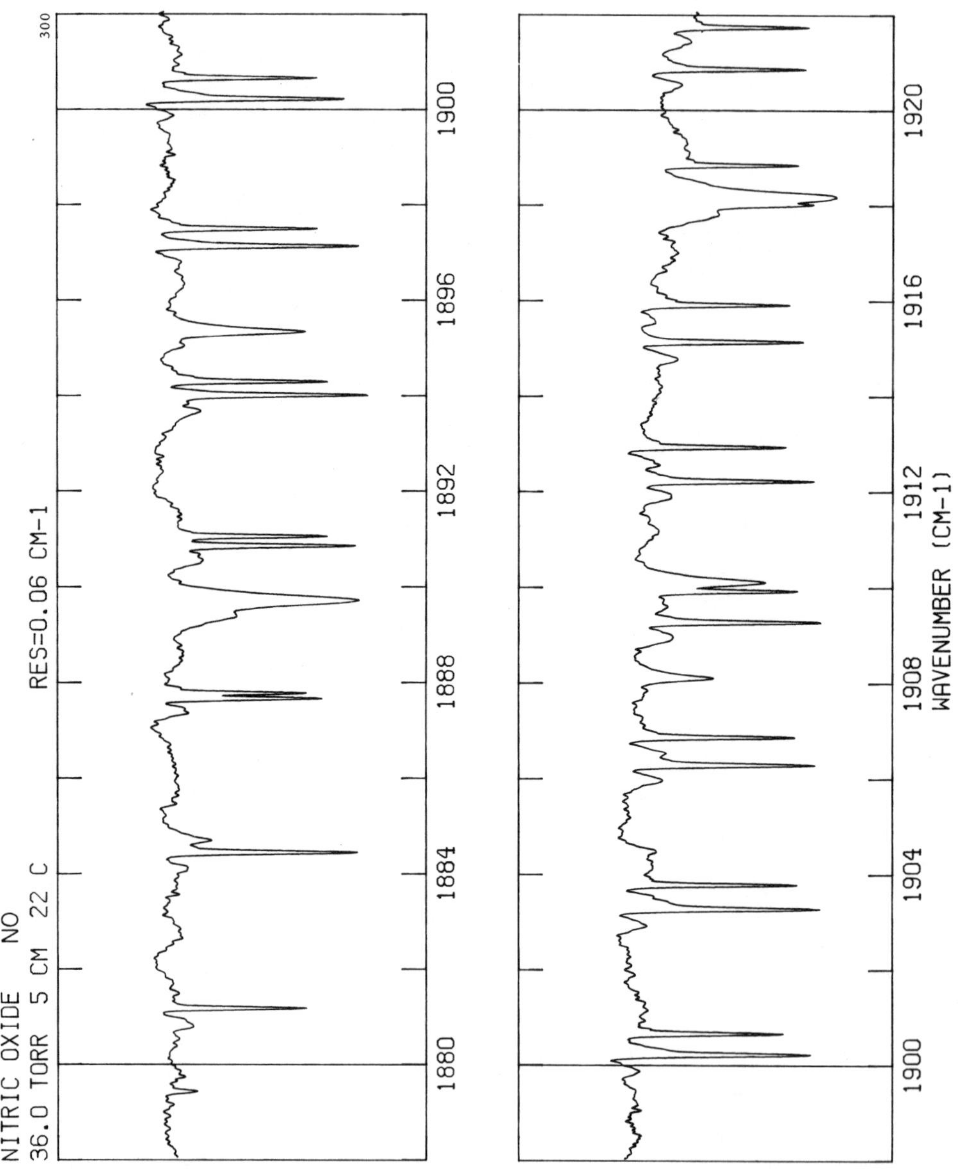
NITRIC OXIDE NO
36.0 TORR 5 CM 22 C
RES=0.06 CM-1
300
1880
1884
1888
1892
1896
1900
1900
1904
1908
1912
1916
1920
WAVENUMBER (CM-1)

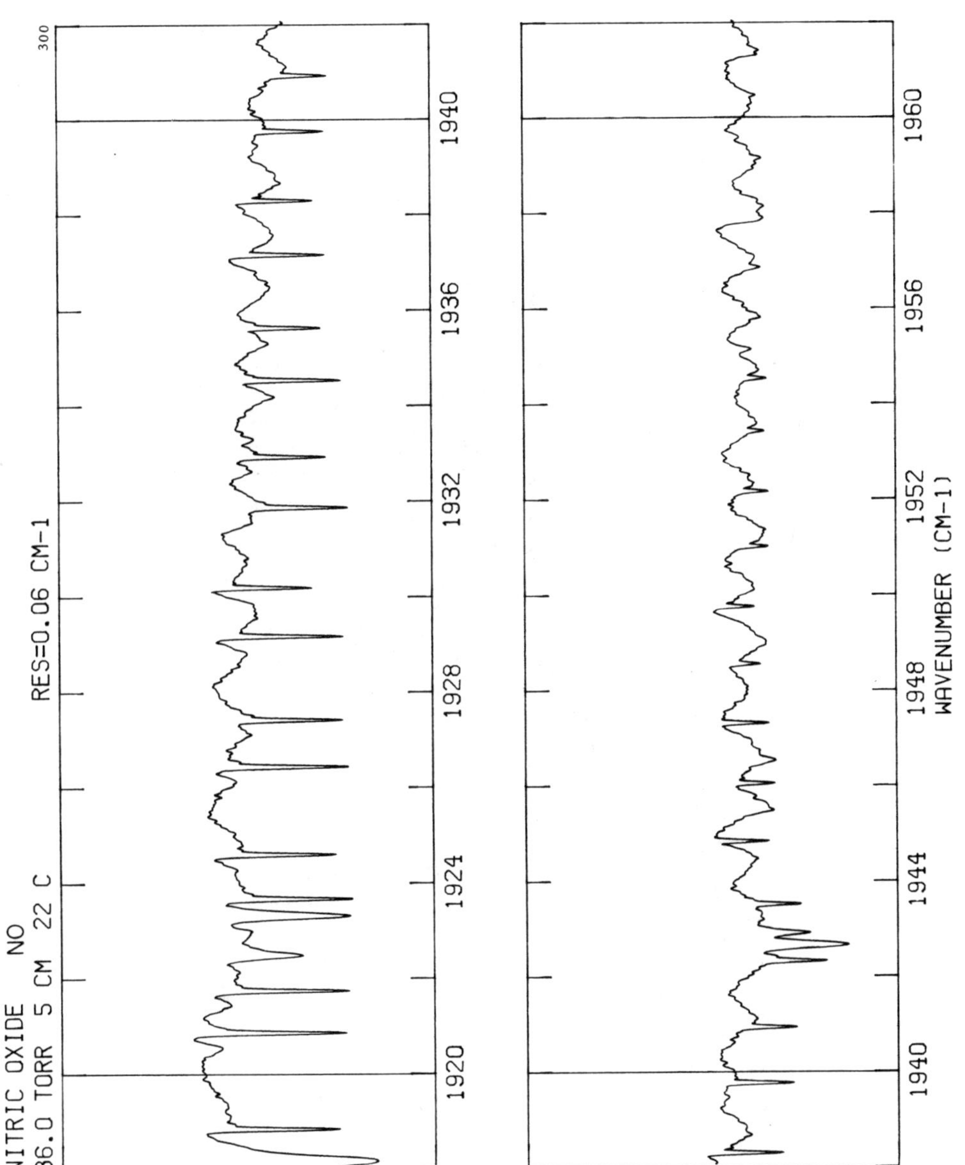
NITRIC OXIDE NO
36.0 TORR 5 CM 22 C
RES=0.06 CM-1
300
1920
1924
1928
1932
1936
1940
1944
1948
1952
1956
1960
WAVENUMBER (CM-1)

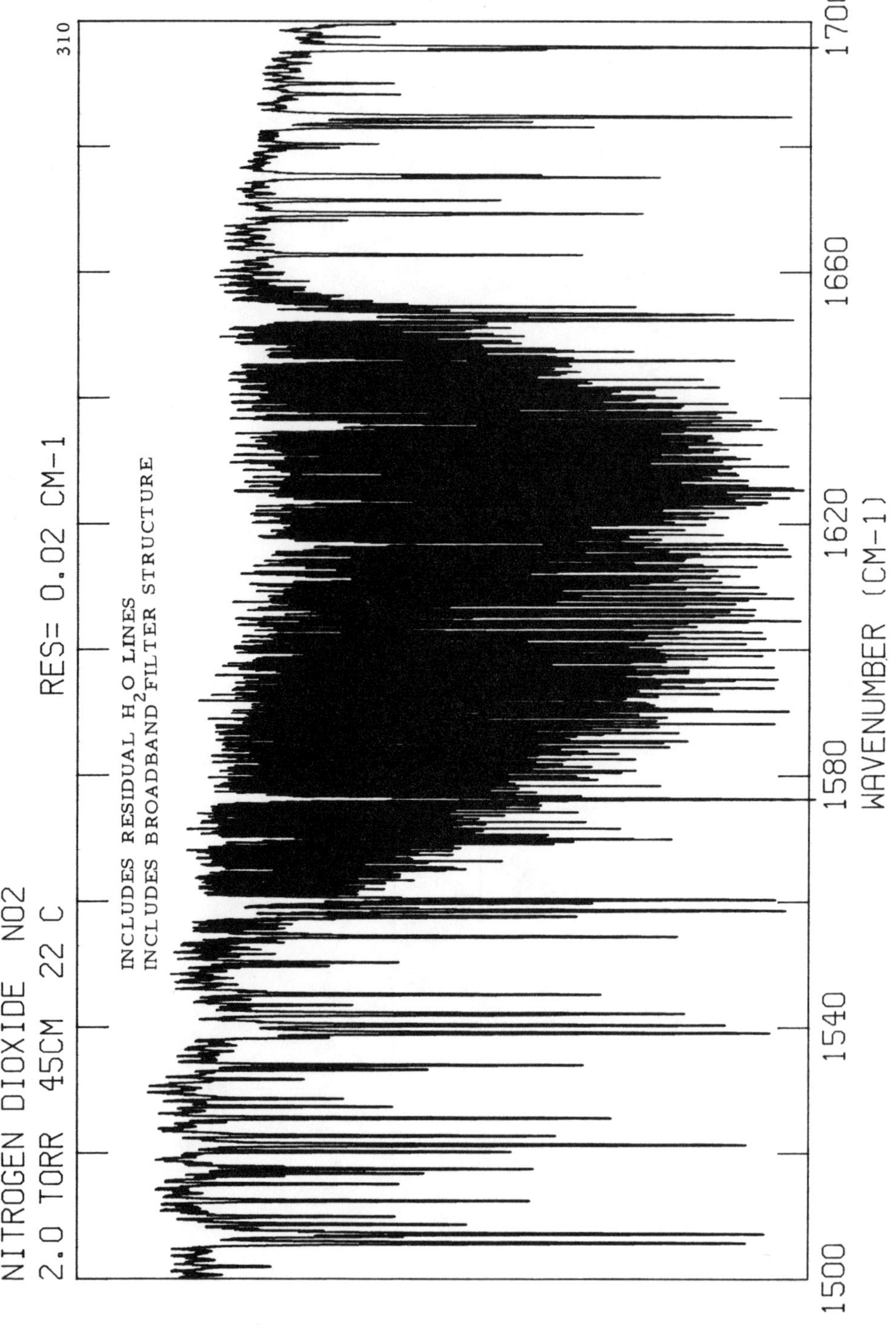

NITROGEN DIOXIDE NO2
2.0 TORR 45CM 22 C
RES= 0.02 CM-1
INCLUDES RESIDUAL H_2O LINES
INCLUDES BROADBAND FILTER STRUCTURE
310
1500
1540
1580
1620
1660
1700
WAVENUMBER (CM-1)

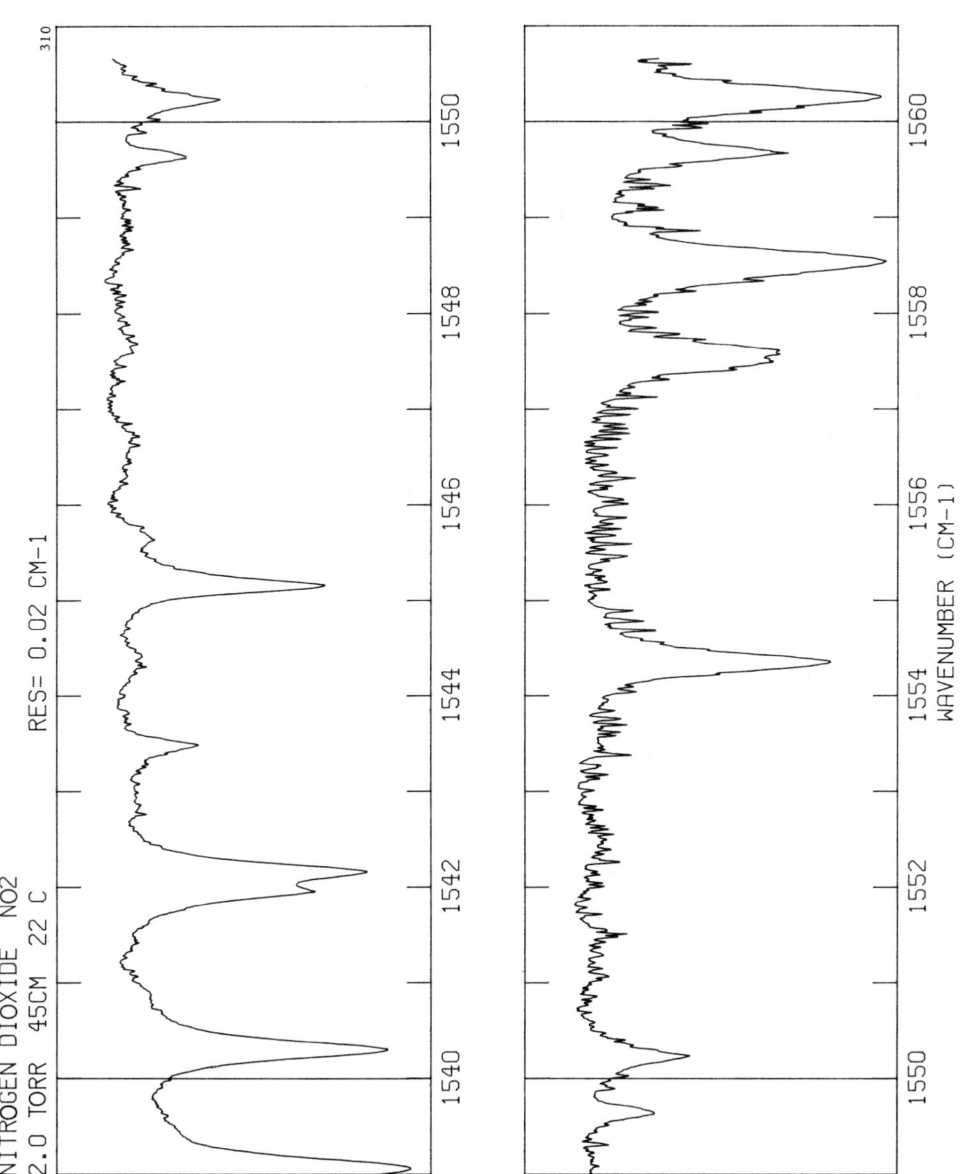
NITROGEN DIOXIDE NO2
2.0 TORR 45CM 22 C
RES= 0.02 CM-1
310
1540
1542
1544
1546
1548
1550
1552
1554
1556
1558
1560
WAVENUMBER (CM-1)

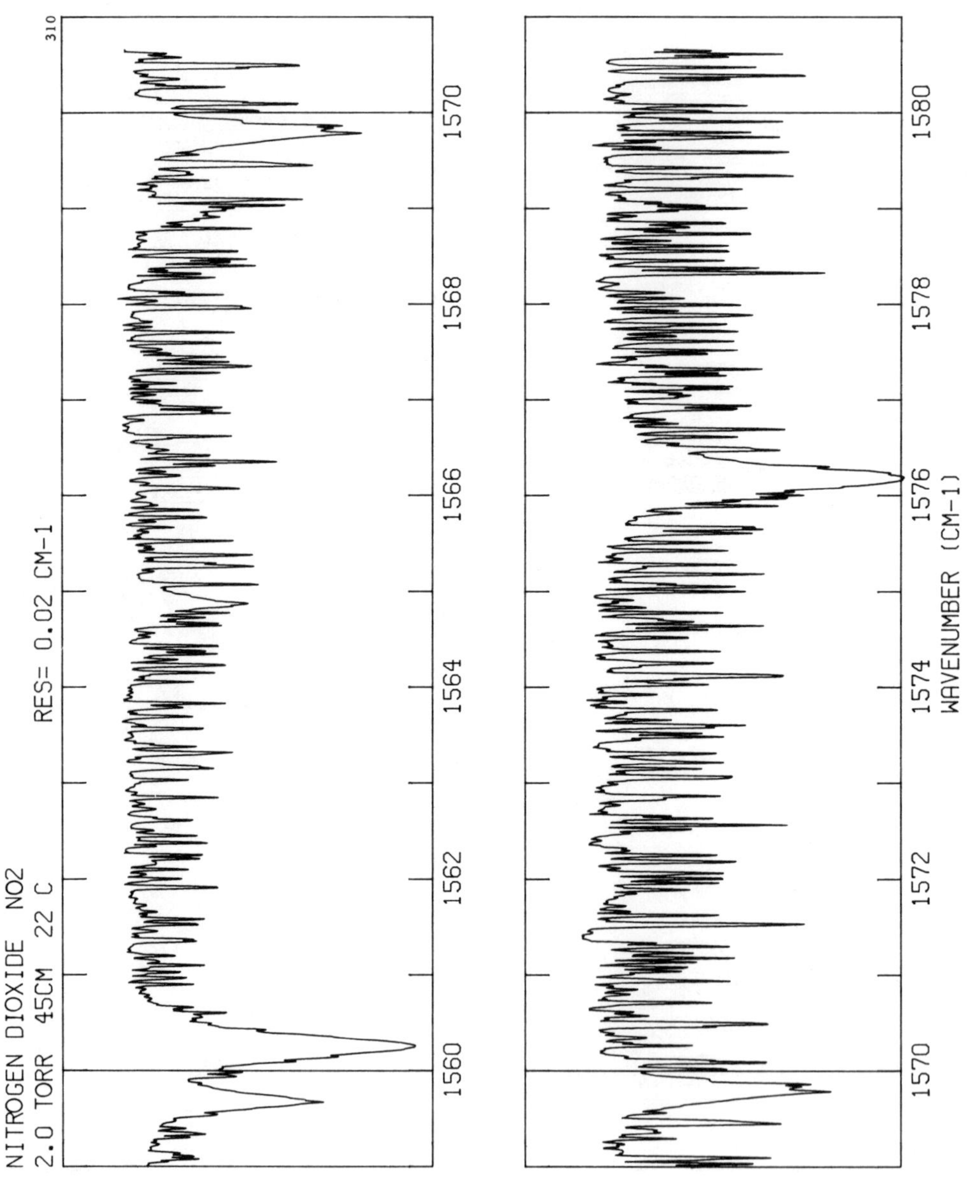
NITROGEN DIOXIDE NO2
2.0 TORR 45CM 22 C
RES= 0.02 CM-1
310
1560
1562
1564
1566
1568
1570
1570
1572
1574
1576
1578
1580
WAVENUMBER (CM-1)

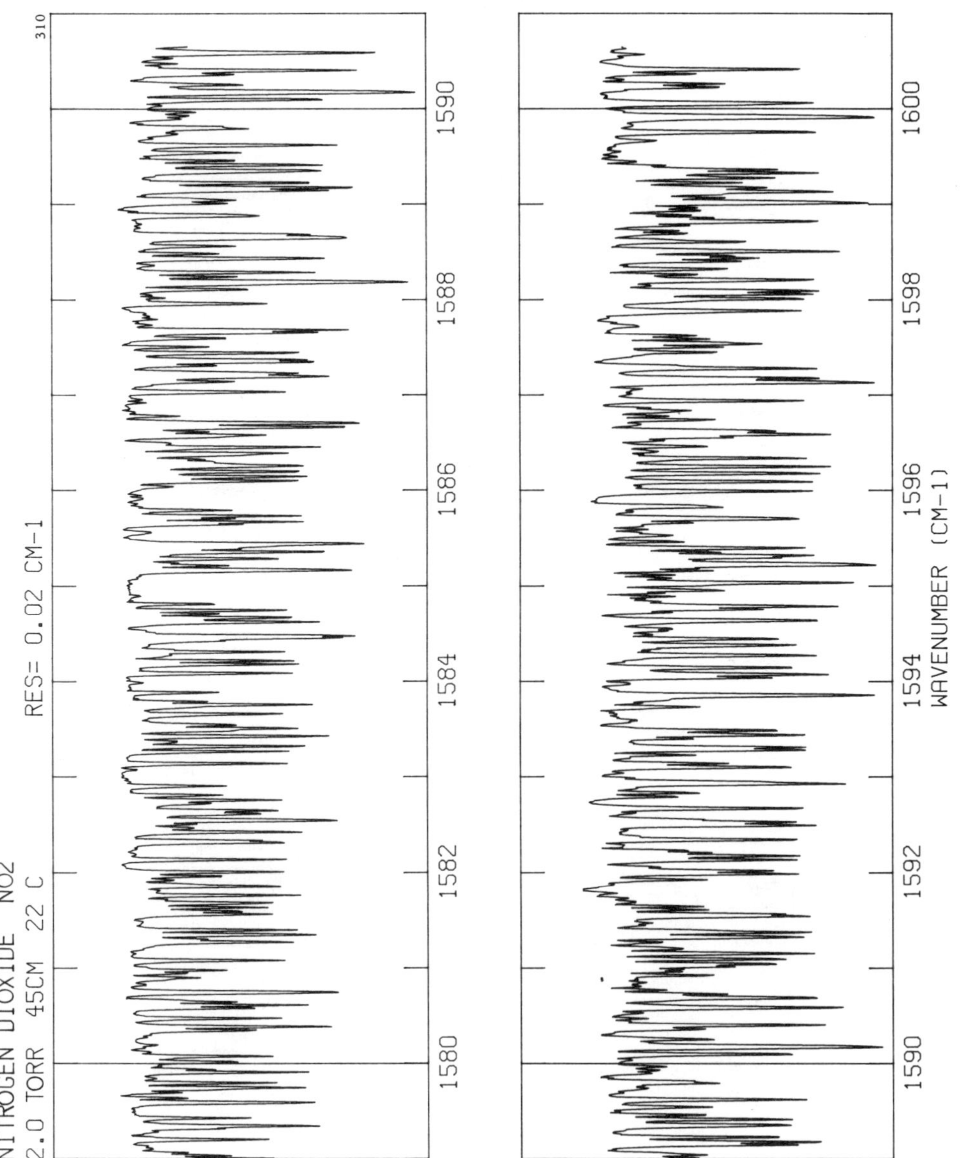
NITROGEN DIOXIDE NO2
2.0 TORR 45CM 22 C
RES= 0.02 CM-1
310
1580
1582
1584
1586
1588
1590
1590
1592
1594
1596
1598
1600
WAVENUMBER (CM-1)

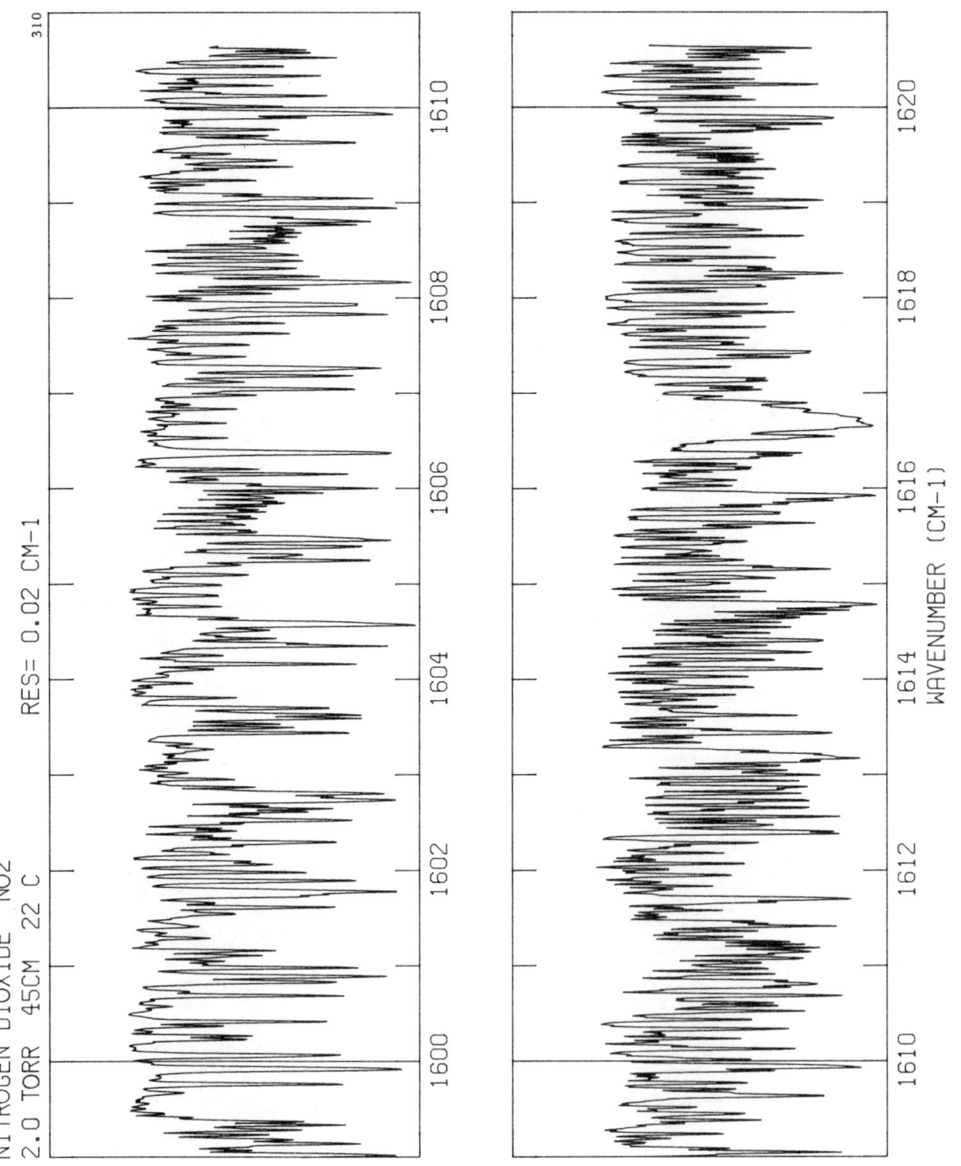
NITROGEN DIOXIDE NO2
2.0 TORR 45CM 22 C
RES= 0.02 CM-1
310
1600
1602
1604
1606
1608
1610
1610
1612
1614
1616
1618
1620
WAVENUMBER (CM-1)

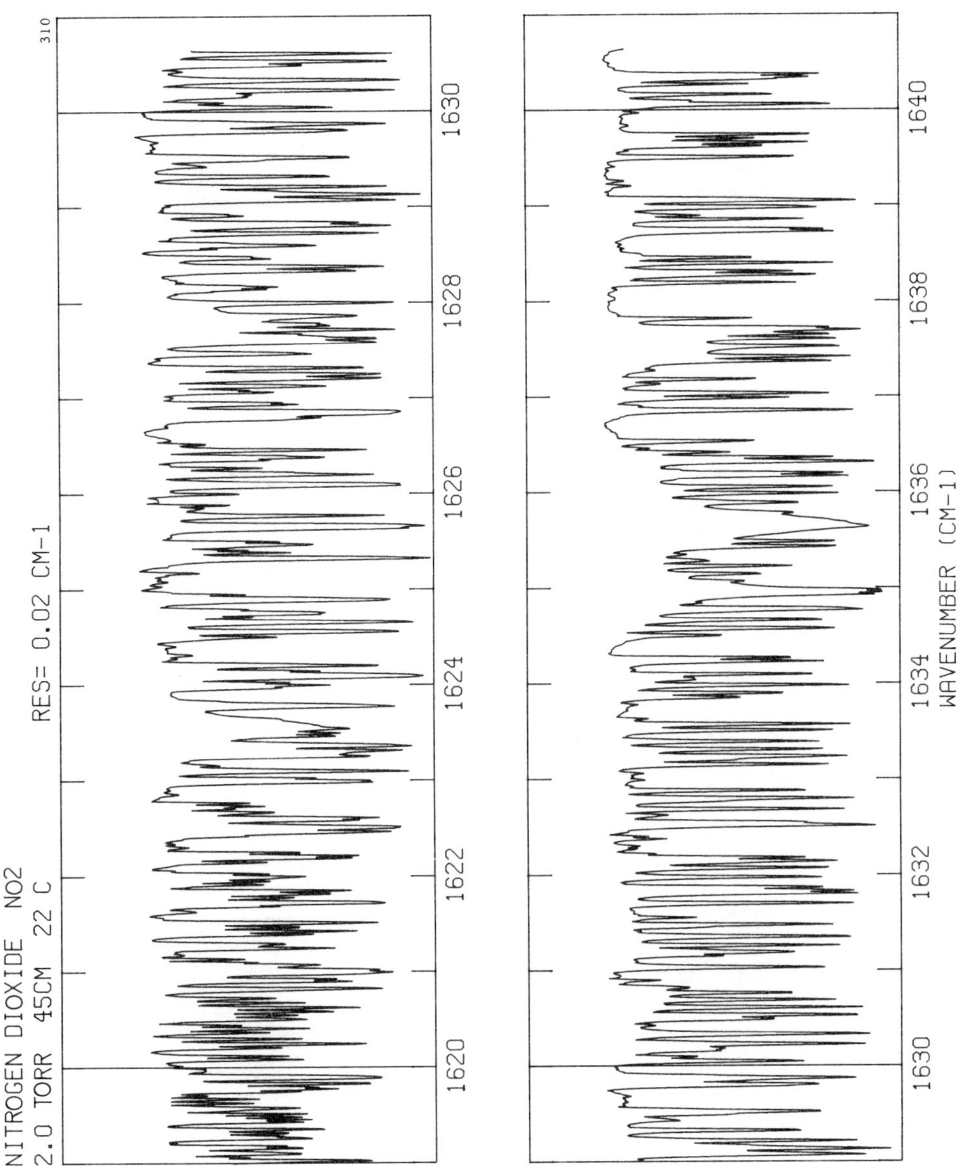
NITROGEN DIOXIDE NO2
2.0 TORR 45CM 22 C
RES= 0.02 CM-1
310
1620
1622
1624
1626
1628
1630
1630
1632
1634
1636
1638
1640
WAVENUMBER (CM-1)

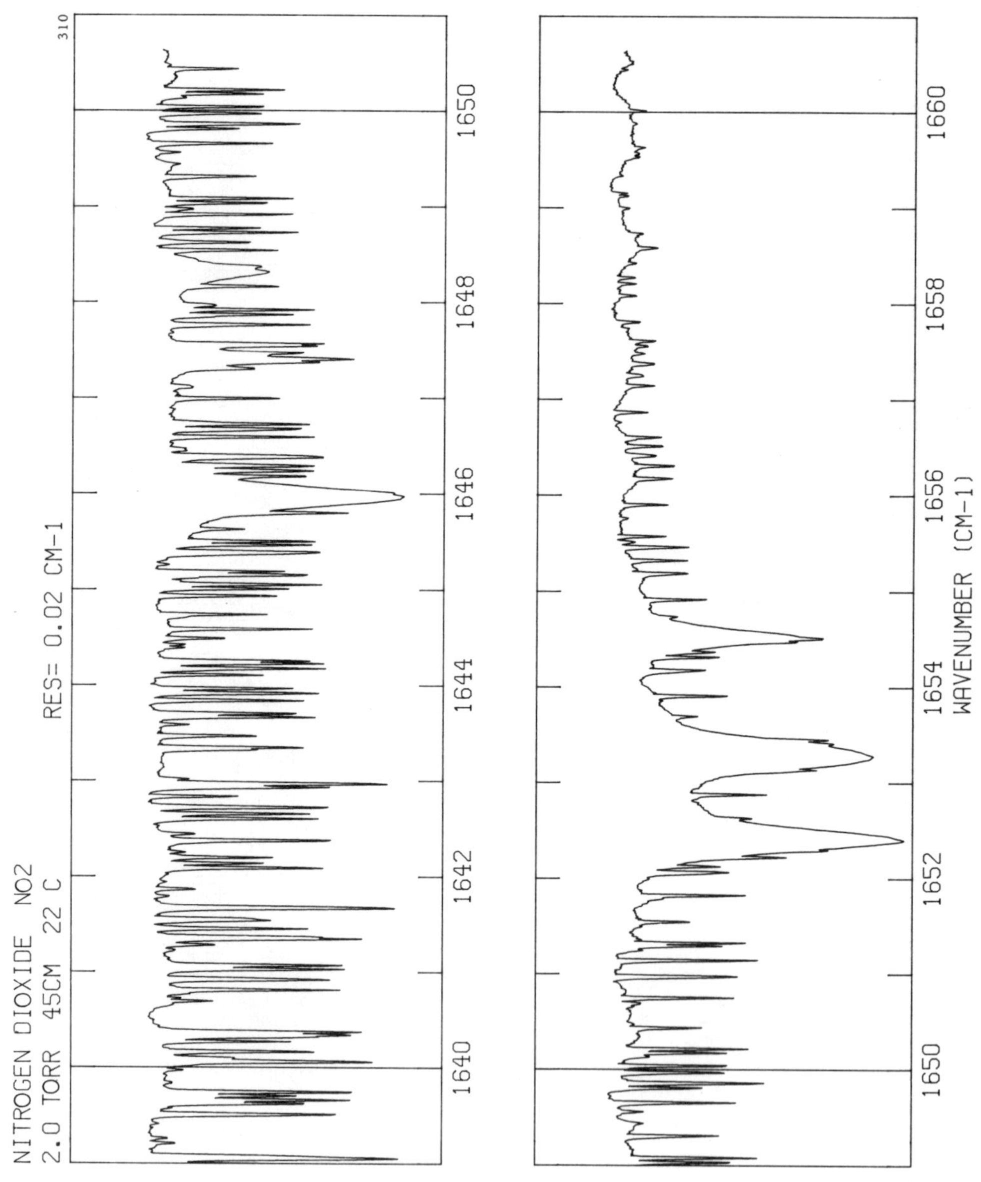
NITROGEN DIOXIDE NO2
2.0 TORR 45CM 22 C
RES= 0.02 CM-1
310
1640
1642
1644
1646
1648
1650
1650
1652
1654
1656
1658
1660
WAVENUMBER (CM-1)

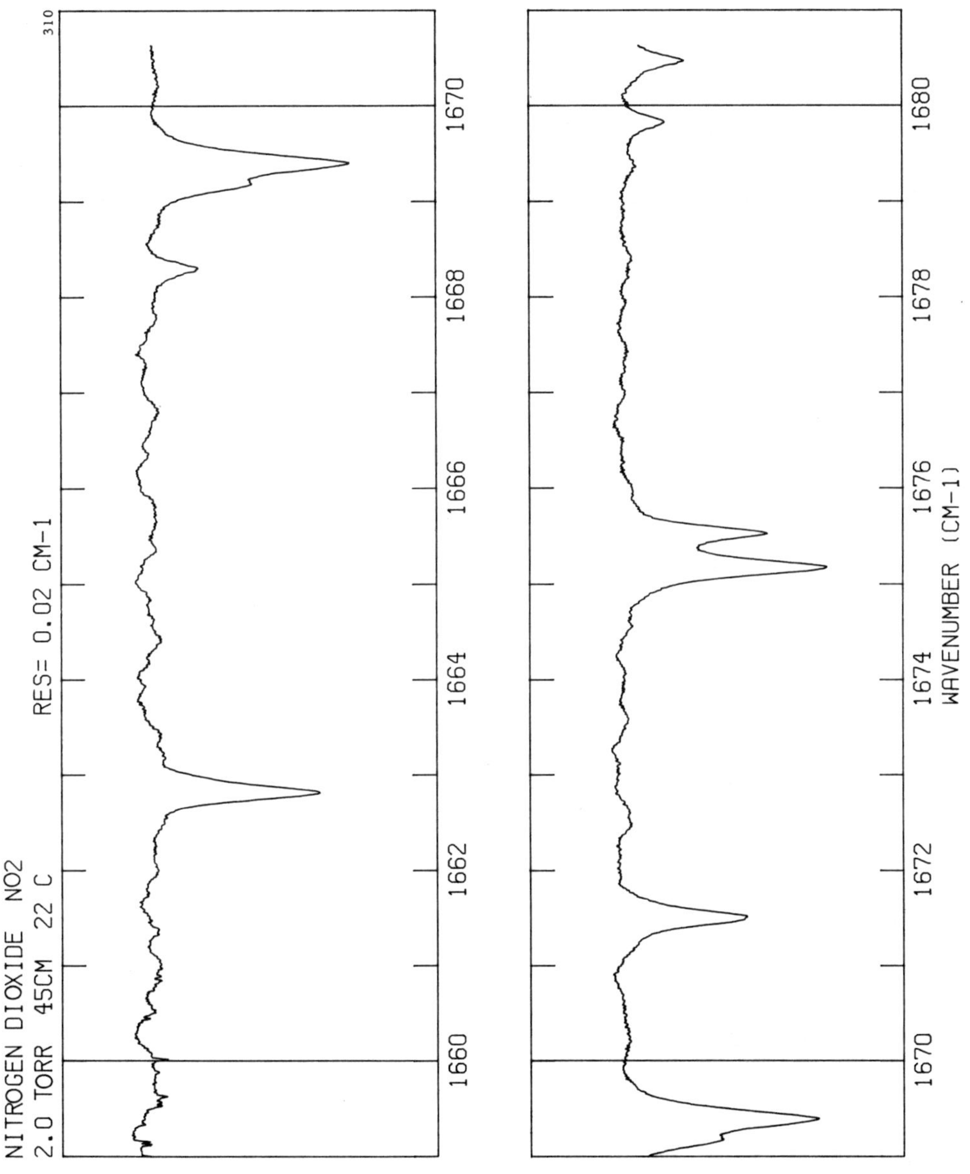
NITROGEN DIOXIDE NO2
2.0 TORR 45CM 22 C
RES= 0.02 CM-1
310
1660
1662
1664
1666
1668
1670
1670
1672
1674
1676
1678
1680
WAVENUMBER (CM-1)

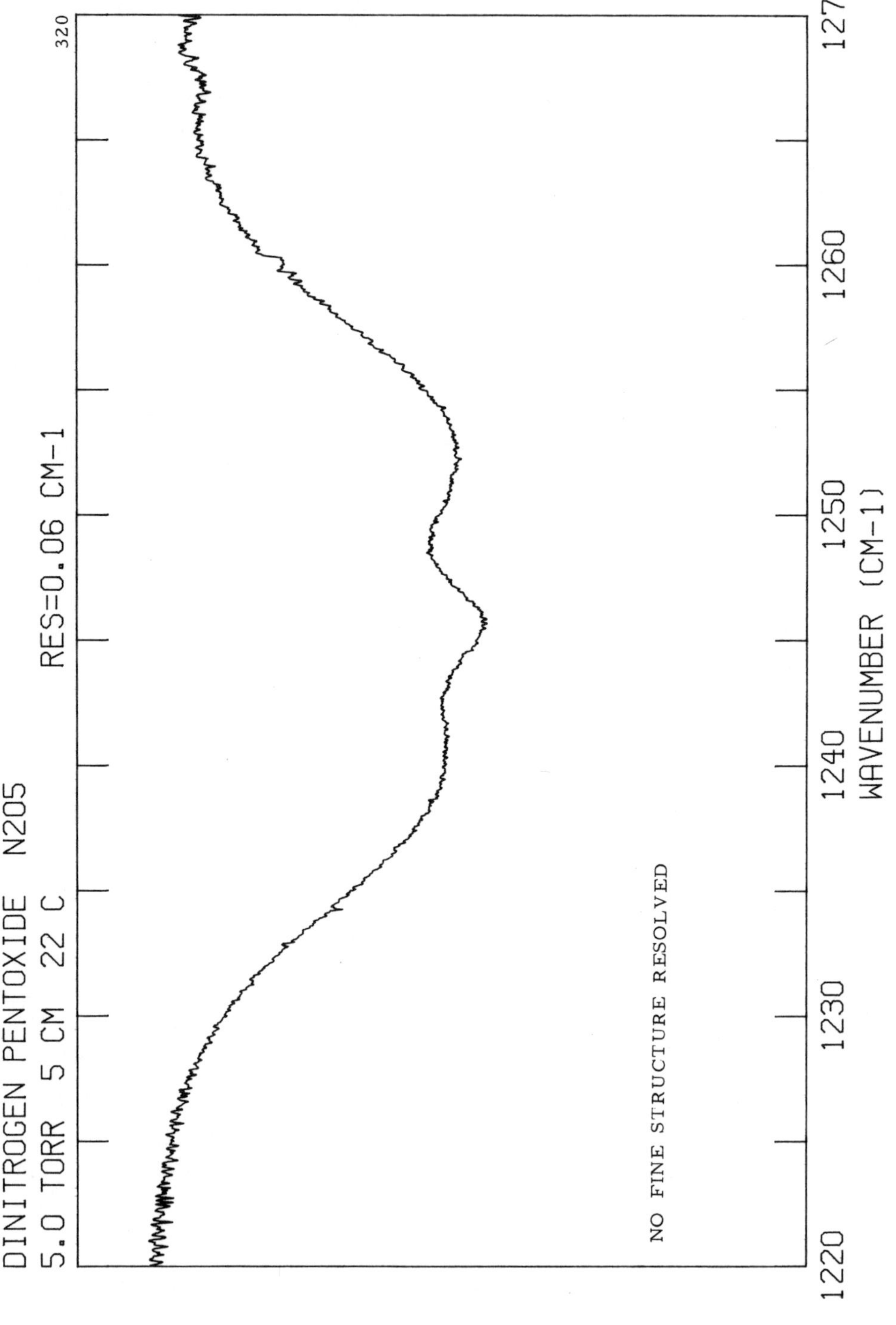
DINITROGEN PENTOXIDE N2O5
5.0 TORR 5 CM 22 C
RES=0.06 CM-1
320
NO FINE STRUCTURE RESOLVED
1220
1230
1240
1250
1260
1270
WAVENUMBER (CM-1)